Geoheritage and Geotourism Resources: Education, Recreation, Sustainability II

Geoheritage and Geotourism Resources: Education, Recreation, Sustainability II

Editors

Hara Drinia
Panagiotis Voudouris
Assimina Antonarakou

Basel • Beijing • Wuhan • Barcelona • Belgrade • Novi Sad • Cluj • Manchester

Editors

Hara Drinia	Panagiotis Voudouris	Assimina Antonarakou
National and Kapodistrian	National and Kapodistrian	National and Kapodistrian
University of Athens	University of Athens	University of Athens
Athens	Athens	Athens
Greece	Greece	Greece

Editorial Office
MDPI
St. Alban-Anlage 66
4052 Basel, Switzerland

This is a reprint of articles from the Special Issue published online in the open access journal *Geosciences* (ISSN 2076-3263) (available at: https://www.mdpi.com/journal/geosciences/special_issues/geoheritage_geotourism_resources_II).

For citation purposes, cite each article independently as indicated on the article page online and as indicated below:

Lastname, A.A.; Lastname, B.B. Article Title. *Journal Name* **Year**, *Volume Number*, Page Range.

ISBN 978-3-7258-0255-5 (Hbk)
ISBN 978-3-7258-0256-2 (PDF)
doi.org/10.3390/books978-3-7258-0256-2

Contents

About the Editors

Hara Drinia

Hara Drinia is a Professor of Palaeoecology and Sedimentology at the Department of Geology and Geoenvironment at the National and Kapodistrian University of Athens. She teaches a wide range of undergraduate and postgraduate courses including Sedimentology, Palaeoecology, Stratigraphy, Marine Ecosystems, Environmental Education, Geoscience Teaching, Geological Heritage, and Geoconservation. Her academic performance, combined with her scientific research, is strong evidence of her ongoing national and international research activity and contributions to the science of geology. She has been actively involved in many international scientific conferences and in regional, national, and international meetings. She has a strong record of participating in research programs and initiatives and has been the Principal Investigator of several research projects. Regarding her wider contribution to the geological community, it is worth noting that Prof. Drinia is an active member of eight geosciences committees and has been a member of the organizing committee of eight international and national geological conferences. She is also been actively involved in the Socrates/Erasmus educational exchange programs in Switzerland, Germany, and Italy.

Panagiotis Voudouris

Panagiotis Voudouris is a Professor of Mineralogy and Petrology at the Department of Geology and Geoenvironment, National and Kapodistrian University of Athens (Greece). He graduated in 1986 and earned his Doctorate (1993) in Mineralogy and Economic Geology at the University of Hamburg (Germany), where he also served as a Visiting Professor. His main research interests are in the fields of ore mineralogy, the genesis of magmatic–hydrothermal ore deposits, the mineralogy and genesis of gemstones, and the promotion of mineralogical and mining heritage in Greece. He is a member of the Editorial Board of *Minerals* and Director of the Mineralogy and Petrology Museum, National and Kapodistrian University of Athens. He has published 99 scientific papers in international refereed journals, is the co-author of four IMA-approved new mineral species, and has co-authored five books. The International Mineralogical Association adopted a new mineral name in his honor (Voudourisite).

Assimina Antonarakou

Assimina Antonarakou is a Professor of Marine Geology–Micropaleontology–Didactics on Geosciences, and is the Vice President of the Faculty of Geology and Geoenvironment at the University of Athens. Her PhD thesis dealt with Miocene cyclic sedimentary successions of the eastern Mediterranean in terms of orbital periodicities and paleoclimatic variations based on planktonic foraminiferal assemblages. Her main research topics are summarized as follows: planktonic foraminiferal eco-biostratigraphy, geobiology, and paleoceanography; astronomical frequencies in paleoclimates; extreme geological events; marine environmental monitoring; ocean dynamics and sea-level changes; natural and human environmental stressors; and foraminiferal trace metals and stable isotopes. She has participated in several national and international projects focused on multiproxy ecosystem responses to past and present environmental events, and she is the co-author of more than 50 peer-reviewed publications in international journals.

 *geosciences*

Editorial

Geoheritage and Geotourism Resources: Education, Recreation, Sustainability II

Hara Drinia *, Panagiotis Voudouris and Assimina Antonarakou

Department of Geology and Geoenvironment, National and Kapodistrian University of Athens, 157 72 Athens, Greece; voudouris@geol.uoa.gr (P.V.); aantonar@geol.uoa.gr (A.A.)
* Correspondence: cntrinia@geol.uoa.gr

Citation: Drinia, H.; Voudouris, P.; Antonarakou, A. Geoheritage and Geotourism Resources: Education, Recreation, Sustainability II. *Geosciences* **2023**, *13*, 350. https://doi.org/10.3390/geosciences13110350

Received: 18 September 2023
Accepted: 16 October 2023
Published: 17 November 2023

In recent years, the world's fascination with the geoenvironment has experienced a remarkable surge. Concepts such as "geosites", "geoparks", and "geodiversity" have become pivotal in the cultural and economic resurgence of various regions. These terms are not mere geological jargon but are intrinsically linked with the revitalization of local communities, fostering newfound growth and sustainability. At the heart of this burgeoning interest lies the exchange of information, skills, experience, and personnel among the world's significant geosites, which are recognized globally for their geopark activities.

Central to this movement is the concept of geotourism, an alternative form of tourism that enables visitors to delve into the geological wonders of the areas they explore while immersing themselves in the diverse tapestry of natural and human resources. Geoeducation, another pivotal element, finds its nexus in the operational heart of geoparks, as they are deemed ideal settings for educational endeavors.

The content of this Special Issue, entitled "Geoheritage and Geotourism Resources: Education, Recreation, and Sustainability II", covers a diverse range of topics related to geology, geoeducation, geotourism, and the preservation of geological heritage. This Special Issue highlights the growing interest in the geoenvironment and its significance in various aspects of local economic and cultural development. It emphasizes the importance of geoeducation in geoparks and the promotion and preservation of geological heritage and geoethical values.

In this second volume, we continue our journey to assemble a comprehensive collection of papers that elucidate the paramount role of the promotion and conservation of geoheritage in the nurturing of geoethical values and the enhancement of sustainability. Beyond research articles encompassing a wide array of geological heritage approaches, researchers responded to our call to submit papers focusing on novel cases and shedding light on contemporary challenges or unexplored knowledge gaps. Additionally, we welcomed long-term review articles that provide an insightful panorama of the evolution within this field.

Fourteen papers explore a wide array of subjects, including scientific studies of geoheritage, techniques for enhancing geoheritage, the significance of geoheritage in human life, geosciences education for sustainability, sustainable geotourism, the sustainable use of natural and cultural heritage, and the valorization of geoheritage for sustainable and innovative tourism development. These papers present research findings, case studies, and methodologies that contribute to a deeper understanding of how geological heritage can be integrated into education, tourism, and sustainable development.

In their review, Mosios et al. [1] conducted a comprehensive assessment of geoethics in Greece, with a particular emphasis on its presence within the educational system and the Greek geoscientific community. They found that although concerns about geoethics were on the rise in Greece, the incorporation of geoethics into educational curricula was limited across all educational levels. Furthermore, the review exposed a notable absence of initiatives aimed at fostering geoethical thinking and instilling the values associated with geological heritage and its preservation. This lack of emphasis on geoethics highlighted the need for a more concerted effort to promote ethical considerations and responsible practices within both the geoscientific community and broader society in Greece, particularly concerning the safeguarding of geological heritage.

In their comprehensive review, Quesada-Valverde et al. [2] conducted a systematic analysis of global research in geoconservation and geotourism, spanning the years 2011 to 2021. Through an in-depth examination of 169 research papers, their study illuminated several key findings. Notably, it showcased a burgeoning interest in geoconservation and geotourism research on a global scale, with Italy, Poland, Brazil, Russia, and China emerging as leading contributors. The study highlighted a diverse range of geomorphological environments under investigation and identified fundamental methods for assessment and promotion, such as geomorphological mapping, economic valuation, fieldwork, geoheritage management, and documentation. Moreover, the review identified critical resources, including UNESCO Global Geoparks, educational programs, digital tools, geomanagement strategies, and geoitineraries. This pioneering research provides a valuable resource for countries embarking on geoheritage studies, offering insights into successful methods and resources to advance geoconservation and geotourism efforts worldwide.

In their article, Valentini et al. [3] emphasize the significance of geoheritage as a valuable geotourism resource that holds the potential to contribute significantly to the sustainable development of society. They highlight that geoheritage serves multiple purposes, including enhancing knowledge about geological and geomorphological characteristics, facilitating education, raising public awareness about geoconservation, and aiding in landscape protection efforts. However, they note that despite its importance, the understanding of this cultural wealth remains limited, with it often being confined to specialized information accessible to a select few. Recognizing the need for more effective communication, the research group embarked on a comprehensive study focusing on central Italy. Their work involved exploring various communication methods, including publications, social media announcements, conferences, live performances, and field excursions, all designed to emotionally engage the public. The study also involved assessing the emotional response of the audience through a digital survey. The researchers found that utilizing diverse forms of art to narrate the landscape established a personal connection with the audience, effectively conveying scientific and cultural themes such as the origin of geological landforms, the cultural traditions associated with the site, and the challenges related to its protection and conservation. This approach demonstrated the effectiveness of emotional engagement in disseminating knowledge about geoheritage and promoting geoconservation among a broad and diverse audience, transcending social, cultural, and age boundaries.

Herrera-Franco et al. [4] underscore the often-overlooked importance of geodiversity in the broader context of biodiversity, emphasizing that geodiversity, which encompasses geological and landscape variety within a region, plays a pivotal role with far-reaching social, economic, educational, scientific, and technological impacts. They focus their study on Guayaquil, a port city in Ecuador, with the objective of assessing the geoheritage values of the region and proposing guidelines and strategies for raising awareness and promoting the understanding of this geoheritage. The researchers employed a comprehensive methodology that encompassed a historical review of Guayaquil's landmarks, surveys to gauge the local population's perception of the city's geoheritage, data analysis, and the development of geodiversity strategies using computer tools. Their findings revealed that individuals initially engage with areas of interest in Guayaquil due to their biodiversity, but once there, they could acquire knowledge about the city's geology, geodiversity, and urban geotourism. Consequently, the study highlights the imperative role of geoheritage in shaping educational plans, initiatives, and promotion strategies, especially regarding using geotourism as a sustainable means of showcasing the city's historical and scientific significance in the context of sustainability. This research underscores the importance of recognizing and preserving a city's geological heritage and its socio-economic and educational value within the broader framework of conservation and sustainable development.

In their article, Drinia et al. [5] underscore the unique geological heritage of Athens, the capital of Greece, which often takes a backseat to its archaeological and historical wealth. The authors specifically focus on the geological aspects of the Acropolis monuments, including the iconic Acropolis Rock. While the Acropolis is celebrated globally for its cultural and historical significance, it is also a remarkable geological monument. Drinia et al. aimed to quantify and recognize the geological features of such monuments, evaluating their educational, touristic, cultural, and accessibility value. They stress the importance of highlighting these geological aspects for both public and scientific communities, emphasizing their potential for geoeducation and geotourism. The article emphasizes the substantial economic and educational benefits that can be derived from promoting these geosites and raising visitor awareness about environmental, geological, and geoconservation issues. Their study also emphasizes the critical need for better integration of geological heritage with cultural heritage and the importance of strategic educational planning and incorporating geo-environmental education into school curricula.

In their study, Fedorov et al. [6] address the crucial factor of accessibility in the utilization of geoheritage resources. They propose an innovative multi-criterion, score-based approach for evaluating the large-scale accessibility of areas abundant in geoheritage. This approach takes into consideration a range of factors, including external and internal public transportation, road infrastructure, local services (including accommodation options), and general settings, to comprehensively assess accessibility. The researchers applied this method to the Russian South, a region featuring three geoheritage-rich areas: the Lower Don, Abrau, and Mountainous Adygeya. The results of their analysis revealed varying levels of accessibility across these regions, with the Lower Don demonstrating excellent accessibility, while Abrau and Mountainous Adygeya showed a moderate level. Importantly, the study highlights the fact that the mere coexistence of geoheritage-rich areas and popular tourist destinations does not guarantee excellent accessibility. These findings hold significance for shaping effective geoheritage resource policies, as well as for planning research and educational endeavors, such as ongoing geochemical investigations and field educational campaigns in the Russian South. The research serves as a valuable contribution to enhancing the utilization of geoheritage resources while considering practical accessibility challenges.

Farabollini and Bendia [7] emphasize the captivating role of geologists in deciphering the physical landscape, revealing its history through the lens of geological and geomorphological processes that have shaped the Earth's terrain over time. They underscore the critical need to disseminate the extensive knowledge and skills within this field, particularly in the context of promoting geodiversity and advocating for its preservation and promotion. The authors present their work as a means of sharing tangible examples of projects that have come to fruition through collaborative agreements among various public entities in the Apennine region of the Marche Region in Italy. This partnership involves the Geology Section of the University of Camerino, the "Consorzio Frasassi" responsible for managing the Frasassi cave complex, and the municipality of Genga, home to the renowned Frasassi Caves. This joint effort has led to the creation of didactic geological notebooks tailored for school groups visiting the caves and interactive museum laboratories in 3D, which illustrate the geological evolution of the area. This knowledge-sharing initiative is poised to educate communities about the significance of their geological heritage while also serving as an inspiration for similar collaborative endeavors in regions where comparable projects can be replicated, thereby furthering the cause of geological education and conservation.

Ozkaya de Juanas et al. [8] emphasize the significance of accessible palaeontological sites as highly suitable environments for meaningful learning experiences in both formal and non-formal geoscience education. Their project focuses on two closely related Cenomanian–Turonian (Upper Cretaceous) outcrops in the Iberian Peninsula—specifically, the sections of Figueira da Foz in Portugal and Tamajón in Spain. Additionally, they consider the Palaeontological and Archaeological Interpretation Centre of Tamajón (CIPAT) as a key component of this initiative. The researchers leveraged modern technology to create three-dimensional virtual models of fossil samples and the main fossil tracks through techniques such as phase-shift scanning, photogrammetry, and structured light and laser triangulation. These virtual fossils have served as the foundation for developing transdisciplinary didactic activities tailored to various educational levels and the general public. The activities are detailed in file cards, including information about participant age, objectives, multiple intelligences, European Union key competencies, required resources, implementation steps, and additional observations. The overarching goal of this work is to enhance the design and execution of didactic sequences for out-of-school education at these palaeontological sites, fostering the creation of effective transdisciplinary teaching tools and cultivating awareness, values, and responsibility toward geoheritage. This project exemplifies the commitment to promoting geoscience education and the preservation of geological heritage through innovative and accessible educational approaches.

Martínez-Martín et al. [9] underscore the significance of UNESCO Global Geoparks (UGGps) as exceptional regions for fostering educational activities on an international scale. They emphasize the didactic potential, multidisciplinarity, and importance of UGGps in facilitating non-formal and informal educational initiatives. These aspects have garnered significant attention from institutions, organizations, and governments in numerous countries. This interest is so pronounced that the number of UGGps continues to grow, with 177 territories spanning 46 countries currently designated as UGGps. These geoparks diligently work to develop diverse educational activities and proposals aimed at contributing to the attainment of "Quality Education", which is the fourth objective of the Sustainable Development Goals (SDGs) outlined in the United Nations 2030 Agenda. The study focuses on describing the various educational initiatives and activities featured on the official websites of Spanish and Portuguese UGGps, with the goal of assessing their visibility prior to visiting these territories and highlighting their importance within the broader framework of educational development. In essence, the research seeks to shed light on the role of UGGps in promoting quality education and their unique contributions to the field of education through accessible and adaptable educational plans.

Nomikou et al. [10] present Nisyros Geopark, situated in the Southeastern Aegean Sea, Greece, as a candidate for the official UNESCO Global Geoparks designation. The geopark is characterized by remarkable geological, natural, and cultural attributes deeply linked to its volcanic origins. It encompasses an extensive area of 481 square kilometers, comprising the main island of Nisyros, an active volcano, as well as the surrounding islets of Pachia, Strongyli, Pergousa, Kandeliousa, and the marine regions connecting them. Nisyros Geopark boasts 24 geosites and a well-established network of walking trails for visitors. Additionally, the entire area is covered by two internationally designated Natura 2000 areas, highlighting its ecological significance. Beyond its geological wonders, the geopark showcases exceptional archaeological and cultural sites, including fortresses, remnants of ancient settlements, and numerous churches and monasteries, making it a unique destination in the broader Eastern Mediterranean region. The management body and scientific team of Nisyros Geopark have undertaken various initiatives to promote its exceptional geodiversity, including the development of an official website, the creation of a mobile application called "Nisyros Volcano App", the production of informative materials like leaflets and guidebooks, as well as the installation of panels and signs at the geosites, all aimed at enhancing the visitor experience and raising awareness about the geopark's unique features.

Tropeano et al. [11] describe the official nomination of a significant area in Puglia, Southern Italy, as an aspiring UNESCO Global Geopark (aUGGp) by the Italian National Commission of UNESCO in November 2021. This designated area encompasses the northwestern part of the Murge territory, where a Cretaceous sector of the Apulia Carbonate Platform is exposed, as well as a portion of the adjacent Premurge territory, where the southwestward lateral extension of the same platform, flexed toward the Southern Apennines Chain, is overlain by thin Plio-Quaternary foredeep deposits. The exceptional geological uniqueness of the aspiring Geopark, known as Murge aUGGp, lies in the fact that it represents the only in situ remnant of the Adria Plate, an ancient continental plate that has largely been compressed between the Africa and Eurasia Plates. In the Murge area of the Apulia Foreland, Adria's crust remains virtually undeformed, contrasting with other regions of the plate that have been involved in subduction and collision processes. The aspiring Geopark preserves the connection between Adria's crust and mantle, and the Cretaceous evolution of the continent is well-documented in the Murge area through the carbonate succession of the Apulia Carbonate Platform, one of the largest peri-Tethys carbonate platforms. Additionally, the Geopark includes the Premurge area, which showcases the remarkable Plio-Quaternary evolution of the outer Southern Apennines foredeep, characterized by an "anomalous" regional uplift during the middle-late Quaternary period. Despite harboring numerous geological features of international significance, the authors highlight the importance of presenting a regional geological framework for the aUGGp to provide context for visitors before delving into the individual geosites. This approach can enhance the appreciation of the individual geosites within the broader geoevolutionary context of the Murge aUGGp and enrich the geotourism experience.

Triantaphyllou et al. [12] highlight the exceptional geosites of Lemnos, emphasizing that these sites not only possess aesthetic beauty but also hold profound scientific connections to the island's geological history, prehistory, archaeology, mythology, and religious heritage. The richness of Lemnos' geosites, coupled with the abundance of archaeological sites, cultural monuments, and museums, forms the foundation of what the authors term "Geo-Archaeo-Routes." These routes are well-defined geographical paths that can be offered, guided, and followed by tourists. The quantitative assessment of Lemnos' geosites conducted by the authors serves as a valuable decision-making tool, facilitating the sustainable development of Geo-Archaeo-tourism at the local level and laying the groundwork for the creation of these specialized routes. These "Geo-Archaeo-Routes" are particularly suited to environmentally friendly forms of tourism, catering to naturalists, hikers, cultural enthusiasts, and those interested in religious heritage—reflecting the diverse needs of 21st century tourists. The established hiking and road routes on Lemnos Island offer a unique touristic product that combines "nature with culture", providing a holistic geotouristic experience that blends natural entertainment with cultural exploration. In essence, these routes offer a distinctive way to appreciate the geological, archaeological, and cultural heritage of Lemnos while embracing sustainable and eco-friendly tourism practices.

Melelli et al. [13] present a method for revitalizing abandoned mines, whose existence had faded from both the physical landscape and collective memory. They illustrate this approach with two case studies in the Umbria region of central Italy. The selected sites, located in the Upper Tiber River Valley in northern Umbria, were once lignite mines that had been completely erased over time. Given their rural locations, recovering the memory of these places, and repositioning them as geosites posed an intriguing challenge. The authors embarked on a comprehensive process to restore and valorize these abandoned mines. They began by recovering and preserving historical documents related to the Caiperino–Terranera and Carsuga lignite mines, subsequently digitizing these materials. This digital archive served as the foundation for the creation of a geolocalized database within a geographic information system (GIS) environment. Building on this framework, Melelli et al. [13] developed a digital media app enriched with multimedia elements, including video content, 3D models, and augmented reality (AR) features. This innovative app enhances the touristic and educational experiences, offering a dynamic and engaging way to explore these "ghost places" while promoting the cultural heritage of the region. In essence, this method not only revitalizes abandoned mines but also reintroduces them into the collective consciousness, transforming them into valuable geosites with historical and educational significance.

In their article, Iranzo-García et al. [14] conducted research that focused on the intersection of environmental concerns, the preservation of geological heritage, and local development in Spanish geoparks. Iranzo-García et al. recognized the growing interest in environmental problems and the need to protect and manage sites to ensure that geoecological processes remain unaffected by human activities. They conducted research to address these concerns and understand how geoparks could play a role in safeguarding geological heritage. Moreover, the researchers adopted an inductive and qualitative research approach which involves collecting and analyzing data to develop theories and insights rather than testing predefined hypotheses and allowed them to explore the complex relationships between geoparks, private initiatives, and local development in a nuanced manner. The researchers identified and analyzed 48 land stewardship initiatives within 11 of the 15 Spanish geoparks. This involved studying how various organizations and individuals were working to protect and manage the geological and cultural heritage within these geoparks. The study also touched on the presence of nature-based schools within geoparks. It noted that such schools were relatively rare within geoparks, with a notable exception in Central Catalonia. This observation prompted discussions about the potential for integrating nature-based education into geopark operations. They emphasized the positive impact of early years education in natural environments in fostering pro-environmental skills and behaviors that extend into adulthood.

We hope that this Special Issue will serve as a valuable resource for scholars, educators, and professionals interested in the intersection of geology, education, tourism, and conservation, offering insights and practical approaches for harnessing the potential of geological heritage for the benefit of both local communities and the broader public.

Conflicts of Interest: The authors declare no conflict of interest.

References

1. Mosios, S.; Georgousis, E.; Drinia, H. The Status of Geoethical Thinking in the Educational System of Greece: An Overview. *Geosciences* **2023**, *13*, 37. [CrossRef]
2. Quesada-Valverde, M.E.; Quesada-Román, A. Worldwide Trends in Methods and Resources Promoting Geoconservation, Geotourism, and Geoheritage. *Geosciences* **2023**, *13*, 39. [CrossRef]
3. Valentini, L.; Guerra, V.; Lazzari, M. Enhancement of Geoheritage and Development of Geotourism: Comparison and Inferences from Different Experiences of Communication through Art. *Geosciences* **2022**, *12*, 264. [CrossRef]
4. Herrera-Franco, G.; Apolo-Masache, B.; Escandón-Panchana, P.; Jácome-Francis, K.; Morante-Carballo, F.; Mata-Perelló, J.; Carrión-Mero, P. Perception of the Geological-Mining Heritage to Promote Geotourism in Guayaquil, Ecuador. *Geosciences* **2022**, *12*, 322. [CrossRef]
5. Drinia, H.; Tripolitsiotou, F.; Cheila, T.; Zafeiropoulos, G. The Geosites of the Sacred Rock of Acropolis (UNESCO World Heritage, Athens, Greece): Cultural and Geological Heritage Integrated. *Geosciences* **2022**, *12*, 330. [CrossRef]
6. Fedorov, Y.A.; Mikhailenko, A.V.; Ruban, D.A. Large-Scale Accessibility as a New Perspective for Geoheritage Assessment. *Geosciences* **2022**, *12*, 414. [CrossRef]
7. Farabollini, P.; Bendia, F. Frasassi Caves and Surroundings: A Special Vehicle for the Geoeducation and Dissemination of the Geological Heritage in Italy. *Geosciences* **2022**, *12*, 418. [CrossRef]
8. Ozkaya de Juanas, S.; Barroso-Barcenilla, F.; Berrocal-Casero, M.; Callapez, P.M. Virtual Fossils for Widening Geoeducation Approaches: A Case Study Based on the Cretaceous Sites of Figueira da Foz (Portugal) and Tamajón (Spain). *Geosciences* **2023**, *13*, 16. [CrossRef]
9. Martínez-Martín, J.E.; Ester Mariñoso, P.; Rosado-González, E.M.; Sá, A.A. Prospective Study on Geosciences On-Line Education: UNESCO Global Geoparks in Spain and Portugal. *Geosciences* **2023**, *13*, 22. [CrossRef]
10. Nomikou, P.; Panousis, D.; Nikoli, E.; Antoniou, V.; Emmanouloudis, D.; Pehlivanides, G.; Agiomavritis, M.; Nastos, P.; Cieslak-Jones, E.; Batis, A. Nisyros Aspiring UNESCO Global Geopark: Crucial Steps for Promoting the Volcanic Landscape's Unique Geodiversity. *Geosciences* **2023**, *13*, 70. [CrossRef]
11. Tropeano, M.; Caldara, M.A.; De Santis, V.; Festa, V.; Parise, M.; Sabato, L.; Spalluto, L.; Francescangeli, R.; Iurilli, V.; Mastronuzzi, G.A.; et al. Geological Uniqueness and Potential Geotouristic Appeal of Murge and Premurge, the First Territory in Puglia (Southern Italy) Aspiring to Become a UNESCO Global Geopark. *Geosciences* **2023**, *13*, 131. [CrossRef]
12. Triantaphyllou, M.V.; Firkasis, N.; Tsourou, T.; Vassilakis, E.; Spyrou, E.; Koukousioura, O.; Oikonomou, A.; Skentos, A. "Geo-Archaeo-Routes" on the Island of Lemnos: The "Nalture" Experience as a Holistic Geotouristic Approach within the Geoethical Perspective. *Geosciences* **2023**, *13*, 143. [CrossRef]

13. Melelli, L.; Palombo, M.; Nazzareni, S. Ghost Mines for Geoheritage Enhancement in the Umbria Region (Central Italy). *Geosciences* **2023**, *13*, 208. [CrossRef]
14. Iranzo-García, E.; Hueso-Kortekaas, K.; Fansa-Saleh, G. Conservation and Education in Spanish Geoparks: Exploratory Analysis of Land Stewardship Experiences and Valuation Proposal through Outdoor Education. *Geosciences* **2023**, *13*, 276. [CrossRef]

Article

Enhancement of Geoheritage and Development of Geotourism: Comparison and Inferences from Different Experiences of Communication through Art

Laura Valentini [1,*], Veronica Guerra [2] and Maurizio Lazzari [3]

[1] Department of Biomolecular Sciences, University of Urbino Carlo Bo, 61029 Urbino, PU, Italy
[2] Department of Pure and Applied Sciences, University of Urbino Carlo Bo, 61029 Urbino, PU, Italy; veronicaguerra@live.com
[3] Institute of Cultural Heritage Sciences, CNR-ISPC, C/da S. Loja, 85050 Tito Scalo, PZ, Italy; maurizio.lazzari@cnr.it
* Correspondence: laura.valentini@uniurb.it

Abstract: Geoheritage is a geotourism resource that could lead to the sustainable development of society, and could contribute to information on geological and geomorphological characteristics, education, public awareness on geoconservation, and landscape protection. Understanding this cultural wealth is still limited, since the information available is often specialized. Communication addressed to the emotional sphere can profoundly engage people, and technical information can be transmitted more effectively. For several years, our research group dealt with geoheritage and geotourism in central Italy. This work aimed to investigate different communication methods, such as publications, announcements through social media, conferences, live performances, and field excursions aimed at the emotional involvement of the public. Furthermore, the emotional response of the public was explored through a digital survey. These communication methodologies enabled a broad public reach, without any distinction based on social origin, cultural background, or age. Narrating the landscape through different forms of art creates a personal relationship with the audience. The emotional participation of the people demonstrates the effectiveness of the method in transmitting scientific and cultural themes, such as the origin of the geological landforms, the culture and the traditions that the site generated, and the problems pertaining to its protection and conservation.

Keywords: geoheritage; communication methods; art; public emotional involvement; geotourism

Citation: Valentini, L.; Guerra, V.; Lazzari, M. Enhancement of Geoheritage and Development of Geotourism: Comparison and Inferences from Different Experiences of Communication through Art. *Geosciences* **2022**, *12*, 264. https://doi.org/10.3390/geosciences12070264

Academic Editors: Hara Drinia, Panagiotis Voudouris, Assimina Antonarakou and Jesus Martinez-Frias

Received: 17 May 2022
Accepted: 28 June 2022
Published: 30 June 2022

Publisher's Note: MDPI stays neutral with regard to jurisdictional claims in published maps and institutional affiliations.

1. Introduction

For the last few decades, there was a growing scientific interest in issues related to geoheritage, and many initiatives emerged worldwide. Geoheritage is an important natural resource that feeds the spread of geotourism around the world, and is rightly defined by Brocx and Semenniuk [1] in this way: "Geoheritage encompasses global, national, statewide, and local features of geology, at all scales that are intrinsically important sites or culturally important sites offering information or insights into the evolution of the Earth; or into the history of science, or that can be used for research, teaching, or reference".

Geoheritage can represent the first step towards the application of strategies useful for geoconservation and the enhancement of this inexhaustible natural resource, leading to the sustainable development of society [1–7]. Geodiversity is fundamental in determining the characteristics of habitats, and contributes to shaping the scenic character of landscapes, which, most times, is the one that is more easily retained by our memory [8]. An understanding of geoheritage could contribute to knowledge on geological and geomorphological characteristics, education, and public awareness on geoconservation and landscape protection issues. Moreover, the study of geosites promotes the process of developing laws on this topic, and the creation of databases useful for a better knowledge and conservation of geological heritage [9,10].

Unfortunately, awareness of this cultural wealth is still limited because the information available to the public is often specialized, and lacks significant emotional power. We should be aware that scientific approaches and languages are poorly understood by the general public because they are too technical, and the messages they convey are not absorbed. On the other hand, communication methods that directly address the emotional sphere seem to engage the public in a deep and meaningful way. Innovative strategies using the arts, gaming, theatrical improvisation, and interactive exhibitions are being developed [11–14]. If people are sensitized, technical information can be communicated more effectively. This work is aimed at testing alternative and innovative ways to engage non-expert audiences.

For several years, our research group worked on geoheritage and geotourism, following different communicative methods in areas of central Italy, particularly in the Marche and southern Emilia-Romagna regions. These areas were selected for their undoubted scientific and perceptive aesthetic value, and for the occurrence of many geosites of geomorphological interest, i.e., geomorphosites still untouched and characterized by various genetic and evolutive processes, and by an extraordinarily rich cultural and ecological context [15–18]. We emphasize that the term geomorphosite (contraction of "geomorphological site") is used in the meaning of "landforms, active or inherited, having particular importance for the comprehension of the history of the Earth's and its present or future evolution" [19–22]. The cultural landscape has to be considered as an additional value, defined by the multiplicity of elements that characterize a place, such as cultural, ecological, economic, and aesthetic criteria [18,23–25]. In the direction of sustainable development of geotourism, the accessibility of the sites, and the presence of accommodation, protected areas (natural reserves or parks), and teaching facilities were also investigated. Finally, vulnerability and the need for protection of geosites were underlined.

In this work, different communication methods, such as using forms of arts employed in popular publications, announcements through social media, conferences, live performances, and several modes of field excursions, always appealing to the emotional involvement of the public, were illustrated and compared. However, it should be noted that a quantitative monitoring of the incidence of these communication tools was carried out only during normal excursions, through on-site and online questionnaires, and interviews carried out with geotourists to assess their emotional, aesthetic–perceptive, and visual impact. Future forms of monitoring, not present in this paper, will consider the quantitative evaluation of the incidence of the coupling of musical, poetic, and serious game arts to traditional geotourist excursions. A touristic proposal that uses natural landscapes in a didactic or entertaining way, based on imagination and emotion about Earth's history may, in fact, provide interesting developments [26], especially considering that tourists and day visitors are attracted to crucial features such as environment and nature, educational tourism, culture, history, events, entertainment, and fun [27]. There is significant evidence that emotional events can be stored more profoundly, and memorized for a longer time, than neutral events. In addition, this research concerns the identification of emotional activations of the public according to the geological process, through a digital survey.

The purpose is to combine and stimulate an interdisciplinary debate on various interconnected themes to the investigated sites, and reflect on the different interactions between humans and the environment, which, over time, favored and induced local and global changes in the physical landscape.

The main goal is to educate by creating a new perception of geological landscapes, starting with their physical beauty, then building on scientific research in cooperation with the arts, improving what we know about their problems and weaknesses, and addressing their culture and other strengths.

In addition, this work could help to inspire those geoscientists who have never considered collaborating with artists. The development of a unified culture would help to involve everybody in a more profound knowledge of the landscapes of the Earth and their delicate and complex mechanisms, preserving the wealth of our planet for future generations.

2. Investigated Areas

The Marche and southern Emilia-Romagna regions (central Italy) represent a restricted area between the Adriatic coast and the Apennine chain (Figure 1), which boasts, in a limited space, a wide diversity of landforms: low sandy coasts, promontories overlooking the sea, rounded and greenish hills, fluvial terraces, spectacular gorges, rugged mountains dislocated by active fault systems, hot springs with thick travertine deposits, and tectonic thrusts. This variety is a consequence of the geological history of the Apennine chain [28], which produced extraordinary contrasts of physical forms in a restricted space.

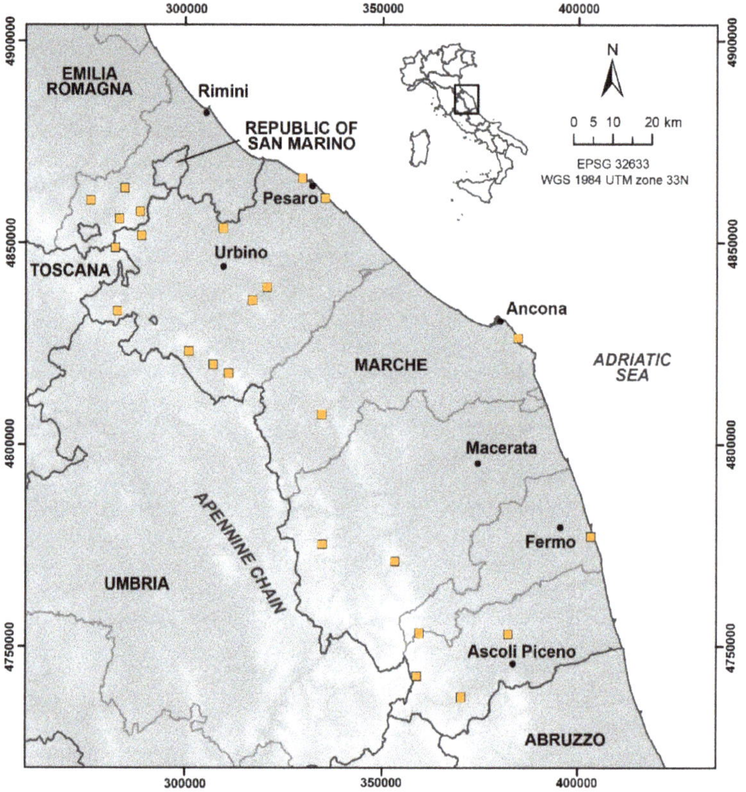

Figure 1. Study area and location (yellow squares) of the investigated sites (base map DEM from [29]).

Moreover, located on the border between the Marche and Emilia-Romagna regions, the Valmarecchia and Montefeltro areas are well-known by geologists all over the world for the origin and evolution of the Valmarecchia nappe, which consists of stacked slices of Ligurian and Epiligurian rocks over-thrusting Tuscan and Umbro–Marchean units, and producing unique geological landscapes [17,30,31]. Historians and artists also know these areas: they were sites of significant human settlements, in some cases since the prehistoric age, leaving us with testimonies of great cultural interest.

The altitude of this varied landscape (coasts, hills, and mountains, reaching a maximum height of 2476 m a.s.l. in about 9500 km^2) also produced a significant floristic and faunal heritage. These regions, moreover, still show intact areas where the landscape, culture, and traditions provide an excellent opportunity for didactic and touristic development [15,32,33].

The territory is characterized by landforms whose formation is due to the complex interaction between endogenous (tectonic) and exogenous (erosive) processes, which, since the Neolithic period, were decisively overlaid by anthropogenic action of landscape shaping and transformation.

The map in Figure 1 shows the study area and the location of the investigated geosites.

3. Communication Methods

Geosciences are traditionally conveyed during conferences or dedicated events as formal, one-way presentations. The official scientific research, as presented in international scientific conferences and journals, is indispensable, but this approach remains mainly addressed to academic or specialistic audiences.

Despite this, one of the main objectives of the scientific research, especially in topics where the interaction with human activities is strong, as in the case of geological and environmental issues, is to spread interest in important scientific themes, trying to reach people of all ages and social backgrounds. Simplifying scientific concepts, and making them attractive to the public, is essential to face challenges around the territory and the environment. In fact, geoscientists use too much specialized language, while non-geoscientists are not predisposed to understand issues that they consider to be the domain of specialists.

Some studies try to understand how geoscientists and non-geoscientists perceive geological concepts and activities, and their cognitive and affective responses. They conclude by remarking that geoscientists are increasingly required to incorporate emotions into communication efforts engaging with non-geoscientists [34,35].

In this view, it appears to be very useful for scientists to collaborate with artists, implementing new tools that can positively influence the emotional sphere, and capture people's attention. During the last decades, there was great attention focused on the communication of even complex content to the general public in the scientific world, to raise awareness of important scientific and environmental topics [11,18,34,36,37]. Art is an essential vehicle of content, involves people profoundly and directly, and has been used for a long time with significant impact.

This paper presents the experiences carried out by the research team in the last decade regarding different communication methods and different approaches to addressing the public (Figure 2). Alongside the scientific approach, we propose a method to communicate using art forms, such as music, poetry, and theatrical improvisation techniques, to encourage people to learn more about landscapes by integrating their origins and physical aesthetics with their natural, cultural, and artistic heritage.

The former idea was to cross the different disciplines of Earth science, poetry, and music, applying them together to specific sites, and producing emotional experiences where encounters and interplay between the different languages became an expression of the place. Since the beginning of our project, our aim was to work on the emotional involvement of the audience, but in recent years we focused on collecting feedback data. From 2021 onwards, we asked the public who actively participated in the themed excursion events to fill in questionnaires for a more accurate analysis.

This experience started in 2014 with a team named *TerreRare*, which means rare earth elements but also rare landscapes, composed of researchers and artists with different skills: two geologists, a writer/poet, a musician/musicologist, an actor, and a video-maker. Live events were initially proposed in central Italy and, parallel to these, the idea of the project was disseminated in national and international scientific conference [38–41].

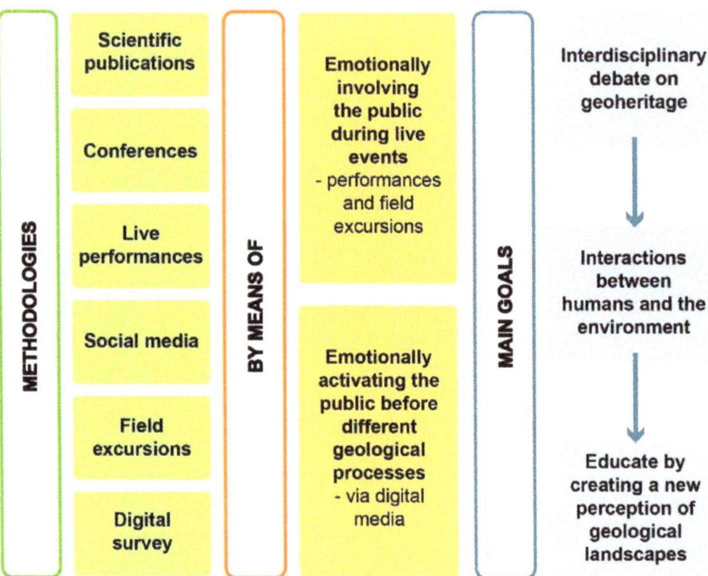

Figure 2. Methodological flow chart.

In the meantime, a project dedicated to 20 geosites in the Marche Region (central Italy) was developed. The project results are included in a book in the native language [15], with one DVD and two CD attached. Starting from the geological and morphological features of these 20 sites, the fundamental concepts of their genesis and evolution were identified. A simple scientific language, conceived for lovers of the territory from all cultural backgrounds, was used. An itinerary with some stops with significant site views was indicated for each location. The layout of the landscape is the result of complicated, unpredictable, and often disastrous events, and originates not only from geological processes, but also from the climate, and human intervention, underlying the delicate fragility of the environment. In a second step, the fundamental concepts previously identified were synthetized in a few key words translated into the poetic and musical languages, trying to activate emotional involvement and paths, ultimately stimulating a deeper understanding of the landscape. The poet suggests original poems, while the musician reproduces the site's emotional impact by searching for a suitable piece of ancient music. Finally, for each site, some aspects of natural, historical, or cultural interest are suggested. Some videos, realized in these places, are contained in the enclosed DVD (the book is also available in an interactive form). The pieces of music, recorded for the project, can be found on the tied CDs [16,18]. A summary of the book's contents is also presented on a website (https://www.terreraremarche.it, accessed on 28 February 2022). The 20 geosites can be visited virtually, and the geological history of these places, the information on the morphologic features, and the cultural aspects around these places can be explored. The poems and the pieces of music are also the soundtracks of videos that, by using the best views of these places, creatively interpret science and nature through art.

The project was presented several times in popular events, always arousing participation and interest, but this work's most effective communicative method is through live events. Since the beginning of this project, the public were addressed directly through shows that combine scientific communication in popular language with the reading of poems, and the live performance of music. A total of ten live performances were organized, sometimes near one of the proposed geosites, sometimes outside the region, with the aim of promoting the richness of the Marche region. The shows are usually dedicated to five or six places and their main geological and geomorphological aspects. The speaker uses conversational language, focusing on the most interesting aspect of the genesis, and how the morphology influenced the history and culture of the site. Finally, the speaker guides the public, and identifies keywords that represent the link between science, music, and poetry. The musical and poetic performances are conducted in front of a large screen with projections of images and videos of the places (Figure 3). The result is the total emotional participation of the public in the places.

Figure 3. Images from one of *TerreRare* live events that combine scientific communication with the reading of poems, and the live performance of music. The musical and poetic performances are conducted in front of a large screen with projections of images and videos of the places, resulting in a total emotional participation of the public.

Communication to the public through social media was not overlooked: the widespread interest in this project is testified by a total of about 30,000 views (see the number of views for each video at https://www.facebook.com/search/videos/?q=terrerare, accessed on 28 February 2022), and 950 followers of the Facebook page.

From 2020, the project *Paesaggeo* (from contraction of the Italian words *paesaggio* and *geologico*, i.e., geological landscape) was founded, for experimenting with online and live actions concerning the emotional involvement of the public. To sum up, the aim was to investigate the involvement of users with conventional and newly proposed means, such as music, poetry, and field experiences. *Paesaggeo*'s trekking proposal was realized during the summer of 2021, including four different approaches to disseminating geoheritage values (i.e., classic guided tours, exploring experiences, direct interaction with the public with narrative and theatrical improvisation, and the emotional engagement of users through music and art). *Paesaggeo*'s experiences were carried out as part of a doctoral research at the University of Urbino, which selected the Valmarecchia and Montefeltro areas as test locations to experiment with different communicative methods, engaging directly with the public. This method involved participants from several perspectives, interacting with their scientific curiosity, and inspiring a sense of wonder and joy or other emotions.

An online survey was also created within the project. It aimed to better understand the existing relationship between geological landscapes and human emotions, or the perception of geological landscapes. The survey was conceived with the help of a psychologist—an expert in ecopsychology—to collect people's emotive responses regarding photographs of different geological landscapes. The answer can be selected from the following multiple options:

a. Desire, hope (joy/serenity/interest)—positive activation of the user;
b. Enthusiasm, optimism (love/gratitude/freedom/passion)—positive activation of the user;
c. Boredom (indifference/detachment)—deactivation of the user;
d. Nostalgia (melancholy/resignation/disappointment)—deactivation of the user;
e. Worry, concern (anxiety/sadness/frustration/overwhelm)—negative activation of the user;
f. Pain, anger (discomfort/fear/helplessness)—negative activation of the user.

For each photograph, another question was asked: "Would you like to see it live?" After viewing 14 geological landscapes, some general information was asked to obtain basic knowledge about people responding (age, origin). The first means of popularization of the online survey was the creation of an Instagram profile (@paesaggeo https://www.instagram.com/paesaggeo/, accessed on 28 February 2022), with the intent to illustrate and promote the geological landscape of the Valmarecchia and Montefeltro areas. A unique graphic design was conceived to give the project identity, thus, making it recognizable and facilitating its positioning in users' minds. The profile was also used, among others, to promote live events performed throughout the summer of 2021, as experiences of enhancement of the geological landscape directly in the field. During these events, the effectiveness of this communicative method through different kinds of experiences is proven. There are, in fact, several types of users who like to get involved in guided tours in nature; a total of 85 trekkers were engaged to promote four different kinds of geologically themed treks that could stimulate the emotional field of participants differently. People were brought to have a first-hand experience of the geological heritage thanks to the support of the "Sasso Simone and Simoncello Park", who promoted 16 appointments as part of his hiking calendar.

The titles and contents of the four kinds of appointments are listed as follows:

- *Let's explore a Geolandscape.* The environmental guide brought the users out of the main trails to find geological treasures (such as minerals in badland areas) and hidden processes in the wildest areas. Users are invited to explore, research, and touch by hand the geological heritage, with an approach that encourages curiosity and drives them to experiment their limits.

- *A story in the Geolandscape*, in which gamification dynamics characterize the experience. This modality is a sort of time travel, through which the users live a story in the first person, set in climatic and historical-cultural contexts of the past, surrounded by realistic characters. Participants are called upon to interact with the story, and the characters involve the audience with theatrical improvisation techniques. This experience was realized in collaboration with the informal group *Malafeltro*, which deals with the unconventional valorization of territories through games and interacting with the public. Each site was narrated through a verisimilar story, based upon historic knowledge, about the interaction between the geological landscape and human settlements.

- *Let's listen to the Geolandscape*, in which the geological heritage is expressed through the languages of music and poetry. The poems were chosen from famous or, when possible, local authors, and prose testimonies or chronicles related to the geological events involved in the excursion were also read. The musical pieces were expressly composed by a local musician, a young guitarist who was previously brought to the selected sites and made aware of their geological features and specific content to translate into music. He then composed and performed the pieces live, halfway through the hike.

- *Trekking in a Geolandscape* is a classic excursion in nature trails, accompanied by an environmental guide, which explains in a traditional way the salient geological features of the various landscapes, accompanied by explanations and examples. This was conceived to be the least emotionally engaging experience, or a "control" modality, to compare the other three kinds of excursion. This expectation is proven wrong, as is further discussed.

At the end of the treks, people were asked to answer a survey about their feelings during the experience. The collection of these additional surveys allowed us the chance to compare how the four different kinds of experiences engaged with the emotional field of the participants.

The promotion of the live events was performed via social media (@paesaggeo and @parcosassosimonesimoncello Instagram accounts, and Parco Sasso Simone e Simoncello Facebook page, which counts more than 9000 followers), and through the official website and newsletter of the Natural Park (counting more than 4000 subscribers). The events were also promoted with monthly paper flyers distributed in the local provinces (Rimini and Pesaro-Urbino).

4. Discussion

This work deals with multidisciplinary and interactive approaches to promote the communication of rigorous and complex scientific content related to geoheritage, combining traditional scientific communication with other languages, such as poetry, music, and theatrical improvisations, throughout popular publications, announcements through social media, conferences, live performances, and across different modes of guided experiences in field excursions.

Communication that addresses the emotional sphere is recognized as being much more effective than traditional communication methods, since it can engage the observer in a profound and passionate way. This experience with the emotional involvement of the public in geoheritage promotion produces encouraging results. A story about the origins of a landscape and its fragility through different forms of art often results in an intimate relationship between the public and the place.

We followed this route by proposing the 20 geosites included in the book previously described [15], as well as in the attached DVD, where music, poems, and videos are combined to illustrate a place. The content of the book, summarized in a website, was reached by visitors who sometimes contacted us to visit these places or for more information about the sites.

The live performances attracted a wide and varied audience, mostly without a scientific background, and highlight the great interest of the people in the geological heritage, and its relationships with the environment, human activities, and culture. These performances are arranged following the multidisciplinary communication method described above, combining science, projections of images and videos, readings of poems, and live musical executions (Figure 3).

The response from the public (audience feedback) is always high, although not easy to quantify, as the early stages of this research do not include the collection of quantitative data on the kind of audience and the relative perception achieved.

Paesaggeo was then conceived, with great attention paid to collecting information on the effectiveness of the method. As a first step, we tried to identify the emotional activations of the public in relation to the geological process, through a digital survey. This online survey about geological landscape perception was promoted through Instagram only, from April to December 2021, an activity that led to only 31 answers. Most of the answers we collected were instead obtained after sending the poll personally to all people who took part in the summer excursions, and to friends and colleagues, until reaching a total of 165 responses.

The results of the survey have been compared by summarizing them in four image groups, listed as follows:

1. Warm-colored geological landscapes: Casteldelci cliff; Colorìo marls; Sasso Simone cliff; Tausano rock-fall; Mt. Pincio, and Aquilone (Figure 4);
2. Cold-colored geological landscapes: honeycomb erosion of Mt. Perticara; San Leo and the Tausani cliff; Maiolo cliff and badlands; Costa dello Speco and Marecchia River; Pratieghi marls (Figure 5);
3. Landslides: small landslide with a tree; active rock-fall of Maiolo; moving boulders of Sasso Simone (Figure 6);
4. Quarry sites: San Giovanni in Galilea quarry; Mt. Ceti quarry (Figure 7).

Figure 4. Warm-colored landscapes of *Paesaggeo* online survey: (**A**) Colorìo marls; (**B**) Casteldelci cliff; (**C**) Tausano rock-fall, and Mt. Pincio and Aquilone; (**D**) Sasso Simone cliff.

The last two groups (landslides and quarry sites) were the first to be established, as they represent very focused portraits of two different, but impressive, phenomena. The other images represent wider areas, or other types of processes (i.e., a river meander, honeycomb erosion, cliffs), and they are divided into two groups by estimating the chromatic characters of the images.

The results are presented here in a merged mode, obtained by summing and putting into percentage the results obtained for each group, and graphically shown in Figure 8:

1. Warm-colored geological landscapes produce the highest positive activations in users, with 80.61%. Deactivation reaches 12.12%, while negative activation is equal to 7.27%. A total of 91.06% of the users would like to see those kinds of geological landscapes, 4.39% of users declare they do not want to, and 4.55% say they do not know.
2. Cold-colored geological landscapes have a slightly lower percentage of positive activations, with 72.49%. Deactivation increases to 16.36%, and negative activation is perceived by 11.15% of the compilers. A total of 84.12% of the users would like to see these geological landscapes live, 7.27% would not, and 8.61% do not know.
3. The images picturing landslides obtain a total of 47.68% of positive activations in users, 18.38% cases of deactivations, and 33.94% of negative activation. Nevertheless, 74.34% of compilers declared that they would like to see that geological landscape live, 15.15% do not want to, while 10.51% do not know.
4. Quarry sites are the most negatively perceived geological landforms in the survey. They receive 43.64% for negative activations, 25.76% for deactivations, and just 30.6%

for positive activations. Overall, 60.3% of users would, however, like to see them live, 27.88% said they do not, and 11.82% said they do not know.

Figure 5. Cold-colored landscapes of *Paesaggeo* online survey: (**A**) honeycomb erosion of Mt. Perticara; (**B**) Pratieghi marls; (**C**) Maiolo cliff and badlands; (**D**) Costa dello Speco and Marecchia River; (**E**) San Leo and the Tausani cliff.

Figure 6. Landslides of *Paesaggeo* online survey: (**A**) active rock-fall of Maiolo; (**B**) moving boulders of Sasso Simone; (**C**) small landslide with a tree.

Figure 7. Quarry sites landscapes of *Paesaggeo* online survey: (**A**) San Giovanni in Galilea quarry; (**B**) Mt. Ceti quarry.

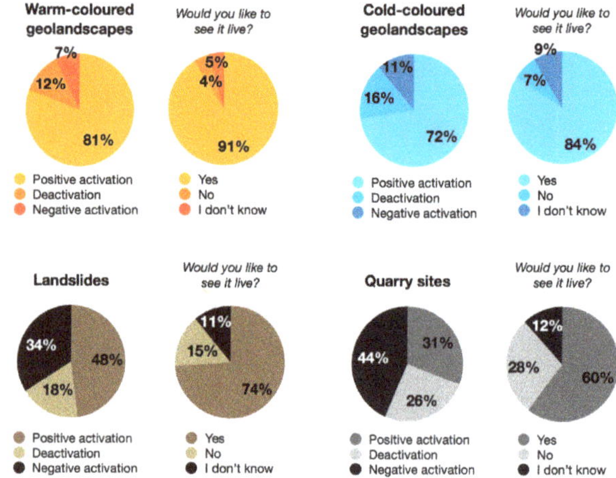

Figure 8. Summarized results from *Paesaggeo*'s online survey.

As our daily lives are littered with pervasive, and sometimes invasive, technologies, the importance of experiencing nature first-hand seems more relevant than ever. It is expressed as a renewed interest on the part of people to live in closer contact with the natural environment. Therefore, it was decided to experiment with disseminating geomorphological heritage concepts through a series of dedicated events. They took place on four weekends from May to August 2021, at Sasso Simone and Simoncello, Mt. Ercole and Mt. San Silvestro, Maiolo cliff and badlands, and Maciano hill and Scavolino palaeo-lake geomorphosites, located in the Valmarecchia and Montefeltro areas.

On Saturdays, *Let's explore a Geolandscape* was proposed in the mornings and early afternoons, while *A story in the Geolandscape* mode took place in the later afternoon and evening. On Sundays, the mornings and early afternoons were dedicated to *Trekking in a Geolandscape*, and the later afternoons and evenings to *Let's listen to the Geolandscape* (Figure 9). Mealtimes were included in all the excursions, to allow the group a moment of sharing, both within the group, and with the guide and the other accompanying persons.

Figure 9. Pictures from *Paesaggeo*'s excursions: (**A**) Trekking in a Geolandscape; (**B**) A story in the Geolandscape; (**C**) Let's explore a Geolandscape; (**D**) Let's listen to the Geolandscape.

At the beginning of each experience, the participants were given a general context of the research, including definitions of geosites, geomorphosites, and geological heritage. They were also asked about the possibility of sending an evaluation survey concerning the emotional involvement experienced during the excursion, and some other contents of the research. The questionnaires (in native language) are available at the following links: Sasso Simone and Simoncello https://forms.gle/mgKjzm157REKUyK99; Mt. Ercole and Mt. San Silvestro https://forms.gle/PJDLx2FrYSXX6cGDA; Maiolo cliff and badlands https://forms.gle/Usoe6v8MF4MwuGa16; Maciano hill and Scavolino palaeo-lake https://forms.gle/oP6vycoHN1ExzWB46 (links accessed on 20 June 2022).

The surveys' outcomes show how the public that engage in such experiences agree on the efficacy of emotional engagement as helpful in remembering geological content. The first four questions aim to obtain the user's general opinion about geosites, and how hiking experiences enhance them. The answers make clear how positive the perception of geological heritage and geologically themed treks is in the groups, and how this kind of experience triggers a curiosity to make similar ones in the future: 100% of the participants think it is important to enhance sites of geological interest through geologically themed excursions; 98.82% believe that the selected geosites should be enhanced through such experiences; 50.59% of the users had never participated in geologically themed tours, led by a hiking guide specialized in geology; and 100% of participants think they will take part in other geological excursions in the future.

From the polls, it follows that 16.47% of participants (14 out of 85) choose the *Trekking in a Geolandscape* mode, and the same percentage choose the *Let's explore a Geolandscape* mode. On the other hand, *A story in the Geolandscape* and *Let's listen to the Geolandscape* appointments reach 34.12% (29) and 32.94% (28 out of 85) of the total participants, respectively, showing how a highly interdisciplinary approach interests more people. All the participants (100%) feel that the trek mode they selected effectively conveys the geological content of that place.

Figure 10 shows how, on a scale from 1 to 5, most of the users feel highly emotionally involved, and how the scores are distributed among the different modes of excursions. Differences are not detectable, as an emotional engagement is probably conveyed simply by spectacular geological landscapes, or the excitement of living experiences in nature, in groups, or in the first person.

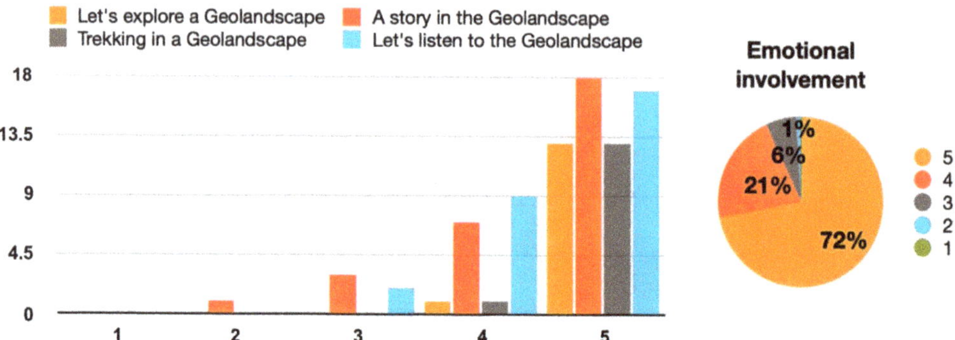

Figure 10. Emotional involvement of participants in *Paesaggeo*'s excursions; the *x*-axis indicates the level of emotional involvement; the *y*-axis shows the number of responses.

Almost all the participants think that enhancing the territory, including emotional engagement, could effectively transmit geological content (98.82%), and the same percentage would be curious to try the other excursion modes (84 out of 85 users).

Finally, some general information is asked to know more about the public age and origin, and how they came to know about the excursions. The results, summarized in Figure 11, show the involvement of different age groups, coming mostly from the Valmarecchia and Montefeltro areas, San Marino, or the surrounding Emilia-Romagna, Marche, and Toscana regions, thus, expressing the involvement mainly of locals and inhabitants, and not tourists. Also, we learned how most of the public came to know about the project: mostly through word-of-mouth, Facebook, the Instagram accounts of Parco Sasso Simone and Simoncello, or a personal invitation by the hiking guide. Lastly, an open question was asked to the users, regarding comments or suggestions: many answers show sincere appreciation, and the hope that similar events could be proposed in the future, also underlying how "the dissemination of scientific content through art is an effective strategy that should be strengthened".

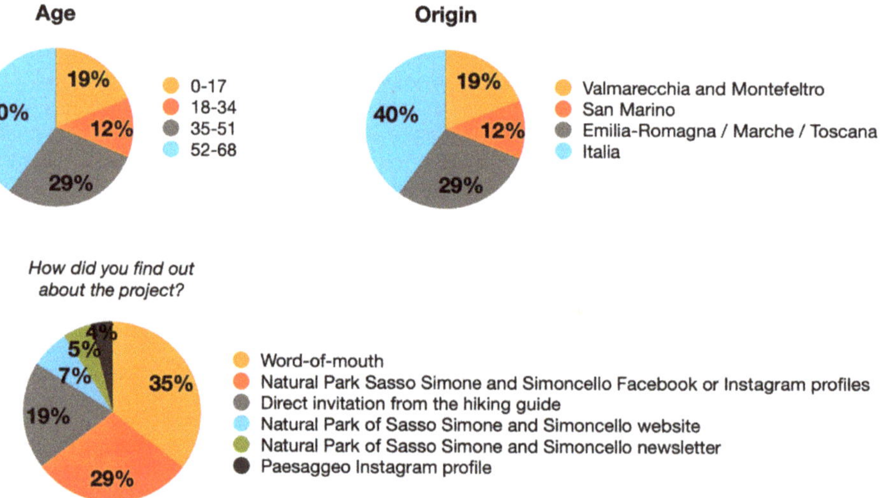

Figure 11. Age and origin of *Paesaggeo*'s excursions participants, and how they learned about the project.

In this paper, it was possible to experiment with the active engagement of the participants both during the excursions of the geotourism proposal *Paesaggeo*, which involved a total of 85 trekkers in the summer of 2021, and thanks to the 165 people who completed the online survey.

Four pictures in the online poll portrayed two geomorphosites enhanced by the live experiences (i.e., Sasso Simone and Simoncello, and Maiolo cliff and badlands), thus, making it possible to show how the same geological landscape can be perceived differently if we compare its digital and live views. In fact, in all the case studies, a live view produces positive feelings different from the digital image, as shown by the following results:

- Maiolo badlands—positive activation from 62.43% (digital) to 94.44% (live); deactivation from 16.36% (digital) to 5.56% (live); negative activation from 21.21% (digital) to 0% (live);
- Southern cliff of Sasso Simone and Simoncello—positive activation from 79.39% (digital) to 91.67% (live); deactivation from 7.88% (digital) to 4.16% (live); negative activation from 12.73% (digital) to 4.16% (live);
- Active rock-fall of Maiolo cliff—positive activation from 69.09% (digital) to 88.88% (live); negative activation from 20.61% (digital) to 5.56% (live); deactivation from 10.3% (digital) to 5.56% (live);
- Boulders of Sasso Simone—positive activation from 56.36% (digital) to 87.5% (live); deactivation from 26.06% (digital) to 12.5% (live); negative activation 17.58% (digital) to 0% (live).

Therefore, it is assumed that a live view of geological heritage results in a general, more positive perception from non-experts, which can increase if the portrayed phenomena are accompanied by explanations of hiking guides, and positive emotional involvement.

5. Conclusions

In the last decade, our experiences with geoheritage communication led us to approach the public in a progressively more conscious way. In this paper, we focus the attention and research activity on improving the scientific communication strategy in a geotourist context. The tools used to make scientific concepts related to the Earth sciences more accessible show that communication through classical scientific language should only be used with an expert audience; otherwise, it is difficult to achieve a proper involvement and understanding of the topics. So, simple and popular language is suitable for events reserved for a heterogeneous and non-expert audience. In particular, in this research, different communication methods, such as publications, announcements through social media, conferences, live performances, and field excursions appealing to the emotional involvement of the public, were adopted. The monitoring of the public emotive answer and agreement was carried out by a questionnaire onsite, and an online survey on Instagram.

The performances with poetry reading and live music (ten live events were organized, sometimes held near the proposed geosites, sometimes in places outside the region) show how the introduction of language linked to art can deeply capture the attention of the public, making even complex scientific content more accessible to a non-expert audience. The live music performances, for example, involved a large and varied audience, sensitive to nature and art, and eager to learn about geology, landscapes, and their relations with human activities.

The immersion in the sites directly in the field (sixteen thematic excursions were organized in some of the case studies) highlight how the perception of a place is much more positive if the experience is lived in the first person, directly on the site. The feedback obtained from the participants in the excursions confirms, once again, the importance of using a multidisciplinary method linked to games, music, acting, reading stories related to the place, and improvised theatrical activity, in the mode of communication.

The emotional involvement of the users highlights the effectiveness of the method, which was used to convey not only the pure geological content, but also the cultural and traditional content, and the possibilities of protection and conservation of the sites.

Comparing these experiences with conventional communication methods, and the communication method used here, developed by integrating it with art forms, it is worth noting that people show a stronger interest in the second case, contacting the organizers after live events to ask for details about the projects, or more information about the geosites or how to visit them. This behavior is reflected in increased interest and appreciation in our social media, especially after the live events.

Experiences with the active engagement of participants during geotourism excursions show how the same geological landscape can be perceived differently when comparing its digital and live views. A live view of the geological heritage results in a more positive overall perception by non-experts, which increases if the phenomena portrayed is accompanied by explanations from hiking guides and positive emotional involvement.

By experimenting with different methods of communication, a wide audience is reached, without distinction in social origin, age, or cultural background, finally achieving the main goal of this project, which is to contribute to the enhancement of our great geological and geomorphological heritage.

Author Contributions: Conceptualization, L.V. and V.G.; methodology, L.V., M.L. and V.G.; field investigation, L.V., M.L. and V.G.; resources, L.V. and V.G.; data curation, L.V., M.L. and V.G.; writing—original draft preparation, L.V. and V.G.; writing—review and editing, L.V., M.L. and V.G.; supervision, M.L.; funding acquisition, L.V. and V.G. All authors have read and agreed to the published version of the manuscript.

Funding: This research was partly supported by the Marche Region. Part of the work derives from Veronica Guerra's doctoral thesis at the University of Urbino "Carlo Bo" and this publication was supported by her personal research funds.

Data Availability Statement: Not applicable.

Acknowledgments: The authors thank the three anonymous referees for contributing to improving the text of the manuscript with their helpful suggestions.

Conflicts of Interest: The authors declare no conflict of interest.

References

1. Brocx, M.; Semeniuk, V. Geoheritage and geoconservation—History, definition, scope and scale. *J. R. Soc. West. Aust.* **2007**, *90*, 53–87.
2. Wimbledon, W.A.P. Geosites—A mechanism for protection, integrating national and international of heritage sites. In Proceedings of the National Meeting "Il Patrimonio Geologico: Una Risorsa da Proteggere e Valorizzare", Sasso di Castalda, PZ, Italy, 29–30 April 2010; pp. 13–25.
3. Gordon, J.E. Geoheritage, Geotourism and the cultural landscape: Enhancing the visitor experience and promoting geoconservation. *Geosciences* **2018**, *8*, 136. [CrossRef]
4. Gray, M. Geodiversity: The backbone of geoheritage and geoconservation. In *Geoheritage: Assessment, Protection, and Management*; Reynard, E., Brilha, J., Eds.; Elsevier: Amsterdam, The Netherlands, 2018; pp. 13–25.
5. Reynard, E.; Brilha, J. *Geoheritage: Assessment, Protection and Management*; Elsevier: Amsterdam, The Netherlands, 2018; p. 450, ISBN 978-0-12-809531-7.
6. Coratza, P.; Bollati, I.M.; Panizza, V.; Brandolini, P.; Castaldini, D.; Cucchi, F.; Deiana, G.; Del Monte, M.; Faccini, F.; Finocchiaro, F.; et al. Advances in Geoheritage Mapping: Application to Iconic Geomorphological Examples from the Italian Landscape. *Sustainability* **2021**, *13*, 11538. [CrossRef]
7. Mariotto, F.P.; Bonali, F.L.; Venturini, C. Iceland, an Open-Air Museum for Geoheritage and Earth Science Communication Purposes. *Resources* **2020**, *9*, 14. [CrossRef]
8. Sibi, P.; Valletta, M. Paesaggio e morfogenesi . . . delle emozioni. Atti del Convegno Dialogo intorno al Paesaggio. *Cult. Territ. Ling.* **2014**, *1*, 314–323.
9. Geremia, F.; Bentivenga, M.; Palladino, G. Environmental geology applied to geoconservation in the interaction between geosites and linear infrastructures in south-eastern Italy. *Geoheritage* **2015**, *7*, 33–46. [CrossRef]
10. Botelho, R.G.M.; Brilha, J. Principles for Developing a National Soil Heritage Inventory. *Geoheritage* **2022**, *14*, 7. [CrossRef]
11. de Mora, L.; Sellar, A.A.; Yool, A.; Palmieri, J.; Smith, R.S.; Kuhlbrodt, T.; Parker, R.J.; Walton, J.; Blackford, J.C.; Jones, C.G. Earth system music: Music generated from the United Kingdom Earth System Model (UKESM1). *Geosci. Commun.* **2020**, *3*, 263–278. [CrossRef]

12. Lanza, T. Introduction: Five years of Earth sciences and art at the EGU General Assembly (2015–2019). *Geosci. Commun.* **2020**, *3*, 127–128. [CrossRef]
13. Locritani, M.; Merlino, S.; Garvani, S.; Di Laura, F. Fun educational and artistic teaching tools for science outreach. *Geosci. Commun.* **2020**, *3*, 179–190. [CrossRef]
14. Venturini, C.; Pasquaré Mariotto, F. Geoheritage Promotion Through an Interactive Exhibition: A Case Study from the Carnic Alps, NE Italy. *Geoheritage* **2019**, *11*, 459–469. [CrossRef]
15. Nesci, O.; Valentini, L. *TerreRare. Le Marche: Scienza Poesia, Musica*; Argalia: Urbino, Italy, 2019; p. 230, ISBN 978-88-89731-20-8.
16. Nesci, O.; Valentini, L. Science, poetry, and music for landscapes of the Marche Region, Ialy: Communicating the conservation of the natural heritage. *Geosci. Commun.* **2020**, *3*, 393–406. [CrossRef]
17. Guerra, V.; Lazzari, M. Geomorphic approaches to estimate short-term erosion rates: An example from Valmarecchia river system (Northern Apennines, Italy). *Water* **2020**, *12*, 2535. [CrossRef]
18. Valentini, L.; Nesci, O. A new approach to enhance the appeal of the Italian territory through art: Three study cases from Marche Region. *Arab. J. Geosci.* **2021**, *14*, 144. [CrossRef]
19. Valentini, L.; Nesci, O.; Carnevali, L.; Baiocchi, S.; Brizigotti, M.; Teodori, S.; Argalia, S. Landscape as a resource: Science, poetry, and ancient music for the enhancement of the Marche Region, central Italy. In *Recent Research on Geomorphology, Sedimentology, Marine Geosciences and Geochemistry. Advances in Scence, Technology and Innovation, Proceedings of the 2nd Springer Conference of the Arabian Journal of Geosciences (CAJG-2), Sousse, Tunisia, 25–28 November 2019*; Springer: Berlin/Heidelberg, Germany, 2022; pp. 107–109. [CrossRef]
20. Grandgirard, V. L'évaluation des gèotopes. *Geol. Insubr.* **1999**, *4*, 66–69.
21. Panizza, M. Geomorphosites: Concepts, methods and examples of geomorphological survey. *Chin. Sci. Bull.* **2001**, *46*, 4–6. [CrossRef]
22. Reynard, E. Geomorphosites: Definitions and characteristics. In *Geomorphosites*; Coratza, P., Regolini-Bissig, G., Eds.; Pfeil: Munchen, Germany, 2009; pp. 9–20, ISBN 978-3-89937-094-2.
23. Reynard, E.; Fontana, G.; Kozlik, L.; Scapozza, C. A method for assessing "scientific" and "additional values" of geomorphosites. *Geogr. Helv.* **2007**, *62*, 148–158. [CrossRef]
24. Pereira, P.; Pereira, D. Methodological guidelines for geomorphosite assessment. *Géomorphol. Relief Processus Environ.* **2010**, *16*, 215–222. [CrossRef]
25. Lazzari, M. Geosites, cultural tourism and sustainability in the Gargano National Park (southern Italy): The case study of the La Salata (Vieste) geoarchaeological site. *Rend. Online Soc. Geol. It.* **2013**, *28*, 97–101.
26. Pralong, J. Geotourism: A new Form of Tourism utilising natural Landscapes and based on Imagination and Emotion. *Tour. Rev.* **2006**, *61*, 20–25. [CrossRef]
27. Morucci, B. La demande touristique: Une approche de son évolution et de ses perspectives. In *Le Tourisme au XXI e Siècle*; Spindler, J., Durand, H., Eds.; L'Harmattan: Paris, France, 2003; pp. 133–155.
28. Alvarez, W. A review of the Earth history record in the Cretaceous, Paleogene, and Neogene pelagic carbonates of the Umbria-Marche Apennines (Italy): Twenty-five years of the Geological Observatory of Coldigioco. In *250 Million Years of Earth History in Central Italy: Celebrating 25 Years of the Geological Observatory of Coldigioco*; Special Paper; Koeberl, C., Bice, D.M., Eds.; Geological Society of America: Boulder, CO, USA, 2019; Volume 542, pp. 1–58.
29. Tarquini, S.; Isola, I.; Favalli, M.; Battistini, A. *TINITALY, a Digital Elevation Model of ITALY with a 10 m-Cell Size (Version 1.0) [Data Set]*; Istituto Nazionale di Geofisica e Vulcanologia (INGV): Rome, Italy, 2007. [CrossRef]
30. Cornamusini, G.; Conti, P.; Bonciani, F.; Callegari, I.; Martelli, L. Geology of the 'Coltre della Val Marecchia' (Romagna-Marche Northern Apennines, Italy). *J. Maps* **2017**, *13*, 207–218. [CrossRef]
31. Guerra, V.; Lazzari, M. Geomorphological mapping as a tool for geoheritage inventory and geotourism promotion: A case study from the middle valley of the Marecchia River (northern Italy). *Géomorphol. Relief Processus Environ.* **2021**, *27*, 127–145, Actes des 21èmes Journées des Jeunes Géomorphologues. [CrossRef]
32. Aringoli, D.; Farabollini, P.; Gentili, B.; Materazzi, M.; Pambianchi, G. Examples of geoparks and geoconservation strategies from the Southern Umbro-Marchean Apennines (central Italy). *Geoacta* **2010**, *3*, 153–166.
33. Nesci, O.; Borchia, R. Landscapes and landforms of the Duchy of Urbino in Italian Renaissance Paintings. In *Landscapes and Landforms of Italy (World Geomorphological Landscapes)*; Soldati, M., Marchetti, M., Eds.; Springer: Berlin/Heidelberg, Germany, 2017; Volume 22, pp. 257–269.
34. Barthel, R.; Seidl, R. Interdisciplinary Collaboration between Natural and Social Sciences—Status and Trends Exemplified in Groundwater Research. *PLoS ONE* **2017**, *12*, e0170754. [CrossRef]
35. Lacchia, A.; Schuitema, G.; McAuliffe, F. The human side of geoscientists: Comparing geoscientists' and non-geoscientists' cognitive and affective responses to geology. *Geosci. Commun.* **2020**, *3*, 291–302. [CrossRef]
36. Illingworth, S. This bookmark gauges the depths of the human: How poetry can help to personalise climate change. *Geosci. Commun.* **2020**, *3*, 35–47. [CrossRef]
37. Menghini, A.; Pontani, S.; Sapia, V.; Lanza, T. ElectroMagnetic Music: A new tool for attracting people's interest in Geosciences, while sensitizing them to planet sustainability. *Geosci. Commun.* **2020**, *3*, 329–341. [CrossRef]

38. Nesci, O.; Valentini, L. Three known Marchean geomorphosites presented using geomorphology, poetry and ancient music. A new perspective for the development of the area. In Proceeding of 5th AIGeo National Meeting: V Convegno Nazionale AIGeo "Geomorphology for Society: From Risk Knowledge to Landscape Heritage", Cagliari, Italy, 28–30 September 2015.

39. Valentini, L.; Nesci, O. Landscapes of Central Italy through science, poetry and music. A perspective for educating to the planet sustainability. *Geophys. Res. Abstr.* **2016**, *18*. Available online: https://ui.adsabs.harvard.edu/abs/2016EGUGA..18.3911N/abstract (accessed on 28 February 2022).

40. Nesci, O.; Valentini, L.; Argalia, S.; Baiocchi, S.; Brizigotti, M.; Carnevali, L.; Ceccaroli, C. Il Paesaggio come risorsa: Scienza Poesia e Musica per la valorizzazione del territorio marchigiano. In Proceedings of the Convegno "Il Patrimonio Geologico: Una Risorsa Scientifica, Paesaggistica, Culturale e Turistica", Bologna, Italy, 7–8 June 2018.

41. Valentini, L.; Guerra, V.; Lazzari, M. Outline of experiences of communication methods on geoheritage: A key tool for public awareness and sustainable development of geotourism. In *ProGEO SW Europe Regional Working Group Virtual Conference on Geoconservation*; Abstracts Book; Pereira, P., Monge-Ganuzas, M., Bollati, I.M., Rouget, I., Eds.; University of Minho: Braga Portugal, 2022; pp. 62–65.

 *geosciences*

Article

Perception of the Geological-Mining Heritage to Promote Geotourism in Guayaquil, Ecuador

Gricelda Herrera-Franco [1,2], Boris Apolo-Masache [3,4,5,*], Paulo Escandón-Panchana [2], Kelly Jácome-Francis [6], Fernando Morante-Carballo [2,3,4], Josep Mata-Perelló [7] and Paúl Carrión-Mero [3,8,*]

1 Facultad de Ciencias de la Ingeniería, Universidad Estatal Península de Santa Elena (UPSE), Avda. Principal La Libertad-Santa Elena, La Libertad 240204, Ecuador
2 Geo-Recursos y Aplicaciones (GIGA), Campus Gustavo Galindo, ESPOL Polytechnic University, Km. 30.5 Vía Perimetral, Guayaquil P.O. Box 09-01-5863, Ecuador
3 Centro de Investigaciones y Proyectos Aplicados a las Ciencias de la Tierra (CIPAT), Campus Gustavo Galindo, ESPOL Polytechnic University, Km. 30.5 Vía Perimetral, Guayaquil P.O. Box 09-01-5863, Ecuador
4 Facultad de Ciencias Naturales y Matemáticas (FCNM), Campus Gustavo Galindo, ESPOL Polytechnic University, Km. 30.5 Vía Perimetral, Guayaquil P.O. Box 09-01-5863, Ecuador
5 Centro del Agua y Desarrollo Sustentable (CADS), ESPOL Polytechnic University, Guayaquil P.O. Box 09-01-5863, Ecuador
6 Independent Consultant, Guayaquil 090101, Ecuador
7 Departamento de Ingeniería Minera y Recursos Naturales, Universidad Politécnica de Cataluña, Manresa 08242, Spain
8 Facultad de Ingeniería Ciencias de la Tierra (FICT), Campus Gustavo Galindo, ESPOL Polytechnic University, Escuela Superior Politécnica del Litoral, ESPOL, Km. 30.5 Vía Perimetral, Guayaquil P.O. Box 09-01-5863, Ecuador
* Correspondence: bhapolo@espol.edu.ec (B.A.-M.); pcarrion@espol.edu.ec (P.C.-M.)

Citation: Herrera-Franco, G.; Apolo-Masache, B.; Escandón-Panchana, P.; Jácome-Francis, K.; Morante-Carballo, F.; Mata-Perelló, J.; Carrión-Mero, P. Perception of the Geological-Mining Heritage to Promote Geotourism in Guayaquil, Ecuador. *Geosciences* **2022**, *12*, 322. https://doi.org/10.3390/geosciences12090322

Academic Editors: Hara Drinia, Panagiotis Voudouris, Assimina Antonarakou and Jesus Martinez-Frias

Received: 6 July 2022
Accepted: 25 July 2022
Published: 29 August 2022

Publisher's Note: MDPI stays neutral with regard to jurisdictional claims in published maps and institutional affiliations.

Abstract: Biodiversity is an essential component of nature, relegating the aspects of geodiversity, which provides geological and landscape variety to a territory. However, the importance of geodiversity and its social, economic, educational, scientific, and technological impact on a region, are not well understood. This article measures the geoheritage values of Guayaquil, a port city in Ecuador, via surveys and analyses of variables, with the aim of proposing guidelines or strategies that promote the knowledge and diffusion of that geoheritage. Our methodology included (i) a review of historical landmarks of Guayaquil and their relationship with geodiversity, (ii) a survey and data tabulation, (iii) an analysis of the local population's perception of the city's geoheritage, and (iv) the development of geodiversity strategies using computer tools. Our results determined that people approach areas of interest because of each site's biodiversity and the available information about the site. Once there, they can obtain knowledge about the city's geology, geodiversity, and urban geotourism. Therefore, geoheritage is an essential consideration in establishing educational plans, initiatives, and promotion strategies. Furthermore, the identification of a city's heritage values following geoeducation, and the recognition by society of the city's geosites and their historical–scientific significance, will provide a basis for using geotourism in a context of sustainability.

Keywords: geodiversity; geoconservation; geoheritage; geosites; SWOT; Ecuador

1. Introduction

The perceptions of a place's heritage play a vital role in assessing the degree of interest in that place, its importance, and/or the reasons for visiting it [1,2]. The establishment of criteria related to the management and conservation of a heritage site or object and the involvement of different stakeholders, such as government authorities, businesses, the local community, and tourists [3–8], are important for the economic, social, and environmental sustainability of a specific place or region [4,9,10].

Among the factors that affect the perception of a place's heritage are the geology and geomorphological processes of the landscape (i.e., the geological heritage). This geological heritage is focused on the sites or areas with geological characteristics that are of scientific, educational, cultural, or aesthetic value [11–18]. In addition, the criteria for geoeducation and geoconservation are linked to human activities, such as mining (i.e., the mining heritage), inclusive of all of the elements of such activities, such as facilities, machinery, structures, and work tools [19,20].

Figure 1 shows the interconnection of the geological heritage and the mining heritage with geosites, geo-resources, natural disasters, and natural diversity in the context of geotourism. These interconnected factors impact an area's sustainability (economic, environmental, and local); they can be managed via geoconservation and geoeducation.

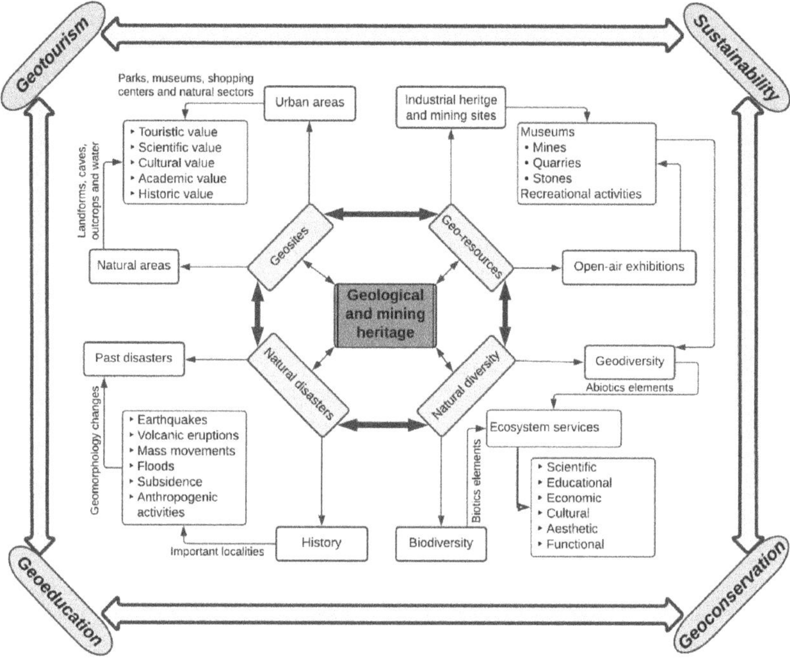

Figure 1. Network of relationships starting with the concepts of geological heritage and mining heritage. Source: adapted from [21–25].

The geological characteristics of a place or region can be analyzed according to geodiversity, which provides an integral framework for managing and evaluating the geological heritage and the mining heritage of the place or region [26–30].

Such an evaluation includes an assessment of the heritage value of sites with geological and mining interest (inside and outside urban areas), which are known as geosites [31–35]. It promotes tourism focused on geology and landscape, also known as geotourism [36]. Unlike conventional tourism, it acquires knowledge and understanding of a site's geology, mining, and geomorphology beyond aesthetic appreciation [37]. This evaluation can be developed from the experts' perspective, who analyze and quantify the intangible value of each sector for its geotourism development [26,38–40]. However, the evaluation of society's expectations allows knowing the perspective of each citizen through surveys and statistical analysis, determining the purpose of their visits and the recommendations to improve each visitor's stay [41,42].

Geoconservation is essential for the preservation and promotion of the development of societies. It is promoted through international programs such as World Heritage Sites (with 1154 heritages) [43] and UNESCO Global Geoparks (with 177 geoparks) [44], which encourage geoconservation and related benefits. Furthermore, geotourism, through geoheritage and geoparks, allows adequate sustainability of the local community of a particular sector [45–49], focusing on rural and urban areas, such as the Hong Kong Geopark (China) [50], where ideal geosites for sustainable tourism based on nature exist within the same city [51–53]. Geotourism and geoheritage can also be promoted by geoscience museums, which should use adequate strategies and tools (including multimedia and interactive ones) to maximise their impact on visitors [54,55].

Ecuador is one of the 17 most megadiverse countries on the planet [56], which has three cultural heritage sites, two natural heritage sites, three intangible heritage sites of humanity, and a geopark endorsed and monitored by UNESCO [57–59]. The Imbabura World Geopark in Ecuador [60] is composed of several geosites that highlight the geological and geomorphological structures of the landscape, such as volcanic complexes, lakes, geothermal sites, mountain ranges, and snow-capped mountains, among others [61]. It favours the promotion of geoeducation, history, and conservation of geoheritage, achieved through sustainable geotourism, for communities' sustainable environmental, social, and economic development [62,63]. The country has several places with abundant geobiodiversity, such as the avenue of volcanoes characteristic of the Ecuadorian Andes due to their composition and geological age [64]. In addition, however, there are urban places with natural sectors, such as the city of Guayaquil (Pearl of the Pacific), characterized by its unique geographical location and geological features that marked its origin, evolution, and development.

Guayaquil city represents one of the geotourism attractions of the coastal part of Ecuador, bordered by the Guayas river and enriched by arms of the sea (salty estuary) [65], which gives it a unique dynamic in the environment of the Gulf of Guayaquil. In addition, it has essential tourist attractions such as parks, museums, shopping centres and natural sectors that allow coexistence with the sector's biodiversity, achieved through natural, active (recreational activities) and cultural tourism [66]. Furthermore, these tourist attractions are connected to several public transportation routes, which facilitates their access, but requires proper disclosure through apps, websites, transport mapping and other means of communication [67–69].

In Guayaquil, the evaluation of 12 geosites has made it possible to quantify and analyse the geological, tourism and educational interests, in addition to the geoconservation index of each geosite, incorporating geotourism as a means for the local, social, and economic development of the city [70]. In addition, there are case studies where higher education institutions promote geotourism and sustainability through different geosite evaluation methodologies [56]. Therefore, the inclusion of the heritage perception by the population allows one to know the society's perspective on geosites around the city and enhance its development.

Based on these premises, the following research question is: What would be the elements of geoheritage that promote the economic and social development of the city and are included as criteria to implement its sustainability? To answer this question, the aim of this work is: to measure the perception or recognition of geoheritage values by citizens, through surveys and analysis of variables, for the proposal of strategies that promote the empowerment of geoheritage. In addition, its natural values should be promoted through dissemination, geo-education, and proper geo-conservation.

2. Materials and Methods

For the promotion of geotourism and geoeducation of geosites or sites of geological and mining interest, it is necessary to know the strengths and weaknesses of these places through a quantitative evaluation, which allows for determining the degree of appreciation and importance of each site [71,72]. Then, knowing the heritage value of geosites, it is possible to promote adequate management of programs and planning for developing tourism and geoconservation of these places [73]. Furthermore, this influences the surrounding sectors, increasing the social benefit of the inhabitants and the geosystem of the geosite; that is what this study is focused on.

2.1. Methodological Approach

Guayaquil has geosites that are visited daily by residents of the sector and foreigners, as well as mining sites that highlight the geology of the place and the exploitation of construction material (Figure 2). Therefore, a quantitative evaluation through surveys analyses the experience and satisfaction of its visitors. Furthermore, it highlights the interest and knowledge in developing activities and management for the benefit of the geosites [38]. This type of study allows us to learn the reasons why people go to these geosites and, through an inferential statistical analysis and interconnection of variables, to assess their perception of the geological/mining heritage of Guayaquil city. The significant variables that explain this measurement are considered predictors that determine the relationships with the knowledge of the geological-mining heritage that the respondents have. These predictors form a perception model through a multiple regression analysis that analyses the incidence of heritage knowledge in geotourism, geodiversity, geology, and tourism in Guayaquil [74,75].

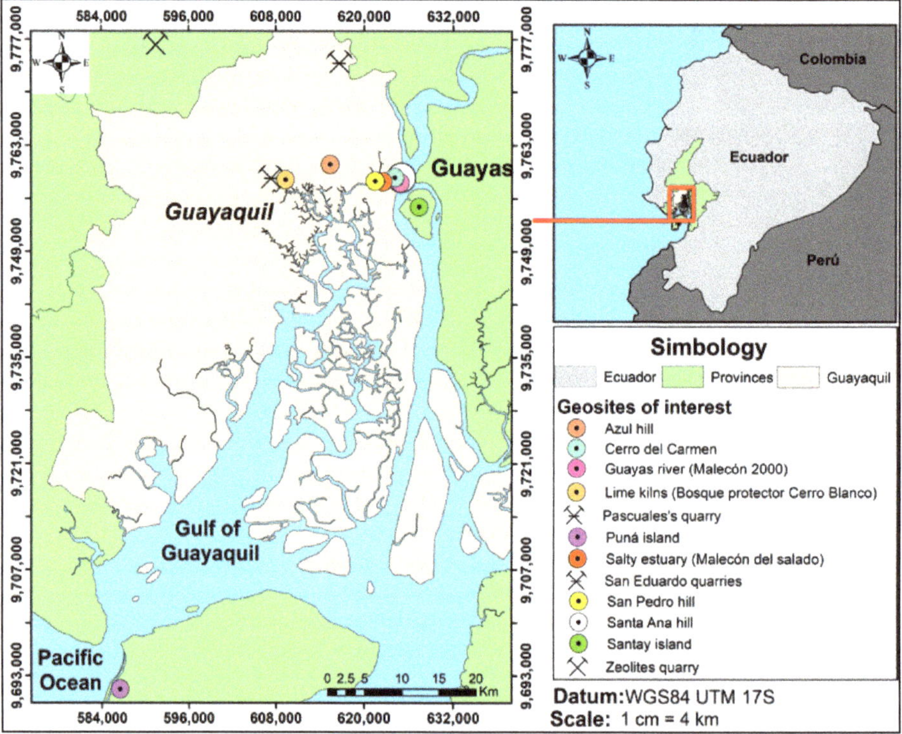

Figure 2. The geographic location of sites of geological and mining interest. Guayaquil, Ecuador. **Source:** Adapted from [76].

The methodological workflow consists of three phases (Figure 3): (i) bibliographic review of the historical milestones of Guayaquil and its relationship with geodiversity, (ii) preparation of a survey and data tabulation, and (iii) analysis of the perception of respondents and the development of key strategies to improve and strengthen the geodiversity of the city.

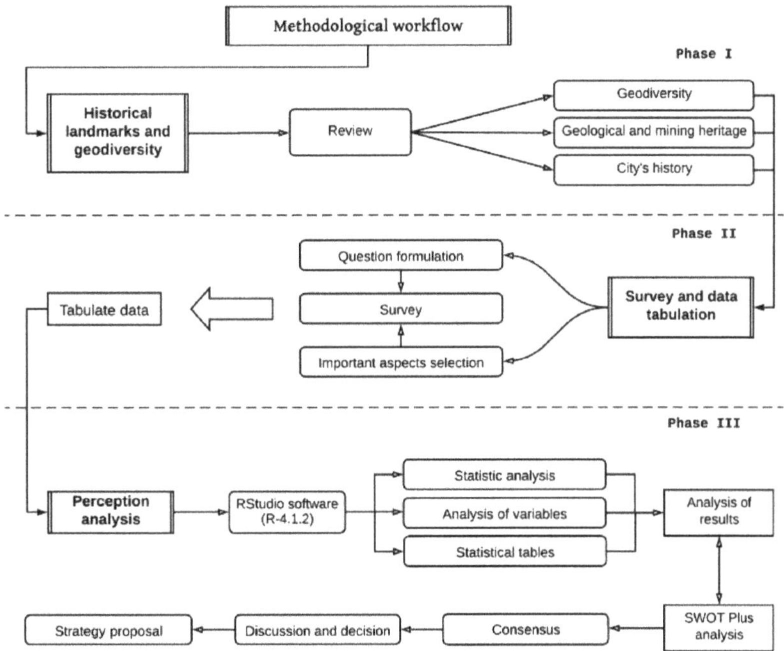

Figure 3. Schematic description of work phases.

2.1.1. Historical Milestones and Geodiversity

Phase I consisted of collecting information through different bibliographic sources, highlighting historical issues of Guayaquil city from its origins. This review allows knowing the milestone events that influenced the development of the current city, in addition to geodiversity and tourism as a source of progress. Furthermore, this phase is fundamental for recognizing places and registering geosites, which are part of the evaluation through surveys directed at society.

2.1.2. Survey and Data Tabulation

Phase II included people with experience and expertise in different areas, aiming to highlight the most important aspects influencing the perception of the city's geological and mining heritage. In addition, the topics discussed allowed for a general survey of the area's inhabitants, and we will analyse its results in Phase III.

The survey carried out for the Guayaquil citizens is based on the survey model implemented by [77], with the purpose of:

1. Sizing the level of knowledge of each interviewee;
2. Identifying the most outstanding tourist areas of Guayaquil;
3. Identifying the level of satisfaction and motivation of the visits;
4. Identifying the interest in promoting geotourism in the evaluated geosites;
5. Identifying the most outstanding issues to improve the degree of citizen satisfaction.

The sample size is obtained using Equation (1) [78,79], considering a probability of success and failure equal to 0.5 ($p = q$), a confidence level of 95% (equivalent to $Z = 2.58$), a sampling error of 5% ($e = 0.05$) [77], and the population size of $N = 2{,}723{,}665$, according to [80]. Equation (2) is considered from the "n" obtained and the population size to optimise the size of this sample:

$$n = \frac{Z^2 * p * q * N}{(\varepsilon^2(n-1)) + (Z^2 * p * q)} = 665.48 \tag{1}$$

$$n_{optimal} = \frac{n}{1 + \frac{n}{N}} = 665.31 \qquad (2)$$

For this work, the analysis determined a minimum of 666 respondents in Guayaquil. Therefore, the questionnaire has three blocks and 34 questions; 31 were designed with closed answers, giving the interviewee a series of alternatives. The remaining three questions allowed the interviewee to formulate their answer. The survey was conducted at the end of September 2021 and can be viewed in the Supplementary Materials Table S1.

2.1.3. Perception Analysis and Strategy Development

Phase III uses the CSV file from phase II, containing the tabulation of the respondents' responses. The data processing and analysis use the statistical program RStudio version R-4.1.2 [81] and the survey questions as categorical variables. The correlation analysis determined the significance of these variables (correlation coefficient R), where the most significant variables explain the perception of the geological and mining heritage [82].

It made it possible to generate a perception model through multiple linear regression, where the dependent variable is the perception of wealth and the independent variables are the set of variables perceived by the inhabitants of the sector (sex, academic training, frequency of use of social networks, international relevance of geosites of Guayaquil, and perception of knowledge in geodiversity, geology, and geotourism). The regression model predicts the perception of the geological and mining heritage through the interaction of each independent variable's estimation coefficients ($\beta_0, \ldots \ldots \beta_n$) [83].

The determination coefficient R^2 explains the proportion of the total variation in the perception of geological-mining heritage in the regression model [84]. The validation of the model focuses on an ANOVA analysis with the significant test statistic (F), which determines if at least one variable explains the perception of the geological and mining heritage [85].

Based on the analysed variables, the matrix identifies the Strengths, Weaknesses, Opportunities, and Threats (SWOT) of the evaluated geosites. Furthermore, in this way, we propose strategies through SWOT analysis to potentially develop geotourism and the local community.

The SWOT matrix is a method with which the appropriate strategies can be established to solve a problem identified in a given topic [86,87]. The generation of a SWOT Plus analysis is an advanced method of the classic SWOT matrix [88,89], focused on the occurrence of existing and potential factors of internal and external origin, as well as their favourable or unfavourable impact on the study of interest [90].

This type of analysis focuses on the significant factors that involve the inhabitants, local authorities, academic institutions, and the intervention of experts [77]. Therefore, through virtual meetings, the internal and external influencing factors are known, mainly based on the respondents' responses on the perception of the geoheritage of Guayaquil city.

3. Results

3.1. History and Geodiversity of Guayaquil

Guayaquil has been the cradle of various settlements of people and their dispossession by locals such as the Huancavilcas. In the pre-Hispanic stage, the ancient cultures that settled in this region lived from maritime trade thanks to various tributaries, such as the Guayas river or the different ramifications of the estuary. Later, with the arrival of the Spaniards in colonial times, this waterway was the largest source of communication and trade [91].

According to legend, the city bears the name of its heroic cacique "Guayas" and his wife "Quil", who encouraged the liberation and fought against their oppressors [92]. Since its foundation, Guayaquil (Pearl of the Pacific), due to its location, became the principal seaport in Ecuador. However, given its geographical position with access to the sea, the city was invaded several times until it achieved its independence, led by the socio-ethnic groups of the time [93]. Its dominant characteristic lies in being an essentially maritime area, flat and close to sea level, crossed by the most prosperous river network on the South American Pacific coast [94]. It has a particular interest in business and tourism due to its commercial activity, event promotion, product export, tourism, or social recreation areas, and other activities that capture people's interest.

In addition, the city has geosites of environmental and geological relevance, highlighting areas with striking geological structures, such as the presence of slumps in Cerro del Carmen, rocky outcrops with almost perpendicular inclination (dips) (Cerro San Pedro), a continuation of the range Chongón Colonche (Cerro Santa Ana, Cerro Azul), rocky material with industrial interest for construction and road filling (San Eduardo, Pascuales and Zeolite Quarries), in addition to geomorphological processes such as the rapid formation of islets in the Guayas River (Santay Island) [70]. The respondents identified several natural sites that highlight the geo-biodiversity around the city, which benefits the tourist interest of the people and the coexistence with the environment, recreational activities and geo-conservation of the urban natural environment. Among these, we have (Table 1):

Table 1. Historical aspects of Guayaquil and its relationship with geodiversity.

Sites of Interest	Historical Aspect	Relationship with Geodiversity	References
Guayas river (Ma-lecón 2000)	Main waterway or trade channel in which they arrived at the boardwalk of the current city. It began as a place of commerce until it became a tourist centre in 2000, enhancing its heritage, tourist, and commercial value.	Two islets stand out, formed by the sedimentation that occurred in the transgression and regression of seawater. In addition, the naval trips allow us to know the natural and historical environment of the river and the boardwalk.	[95,96]
Santa Ana hill	The leading site where the city started. It was one of the colonial bastions for the city's defence against pirate attacks. In addition, it is one of the scenes of the deed of the independence of Guayaquil.	Rocky massif formed by shale with a high degree of fracture and sandstone intercalations.	[91,97]
Cerro del Carmen	It has a viewpoint and a natural beauty that enhances its tourist activity, declared the nation's cultural heritage in 2003, preserving the history and memory of important people, including war heroes, presidents, and artists who contributed to the city's development.	At the city's beginning, it served as quarries to fill and urbanize the swampy areas of Guayaquil. As a result, it has silicified and fractured shale from the Guayaquil formation, which generate block slides (of various sizes), prompting the search for solutions for slope stabilization.	[91,98,99]
Salty estuary (Malecón del estero Salado)	It has broad ramifications throughout the city. However, several of these areas have grown demographically, and the urban area is more extensive in terms of housing.	It is an estuary influenced by the transgression and regression of the sea that overflows with a tangle of rivers, estuaries and lagoons. It has excellent landscape, aesthetic, and recreational value. It is a unique natural ecosystem, home to diverse flora and fauna, such as mangroves settled on delta-type soils.	[94,100]
Quarries	They are suitable places to promote research and development due to their potential as relevant material in different areas of industry.	The San Eduardo formation's quarries have limestone rock to produce cement for the entire country, in addition to the zeolite quarries located in volcano-sedimentary rocks.	[101]
Lime kilns (Bosque protector Cerro Blanco)	The lime kilns were optimal places for tourist overnight camps and their coexistence with nature. They are in Cerro Blanco, a protected forest since 1989 and the most consolidated in the city.	It is a dry forest with abundant geo-biodiversity and endemic species threatened worldwide, in addition to its geotouristic influence without affecting the intangible natural area of the reserve.	[102,103]
Santay island	It is one of the islets formed in the Guayas River. It has a population that has been arriving from different places but in a controlled manner. In 2010, it was declared a RAMSAR site (type I) on the Ecuadorian coast.	Its diversity of flora and fauna entails a critical tourist and investigative interest. In addition, it is in a brackish environment where various aquatic, terrestrial, bird, and halophilic vegetation (mangrove) species meet.	[104,105]

3.2. Perception of Mining Geological Heritage

3.2.1. Demography

The surveys comprise 62.5% men and 37.5% women in the northern, central, and southern parts of Guayaquil (59.4%, 21.1%, and 28.6%, respectively). In addition, 70% of those surveyed are university students, while 13% and 12% have completed their university and college studies, respectively. Most people are informed or communicate more frequently through social networks. This leads to the continuous use and exploitation of these media to publicize various topics of interest to society, current projects, works, and initiatives and communicate different actions in a particular place.

3.2.2. Distribution of Variables of the Perception Model

The construction of the multiple regression model consisted of the statistical analysis between variables, especially those significantly correlated with the perception of the geological-mining heritage, which is achieved based on the respondents' perception. Variables include gender, academic training, frequency of use of social networks, international relevance of geosites in Guayaquil city, and perception of knowledge in geodiversity, geology, and geotourism.

Variables X_1 to X_7 focus on people's perceptions from different perspectives related to the response variable (Y), as seen in Table 2.

Table 2. Description of variables involved in the perception model.

Variable	Name	Description
Y	Heritage perception	Variable response of perception of geological-mining heritage
X_1	Mode_information (1 *, 2 * and 3 *)	Perception of information through news
X_2	Benefit_tourism (0: No and 1: Yes)	Perception of tourism benefit
X_3	Geodiversity (0: No and 1: Yes)	Perception of knowledge of geodiversity
X_4	Geology (0: No and 1: Yes)	Geology knowledge perception
X_5	Around the city geotourism (0: No and 1: Yes)	Perception of geotourism around Guayaquil city
X_6	Urban geotourism (0: No and 1: Yes)	Perception of knowledge of geotourism in the city
X_7	Biodiversity (0: No and 1: Yes)	Perception of visit motivation due to biodiversity

1 *: Read the headline and news content. 2 *: Read the content and verify information in other media. 3 *: Read only the headline of news, photo, or video.

3.2.3. Perception of the Respondents

The respondents' perceptions can be viewed through the mean of the survey responses ($\overline{X}$) according to the significant variables (Table 2). For example, the information mode variable (X_1) has a scale from 1 to 3 (1: Reads the news headline and content; 2: Reads the content and verifies information in other media; 3: Reads only the news headline, photograph, or video). On the contrary, the other variables have a dichotomous value in the perception that depends on the respondents' knowledge (0: No and 1: Yes). In addition, the values of the standard deviation (SD) demonstrate the degree of concentration and coincidence of the average values of perception of the respondents [106].

Table 3 shows the inhabitants' perception of Guayaquil, using the categorical variables with the highest correlation (R = 0.653).

The order of the variables in Table 3 does not represent the degree of significance of each one. Nevertheless, more than 95% of respondents agree that visiting various sites, such as Santa Ana hill, Guayas River on the Malecón 2000, Salty estuary on the 'Malecón del Estero Salado' (acronym in Spanish), and Santay Island, represent beneficial recreational activities for the inhabitants of the city ($\overline{x_2} = 0.97$).

In the context of geoknowledge, a large part of the respondents know the heritage, mainly in recognition of the geodiversity ($\overline{x_3} = 0.98$) and geology ($\overline{x_4} = 0.98$) of geosites, which encourages learning about tourism linked to the natural and the geological. This action is also known as around the city geotourism ($\overline{x_5} = 0.99$). In addition, following the events caused by the COVID-19 pandemic, people have reflected and agreed that open spaces are the best place to be with family and in harmony with nature. Furthermore, natural areas can significantly extend territory for tourism and coexistence with the natural ecosystem within a city ($\overline{x_6} = 0.97$).

Table 3. Perception of the respondents based on significant variables.

N°	Explanatory Variables	Mean ($\overline{X}$)	Standard Deviation (SD)
X_1	Mode_information	0.97; 0.99 and 0.93	0.17; 0.12 and 0.26
X_2	Benefit_tourism	0.63 and 0.97	0.49 and 0.16
X_3	Geodiversity	0.76 and 0.98	0.42 and 0.13
X_4	Geology	0.62 and 0.98	0.49 and 0.15
X_5	Around the city geotourism	0.87 and 0.99	0.34 and 0.10
X_6	Urban geotourism	0.72 and 0.97	0.45 and 0.15
X_7	Biodiversity	0.93 and 0.98	0.25 and 0.12

Focusing on the reasons for visits, we have the variable "Biodiversity" ($\overline{x_7} = 0.98$); in addition, the variable "Mode_information" ($\overline{x_1} = 0.99$) highlights the form of communication in which people choose to read the headline and content of news or topics of collective interest.

3.2.4. Geological-Mining Heritage Perception Model

The estimation coefficients β determine the relative importance and sense of the relevance of the explanatory variables or predictors, which contribute to the measurement of the perception of the geological and mining heritage [107]. However, since no predictor explains the perception of wealth, it takes a constant value (intercept). In addition, the significance level (p) establishes the probability that the perceptions of the geosystem are possible and allow measuring the level of perception of the respondents [82,84].

The estimation model focuses on the respondents' perception of the geological-mining heritage and seven significant predictors. According to the β coefficients and the significance level (p) being less than 0.05, these predictors indicate the degree of importance or contribution to the main focus topic. Based on the results, the analysed sites have a tourist interest. The people who come to these places seek to appreciate some of the biodiversity as a priority. In addition, the survey is prepared to find out the interests in geodiversity, and some people are interested in geology and urban geotourism in the city's environment with its landscapes (Table 4).

Table 4. Multiple linear regression results.

Explanatory Variables	β Coefficients	Standard Error β	Significance Level (* *p*)
Intercept	0.260	0.049	
Benefit_tourism	0.220	0.038	* 0.000
Knowledge Geo			
Geology	0.198	0.034	* 0.000
Urban geotourism	0.162	0.029	* 0.000
Geodiversity	0.127	0.022	* 0.000
Around the city geotourism	0.040	0.015	0.007
Reason for visit			
Biodiversity	0.027	0.010	0.014
Mode_information	0.015	0.007	0.021

* $p < 0.05$ significant variables of the perception model, * $0.000 < 0.001$.

Table 4 presents the parameters that significantly measure the perception of the geological and mining heritage through the β coefficients in the multiple regression model, where the best way to measure the perception of heritage is through tourism ($\beta = 0.220$) in the geosites of the city. This activity allows people to live together in Guayaquil's natural and historical environment.

Interrelated variables, such as "Geology", "Around the city geotourism", and "Geodiversity", are due to the estimators that determine the increase and decrease in heritage perception. The geology (β = 0.198) and geodiversity (β = 0.127) of the geosites influence the perception of tourists, who seek the opportunity for social, cultural, natural, and historical recreation within the city through urban geotourism (β = 0.162). Unlike around the city, geotourism (β = 0.040) deals with large areas and greater biological and geological diversity.

Biodiversity (β = 0.027) is a reason for visiting natural sites inside and outside the city which link humans with the flora and fauna of the various geosystems in a natural environment. Therefore, people tend to obtain information through different media, such as television, social networks, web pages, and magazines (β = 0.015). Therefore, it affects the acceptable estimation of the perception of the geological-mining heritage of the geosites of the city.

3.2.5. Validation of the Perception Model

The coefficient of determination values establishes the goodness of fit of estimation models and the variation in the response variable explained by predictor variables. For example [84,108] determine that R^2 depends mainly on the research area and suggest measurements of 0.67, 0.33, and 0.19 for model validation of substantial, moderate, and weak population perception, respectively. The goodness of fit of this model determined a determination coefficient of 0.426, which indicates that 43% of the variation in the perception of the geological and mining heritage (variable Y, Table 2) can be explored correctly through the seven significant parameters of the multiple regression model (Variables X_1 to X_7, Table 2). Finally, Table 5 shows the results of the ANOVA analysis, considering that the model can be used to explain the relationship between the perception of wealth and the different perception parameters of the respondents.

Table 5. ANOVA analysis of the geological-mining heritage perception model.

Models	Sum of Squares	* DF	* F	Significance Level (* *p*)
Regression	7.613	8		
Residual	10.303	938	43.73	0.0000
Total	17.916	946		
R^2			0.426	

* DF: degrees of freedom; * F: F-statistic; * p < 0.001 highly significant.

The F test statistic was 43.73, representing a highly significant value of (p < 0.001), which indicates that at least one of the independent variables (X_n) acceptably explains the perception of geological and mining heritage.

3.3. Strengths, Weaknesses, Opportunities, and Threats (SWOT) Analysis Plus

This section aims to determine the strategies that promote the geotourism of the sites within Guayaquil. Therefore, a SWOT Plus matrix identifies the significant factors that surround the topic of geotourism in the city, as shown in Table 6.

Based on the identified strengths (existing, favourable internal/external factors) and recognized opportunities (potential, favourable internal/external factors), the following six strategies can mitigate the present weaknesses (existing, unfavourable internal/external factors) and permissible threats (potential, unfavourable internal/external factors):

- Promote geosites as suitable and heritage sectors with abundant geodiversity, landscape areas, recreational areas, and natural infrastructures optimal for increasing tourism and promoting the local security of the place.
- Inculcate and engage local communities in sustainable management of geosites, preserving natural resources and geo-biodiversity of national and international tourist relevance.
- Propose information and security centres to increase geoeducation, disseminate geosite information and improve the city's local security protocol regulation.

- Creation of projects, programs, or activities that encourage the support of government entities to promote geoconservation and economic growth and increase sales and services in the sector through geotourism.
- Encourage alliances in the geoscience area with academic entities to promote the study of natural resources, evidenced impacts and generation of multidisciplinary projects to obtain international support for the preservation and improvement of the geosystem.
- Develop workshops with communities, government entities, and local authorities of the sector to improve the management of geosites and avoid conflicts of interest, which harm the generation of future projects and adequate planning.

Table 6. Matrix SWOT Plus.

Existing Factors	
Favourable internal factors	**Favourable external factors**
<ul><li>Infrastructures of relevance to tourism and nature</li><li>Biodiversity (flora and fauna)</li><li>Recreational areas</li><li>Access routes</li><li>Landscapes or sites with a large landscape area</li><li>Natural heritage</li><li>Job generation</li></ul>	<ul><li>Various entities protecting areas</li><li>Optimal geographical position</li><li>Areas recognized as tourism sectors nationally and internationally</li></ul>
Unfavourable internal factors	**Unfavourable external factors**
<ul><li>Infrastructures with the presence of deterioration</li><li>Low level of security at the local level</li><li>Few information posts throughout the geosites</li><li>No security strategy for visitors</li><li>Few security protocols according to the time</li><li>There is no geo-education material</li></ul>	<ul><li>The average level of education on geodiversity and geotourism</li><li>Low level of preservation of natural resources</li><li>Poor level of security in the city</li><li>Lack of regulation for the security protocol</li><li>There is no culture or legislation for geoeducation</li></ul>
Potential factors	
Favourable internal factors	**Favourable external factors**
<ul><li>Greater knowledge and dissemination through social networks</li><li>Increase in the dissemination of geosites by tourists</li><li>Preservation of natural resources</li><li>Economic growth at the local level</li><li>Increased sales and services</li><li>Support from local authorities in these initiatives</li><li>Great variety of geodiversity</li></ul>	<ul><li>Recognition at the national and international levels</li><li>Tourism and economic growth at the regional level</li><li>Identification of the natural heritage in the region</li><li>Possibility of projects for the development of geodiversity</li></ul>
Unfavourable internal factors	**Unfavourable external factors**
<ul><li>Lack of support from government entities</li><li>Poor management by the authorities</li><li>Lack of rules or regulations to safeguard geosites</li><li>There is no motivation to generate projects</li></ul>	<ul><li>Inefficiency in solving problems (basic needs and population growth)</li><li>Impacts due to climate change</li><li>Conflict of interest between residents, administrators, and private entities</li><li>Lack of financing for projects</li><li>There are no opportunities for multidisciplinary projects</li></ul>

4. Discussion

This work presents three study approaches that focus on the results obtained, focusing on:

1. Guayaquil's history and geodiversity, which allow knowing the evolution of the city and its link with nature [109–111];
2. The identification of variables, identifying those that best measure the perception of the geological and mining heritage based on public opinion [41,112,113];
3. Analysis of the existing and potential factors through a SWOT Plus matrix is necessary in analysing internal and external factors of a particular case for the proposal of strategies that increase the geoconservation and geotourism of the city [114,115].

These approaches are elaborated on below.

4.1. Geoscientific and Historical Literature of Guayaquil City

The literature review allows us to collect historical information about this city and learn about the various events (geological and demographic), which have given re-knowledge to these places that are commonly visited today. This city played an essential role in intercolonial relations through its triple function as the only port of the Audiencia de Quito, the leading shipyard in the American Pacific, and a great producer and exporter of cocoa and wood [116]. Many places in Guayaquil have an impressive wealth of history, culture, and heritage, attracting tourists from all over. However, their citizens do not have sufficient knowledge of the cultural treasure they possess [117].

One of the most iconic places in the city is the so-called mangroves, characteristic of estuarine areas. However, the wealth it keeps is not fully known or protected by its inhabitants [118]. In addition, due to the expansion of the city, several mangrove ecosystems and estuary branches have been lost, connected to pollution problems, such as the discharge of urban and industrial wastewater into unregulated channels, solid and domestic waste from the population, runoff discharges, and clandestine drainage pipes, among others [119].

4.2. Perception of Geological and Mining Heritage

Tourism, geotourism, geology, geodiversity, biodiversity, and information on sites in Guayaquil are essential and significant parameters to explain the perception of geological and mining heritage. These variables determine people's interest in natural and geodiverse sites to carry out human, tourist, and coexistence activities. For this reason, the geological-mining heritage perception model presents an acceptable goodness of fit ($R^2 = 43\%$) in the relationship between these variables. Similarly to the model of [84], they obtained an R^2 of 43.3% relating variables of participation, promotion, and sense of belonging in the perception of sustainable conservation of cultural heritage in Lenggong Valley (Malaysia). Otherwise, the model of [82] obtained an R^2 of 35% using variables such as tourism attraction, facilities, and environment in the satisfaction perception of Safari tourism. Furthermore, [83], in their model of the visual perception of heritage architecture built in a historical city in India, found an R^2 of 26% to 43%, with the variables in this study being heritage characteristics.

The perception model determines that geotourism in cities is growing because people have better access to and preferences for outdoor places in contact with nature. Geotourism, in general, occurs mainly in rural areas. Because people do not have the time or the mobilization facilities for more extensive transfers, they look for options to carry out the said activity within the city, which is called urban tourism, as analysed in the study by [66].

4.3. Approach of Strategies for the Geoconservation of Heritage

According to the historical study and the analysis of surveys, Guayaquil has a significant patrimonial and geodiverse value in several of its geosites. Therefore, its geoconservation is essential to increase the geo-ecosystem benefit it possesses; analysing the strengths, weaknesses, opportunities, and threats (SWOT) is a way to recognise the strategies to promote the geoconservation and geotourism of the city [120,121], being able to solve the problems present in the place through its potential benefits [122,123].

The analysis of the SWOT Plus matrix makes it possible to propose strategies or proposals that benefit the geological, environmental, and cultural area of the sector, focusing on:

1. Promoting the heritage value of the sector;
2. Encouraging geo-education and the dissemination of information;
3. Increasing geotourism and local security;
4. Encouraging support from authorities and academic institutions;
5. National and international recognition through geotourism and geoconservation of geosites.

Optimal strategies for sustainable development through different approaches, cases, and ways of using the SWOT matrix worldwide are discussed below [124–126].

These strategies focus mainly on the sustainability and geoconservation of natural resources and the existing geo-biodiversity in geosites [127], geotourism being an activity that allows achieving this objective and influencing people, managing to improve their coexistence with nature. Moreover, nature increases the sector's social and economic benefits [128–130]. This has been proposed in places such as the Shandong peninsula (China) [131], the Greek island of Gavdos (Greece) [132], the Kanshi mountain basin (Pakistan's the Salt Range) [133], the Caguanes National Park (Cuba) [127], and the ESPOL university campus (Ecuador) [56], which represent places commonly visited.

5. Conclusions

This work allowed the generation of quantitative results from categorical variables with an evaluation of perception through surveys, achieving acceptable goodness of fit of $R^2 = 43\%$ in a sample of 945 respondents in Guayaquil. Based on this study, the significant variables are information mode, tourism benefit, geodiversity, geology, biodiversity, and geotourism around the city and urban form. In addition, the surveys carried out online focused on topics that were easy to understand, had multiple options, and had a lower proportion of free responses, which allowed open access for people without the need for face-to-face counselling due to the bio-safety of COVID-19. The results helped establish that the relevant strategies in a geosystem are geoeducation and heritage disclosure, which generate geoconservation and sustainable development in the city and its inhabitants.

This research identified the variables that best measure the perception of heritage by citizens-highlighting biodiversity, which is a reason for visiting and is present in natural sites, which entails an essential tourist interest to generate a link between the environment and people. However, the respondents have a low degree of geoknowledge (e.g., geology, geotourism, geodiversity), which implies the need to increase information, advertising, and education sites in educational centres so that people know the wide geobiodiversity that exists within the city.

The surveys showed that people look for natural sites to get out of the urban routine, so that they can live together in a healthy, natural environment without danger of contagion. Therefore, the main contribution of this study focuses on recognising geological and mining heritage as an enhancer of natural tourism or geotourism in vast and diverse environments. It is an alternative within cities, allowing sustainable urban development (economic and environmental) within the framework of geobiodiversity through dissemination, geo-education, and adequate geoconservation. In this way, actions could be taken for the correct conservation and management of the city's geological and natural assets, allowing preparation for their enhancement and arousing the interest of people who like sustainable tourism.

Through SWOT analysis, it was possible to provide guidelines to enhance the perception of the heritage in Guayaquil, based on the most representative variables and the existing and potential factors, to minimize the weaknesses and present threats. Among these are:

- Promotion or dissemination of geosites and their heritage value;
- Inculcate sustainable management at the urban, national and international levels;
- Strategic location of geosite information and security centres;
- Encourage the creation of projects or programs that link nature with the city;
- Promote geoscientific alliances to improve the study of natural resources and sustainability;

- Develop workshops as part of a geo-educational program to improve the recognition of natural spaces and reduce conflicts of interest;
- Propose strategies to promote geoeducation in educational centres at all levels for the knowledge of their geoheritage and its conservation.

The limitations of this study are its focus on the digital form through social networks in a planned format and requiring collaboration to carry out the surveys since, due to the confinement stage of COVID-19, the interaction between interviewer and respondent had to be virtual. For future research, the type of surveys (online and face-to-face) could be improved, improving the respondent's understanding, and the characteristics of the study area which generate the reasons for visiting could be verified.

Supplementary Materials: The following supporting information can be downloaded at: https://www.mdpi.com/article/10.3390/geosciences12090322/s1, Table S1: Geological and mining heritage perception survey questions.

Author Contributions: Conceptualization, G.H.-F., B.A.-M., P.E.-P., F.M.-C. and P.C.-M.; methodology, G.H.-F., B.A.-M., P.E.-P., K.J.-F., F.M.-C. and P.C.-M.; software, B.A.-M. and P.E.-P.; validation, B.A.-M. and P.E.-P.; formal analysis, G.H.-F., B.A.-M., P.E.-P., F.M.-C., J.M.-P. and P.C.-M.; investigation, G.H.-F., B.A.-M., P.E.-P., F.M.-C., J.M.-P. and P.C.-M.; data curation, B.A.-M. and P.E.-P.; writing—original draft preparation, G.H.-F., B.A.-M., P.E.-P., K.J.-F., F.M.-C., J.M.-P. and P.C.-M.; writing—review and editing, G.H.-F., B.A.-M., P.E.-P., F.M.-C., J.M.-P. and P.C.-M.; visualization, G.H.-F., B.A.-M., P.E.-P., F.M.-C. and P.C.-M.; supervision, G.H.-F., F.M.-C. and P.C.-M. All authors have read and agreed to the published version of the manuscript.

Funding: This research received no external funding.

Data Availability Statement: Not applicable.

Acknowledgments: This research was supported by the "Registry of geological and mining heritage and its impact on the defence and preservation of geodiversity in Ecuador" research project by ESPOL University (CIPAT-01-2018).

Conflicts of Interest: The authors declare no conflict of interest.

References

1. Gonzalez, V.M.; Coromina, L.; Galí, N. Overtourism: Residents' Perceptions of Tourism Impact as an Indicator of Resident Social Carrying Capacity-Case Study of a Spanish Heritage Town. *Tour. Rev.* **2018**, *73*, 277–296. [CrossRef]
2. Poria, Y.; Reichel, A.; Biran, A. Heritage Site Perceptions and Motivations to Visit. *J. Travel Res.* **2006**, *44*, 318–326. [CrossRef]
3. Korani, Z.; Sam, F. A Comparative Study of Heritage Visitors' Perceptions of Time in the Modern Era. *Curr. Issues Tour.* **2022**, 1–16. [CrossRef]
4. Nair, P.; Singh, D.P.; Munoth, N. "Heritage Village" Designation Impact on the Village: A Case Study on the Perception of the Residents of Kalpathy Village (Kerala). *J. Inst. Eng. Ser. A* **2022**, *103*, 493–500. [CrossRef]
5. de Oliveira, R.A.; Baracho, R.M.A.; Cantoni, L. The Perception of UNESCO World Heritage Sites' Managers about Concepts and Elements of Cultural Sustainability in Tourism. *J. Cult. Herit. Manag. Sustain. Dev.* **2022**. [CrossRef]
6. Kibria, M.M.; Dilshad, T.; Al Asek, A. Policy Implications Based on Stakeholders' Perceptions for Integrated Management of the Halda River: Bangabandhu Fisheries Heritage of Bangladesh. *Water Policy* **2022**, *24*, 517–533. [CrossRef]
7. Herrera-Franco, G.; Carrión-Mero, P.; Aguilar-Aguilar, M.; Morante-Carballo, F.; Jaya-Montalvo, M.; Morillo-Balsera, M.C. Groundwater Resilience Assessment in a Communal Coastal Aquifer System. The Case of Manglaralto in Santa Elena, Ecuador. *Sustainability* **2020**, *12*, 8290. [CrossRef]
8. Reynard, E.; Brilha, J. (Eds.) *Geoheritage: Assessment, Protection and Management*; Elsevier: Amsterdam, The Netherlands, 2018; 450p, ISBN 978-0-12-809531-7.
9. Briones-Bitar, J.; Carrión-Mero, P.; Herrera-Franco, G. Management Practices for a Sustainable Community and Its Impact on Development, Manglaralto-Santa Elena, Ecuador. In Proceedings of the 17th LACCEI International Multi-Conference for Engineering, Education, and Technology: "Industry, Innovation, and Infrastructure for Sustainable Cities and Communities", Montego Bay, Jamaica, 24–26 July 2019.
10. Gray, M. Geodiversity, Geoheritage and Geoconservation for Society. *Int. J. Geoheritage Park.* **2019**, *7*, 226–236. [CrossRef]
11. Lazzari, M.; Aloia, A. Geoparks, Geoheritage and Geotourism: Opportunities and Tools in Sustainable Development of the Territory. *Geoj. Tour. Geosites* **2014**, *13*, 8–9.
12. Panizza, M. Outstanding Intrinsic and Extrinsic Values of the Geological Heritage of the Dolomites (Italy). *Geoheritage* **2018**, *10*, 607–612. [CrossRef]

13. Muda, J. The Geological Heritage Values and Potential Geotourism Development of the Beaches in Northern Sabah, Malaysia. *Bull. Geol. Soc. Malays.* **2013**, *59*, 1–7.
14. Szepesi, J.; Ésik, Z.; Soós, I.; Németh, B.; Sütő, L.; Novák, T.J.; Harangi, S.; Lukács, R. Identification of Geoheritage Elements in a Cultural Landscape: A Case Study from Tokaj Mts, Hungary. *Geoheritage* **2020**, *12*, 89. [CrossRef]
15. Alexandrowicz, Z.; Urban, J.; Miśkiewicz, K. Geological Values of Selected Polish Properties of the UNESCO World Heritage List. *Geoheritage* **2009**, *1*, 43–52. [CrossRef]
16. Morante-Carballo, F.; Herrera-Narváez, G.; Jiménez-Orellana, N.; Carrión-Mero, P. Puyango, Ecuador Petrified Forest, a Geological Heritage of the Cretaceous Albian-Middle, and Its Relevance for the Sustainable Development of Geotourism. *Sustainability* **2020**, *12*, 6579. [CrossRef]
17. Pasquaré Mariotto, F.; Bonali, F.L.; Venturini, C. Iceland, an Open-Air Museum for Geoheritage and Earth Science Communication Purposes. *Resources* **2020**, *9*, 14. [CrossRef]
18. Palladino, G.; Prosser, G.; Bentivenga, M. The Geological Itinerary of Sasso Di Castalda: A Journey into the Geological History of the Southern Apennine Thrust-Belt (Basilicata, Southern Italy). *Geoheritage* **2013**, *5*, 47–58. [CrossRef]
19. Almeida, D.G.; Lacaba, R.G.; Torrens, R.B. La Conservación del Patrimonio Geológico y Minero como Medio para Alcanzar el Desarrollo Sostenible. *Minería y Geol.* **2003**, *19*, 121–128.
20. Bonomo, A.E.; Acito, A.M.; Prosser, G.; Rizzo, G.; Munnecke, A.; Koch, R.; Bentivenga, M. Matera's Old Quarries: Geological and Historical Archives That Need Protection and Valorization. *Geoheritage* **2019**, *11*, 1603–1619. [CrossRef]
21. Brilha, J.; Gray, M.; Pereira, D.I.; Pereira, P. Geodiversity: An Integrative Review as a Contribution to the Sustainable Management of the Whole of Nature. *Environ. Sci. Policy* **2018**, *86*, 19–28.
22. Lazzari, M. Geosites, Cultural Tourism and Sustainability in the Gargano National Park (Southern Italy): The Case Study of the La Salata (Vieste) Geoarchaeological Site. In Proceedings of the 12th European Geopark Conference, Cilento, Italy, 4–6 September 2013; Volume 28, pp. 97–101.
23. Carrión-Mero, P.; Ayala-Granda, A.; Serrano-Ayala, S.; Morante-Carballo, F.; Aguilar-Aguilar, M.; Gurumendi-Noriega, M.; Paz-Salas, N.; Herrera-Franco, G.; Berrezueta, E. Assessment of Geomorphosites for Geotourism in the Northern Part of the "Ruta Escondida" (Quito, Ecuador). *Sustainability* **2020**, *12*, 8468. [CrossRef]
24. Pijet-Migoń, E.; Migoń, P. Geoheritage and Cultural Heritage—A Review of Recurrent and Interlinked Themes. *Geosciences* **2022**, *12*, 98. [CrossRef]
25. Pásková, M.; Zelenka, J.; Ogasawara, T.; Zavala, B.; Astete, I. The ABC Concept—Value Added to the Earth Heritage Interpretation? *Geoheritage* **2021**, *13*, 38. [CrossRef]
26. Carrión-Mero, P.; Montalván-Burbano, N.; Herrera-Narváez, G.; Morante-Carballo, F. Geodiversity and Mining Towards the Development of Geotourism: A Global Perspective. *Int. J. Des. Nat. Ecodynamics* **2021**, *16*, 191–201. [CrossRef]
27. Ferrero, E.; Giardino, M.; Lozar, F.; Giordano, E.; Belluso, E.; Perotti, L. Geodiversity Action Plans for the Enhancement of Geoheritage in the Piemonte Region (North-Western Italy). *Ann. Geophys.* **2012**, *55*, 3.
28. Erikstad, L. Geoheritage and Geodiversity Management—The Questions for Tomorrow. *Proc. Geol. Assoc.* **2013**, *124*, 713–719. [CrossRef]
29. Finzi, Y.; Avni, S.; Maroz, A.; Avriel-Avni, N.; Ashckenazi-Polivoda, S.; Ryvkin, I. Extraordinary Geodiversity and Geoheritage Value of Erosional Craters of the Negev Craterland. *Geoheritage* **2019**, *11*, 875–896. [CrossRef]
30. da Silva, P.; Gomes, R.; Mansur, K.L.; de Souza Ferreira de Castro, A. Consolidação Da Geodiversidade como Patrimônio e o Valor Geológico dos Monumentos do Rio de Janeiro. *Anu. do Inst. Geocienc.* **2020**, *43*, 488–497. [CrossRef]
31. Palacio-Prieto, J.L. Geoheritage within Cities: Urban Geosites in Mexico City. *Geoheritage* **2015**, *7*, 365–373. [CrossRef]
32. Corbí, H.; Martínez-Martínez, J.; Martin-Rojas, I. Linking Geological and Architectural Heritage in a Singular Geosite: Nueva Tabarca Island (SE Spain). *Geoheritage* **2019**, *11*, 703–716. [CrossRef]
33. Reverte, F.C.; Garcia, M.G.M. The Geological Heritage of São Sebastião, SP: Inventory and Potential Use of Geosites with Scientific Value I O Patrimônio Geológico de São Sebastião—Sp: Inventário e Uso Potencial de Geossítios Com Valor Científico. *Geociencias* **2016**, *35*, 495–511.
34. Pilogallo, A.; Nolè, G.; Amato, F.; Saganeiti, L.; Bentivenga, M.; Palladino, G.; Scorza, F.; Murgante, B.; Las Casas, G. Geotourism as a Specialization in the Territorial Context of the Basilicata Region (Southern Italy). *Geoheritage* **2019**, *11*, 1435–1445. [CrossRef]
35. Brilha, J. Geoconservation and Protected Areas. *Environ. Conserv.* **2002**, *29*, 273–276.
36. Dowling, R.; Newsome, D. *Geotourism: The Tourism of Geology and Landscape*; Goodfellow Publishers Ltd.: Oxford, UK, 2010.
37. Hose, T.A. Selling the Story of Britain's Stone. *Environ. Interpret.* **1995**, *10*, 16–17.
38. Bruschi, V.M.; Cendrero, A. Geosite Evaluation; Can We Measure Intangible Values? *J. Quat. Sci.* **2005**, *18*, 293–306.
39. Francesco, G.; Mario, B.; Giuseppe, P. Environmental Geology Applied to Geoconservation in the Interaction Between Geosites and Linear Infrastructures in South-Eastern Italy. *Geoheritage* **2015**, *7*, 33–46. [CrossRef]
40. Wimbledon, W.A.P.; Smith-Meyer, S. *Geoheritage in Europe and Its Conservation*; ProGEO Oslo: Uppsala, Sweden, 2012; Volume 405.
41. Kubalíková, L.; Bajer, A.; Balková, M. Brief Notes on Geodiversity and Geoheritage Perception by the Lay Public. *Geosciences* **2021**, *11*, 54.
42. Chrobak, A.; Ugolini, F.; Pearlmutter, D.; Raschi, A. Thermal Tourism and Geoheritage: Examining Visitor Motivations and Perceptions. *Resources* **2020**, *9*, 58.
43. UNESCO World Heritage List. Available online: https://whc.unesco.org/en/list/ (accessed on 27 April 2022).

44. UNESCO List of UNESCO Global Geoparks (UGGp). Available online: https://www.unesco.org/en/articles/unesco-designates-8-new-global-geoparks (accessed on 27 April 2022).
45. Torabi Farsani, N.; Coelho, C.; Costa, C. Geotourism and Geoparks as Gateways to Socio-Cultural Sustainability in Qeshm Rural Areas, Iran. *Asia Pac. J. Tour. Res.* **2012**, *17*, 30–48. [CrossRef]
46. Herrera Franco, G.; Carrion Mero, P.; Morante Carballo, F.; Herrera Narváez, G.; Briones Bitar, J.; Blanco Torrens, R. Strategies for the Development of the Value of the Mining-Industrial Heritage of the Zaruma-Portovelo, Ecuador, in the Context of a Geopark Project. *Int. J. Energy Prod. Manag.* **2020**, *5*, 48–59. [CrossRef]
47. Rosado-González, E.M.; Sá, A.A.; Palacio-Prieto, J.L. UNESCO Global Geoparks in Latin America and the Caribbean, and Their Contribution to Agenda 2030 Sustainable Development Goals. *Geoheritage* **2020**, *12*, 36. [CrossRef]
48. Bentivenga, M.; Cavalcante, F.; Mastronuzzi, G.; Palladino, G.; Prosser, G. Geoheritage: The Foundation for Sustainable Geotourism. *Geoheritage* **2019**, *11*, 1367–1369. [CrossRef]
49. Hose, T.A. *Geoheritage and Geotourism: A European Perspective*; Boydell & Brewer: Woodbridge, UK, 2016; Volume 19, ISBN 1783271477.
50. Wang, L.; Tian, M.; Wang, L. Geodiversity, Geoconservation and Geotourism in Hong Kong Global Geopark of China. *Proc. Geol. Assoc.* **2015**, *126*, 426–437. [CrossRef]
51. UNESCO Hong Kong Geopark's Top Attractions. Available online: https://www.geopark.gov.hk/en (accessed on 23 April 2022).
52. Guo, W.; Chung, S. Using Tourism Carrying Capacity to Strengthen UNESCO Global Geopark Management in Hong Kong. *Geoheritage* **2019**, *11*, 193–205. [CrossRef]
53. Fung, C.K.W.; Jim, C.Y. Segmentation by Motivation of Hong Kong Global Geopark Visitors in Relation to Sustainable Nature-Based Tourism. *Int. J. Sustain. Dev. World Ecol.* **2015**, *22*, 76–88.
54. Pasquaré Mariotto, F.; Venturini, C. Strategies and Tools for Improving Earth Science Education and Popularization in Museums. *Geoheritage* **2017**, *9*, 187–194. [CrossRef]
55. Venturini, C.; Pasquaré Mariotto, F. Geoheritage Promotion Through an Interactive Exhibition: A Case Study from the Carnic Alps, NE Italy. *Geoheritage* **2019**, *11*, 459–469. [CrossRef]
56. Morante-Carballo, F.; Merchán-Sanmartín, B.; Cárdenas-Cruz, A.; Jaya-Montalvo, M.; Mata-Perelló, J.; Herrera-Franco, G.; Carrión-Mero, P. Sites of Geological Interest Assessment for Geoeducation Strategies, ESPOL University Campus, Guayaquil, Ecuador. *Land* **2022**, *11*, 771. [CrossRef]
57. Carrión-Mero, P.; Morante-Carballo, F.; Herrera-Franco, G.; Maldonado-Zamora, A.; Paz-Salas, N. The Context of Ecuador's World Heritage, for Sustainable Development Strategies. *Int. J. Des. Nat. Ecodynamics* **2020**, *15*, 39–46. [CrossRef]
58. UNESCO Protect Our Natural and Cultural Heritage. Available online: https://whc.unesco.org/ (accessed on 26 April 2022).
59. UNESCO UNESCO World Network of Biosphere Reserves. Available online: https://en.unesco.org/news/18-new-sites-join-unescos-world-network-biosphere-reserves (accessed on 27 March 2022).
60. UNESCO Imbabura UNESCO Global Geopark Ecuador. Available online: https://en.unesco.org/galleries/imbabura-unesco-global-geopark-ecuador (accessed on 28 April 2022).
61. Berrezueta, E.; Sánchez-Cortez, J.L.; Aguilar-Aguilar, M. Inventory and Characterization of Geosites in Ecuador: A Review. *Geoheritage* **2021**, *13*, 93. [CrossRef]
62. Vázquez Taset, Y.M. Imbabura: The First UNESCO Geopark in Ecuador. *Bionatura* **2019**, *4*, 830–831. [CrossRef]
63. González, Á.R.P.; Palacios, J.C.A.; Quelal, L.R.C. Marketing y Turismo Sostenible en el Geoparque Imbabura, Ecuador. *ECA Sinerg.* **2021**, *12*, 97. [CrossRef]
64. Carrión-Mero, P.; Dueñas-Tovar, J.; Jaya-Montalvo, M.; Berrezueta, E.; Jiménez-Orellana, N. Geodiversity Assessment to Regional Scale: Ecuador as a Case Study. *Environ. Sci. Policy* **2022**, *136*, 167–186. [CrossRef]
65. Reynaud, J.-Y.; Witt, C.; Pazmiño, A.; Gilces, S. Tide-Dominated Deltas in Active Margin Basins: Insights from the Guayas Estuary, Gulf of Guayaquil, Ecuador. *Mar. Geol.* **2018**, *403*, 165–178. [CrossRef]
66. Orden-Mejía, M.; Carvache-Franco, M.; Huertas, A.; Carvache-Franco, W.; Landeta-Bejarano, N.; Carvache-Franco, O. Post-COVID-19 Tourists' Preferences, Attitudes and Travel Expectations: A Study in Guayaquil, Ecuador. *Int. J. Environ. Res. Public Health* **2022**, *19*, 4822. [CrossRef]
67. Brunett, A.L.L.; Masache, E.V.; Brunett, A.H.L. Turismo 2.0 como Herramienta para Promocionar los Atractivos Culturales de Guayaquil. *INNOVA Res. J.* **2017**, *2*, 154–163. [CrossRef]
68. Barreth, A.C.; Sang, Y.F.; Quiroga, S.T. Tourist Bus Maps as Contribution to Tourist Movements in Guayaquil. *J. Tour. Herit. Res.* **2020**, *3*, 452–477.
69. Fun Sang Vera, Y.M. Análisis de la Conectividad entre el Transporte Público y los Atractivos Turísticos de Guayaquil para una Propuesta de Mapa Turístico de Buses. Bachelor's Thesis, Universidad de Guayaquil, Guayaquil, Ecuador, 2018.
70. Carrión-Mero, P.; Morante-Carballo, F.; Apolo-Masache, B. Evaluation of Geosites as an Alternative for Geotouristic Development in Guayaquil, Ecuador. *WIT Trans. Ecol. Environ.* **2020**, *241*, 45–56.
71. Pijet-Migoń, E.; Migoń, P. Linking Wine Culture and Geoheritage—Missing Opportunities at European UNESCO World Heritage Sites and in UNESCO Global Geoparks? A Survey of Web-Based Resources. *Geoheritage* **2021**, *13*, 71. [CrossRef]
72. Bazsik, I.; Bujdosó, Z.; Koncz, G. Interrelations between Wine Tourism and Geotourism: A Wine Consumption Survey in Monor (Hungary). *Geoj. Tour. Geosites* **2021**, *39*, 1517–1524. [CrossRef]

73. Gravis, I.; Németh, K.; Procter, J.N. The Role of Cultural and Indigenous Values in Geosite Evaluations on a Quaternary Monogenetic Volcanic Landscape at Ihumātao, Auckland Volcanic Field, New Zealand. *Geoheritage* **2017**, *9*, 373–393. [CrossRef]
74. Ek Styvén, M.; Näppä, A.; Mariani, M.; Nataraajan, R. Employee Perceptions of Employers' Creativity and Innovation: Implications for Employer Attractiveness and Branding in Tourism and Hospitality. *J. Bus. Res.* **2022**, *141*, 290–298. [CrossRef]
75. Robinson, D.; Newman, S.P.; Stead, S.M. Community Perceptions Link Environmental Decline to Reduced Support for Tourism Development in Small Island States: A Case Study in the Turks and Caicos Islands. *Mar. Policy* **2019**, *108*, 103671. [CrossRef]
76. IGM. Capas de Información Geográfica Básica Del IGM de Libre Acceso. Cartas Topográficas Escala 1:50.000, Formato Shp. Available online: http://www.geoportaligm.gob.ec/portal/index.php/cartografia-de-libre-acceso-escala-50k/ (accessed on 24 July 2022).
77. Herrera-Franco, G.; Carrión-Mero, P.; Alvarado, N.; Morante-Carballo, F.; Maldonado, A.; Caldevilla, P.; Briones-Bitar, J.; Berrezueta, E. Geosites and Georesources to Foster Geotourism in Communities: Case Study of the Santa Elena Peninsula Geopark Project in Ecuador. *Sustainability* **2020**, *12*, 4484. [CrossRef]
78. Alcaraz-Quiles, F.J.; Navarro-Galera, A.; Ortiz-Rodríguez, D. Factors Determining Online Sustainability Reporting by Local Governments. *Int. Rev. Adm. Sci.* **2014**, *81*, 79–109. [CrossRef]
79. Murray, R.; Larry, J. *Estadística*, 4th ed.; Mc Grawill: New York, NY, USA, 2009.
80. INEC. Proyecciones Poblacionales Cantonales. Available online: https://www.ecuadorencifras.gob.ec/inec-presenta-sus-proyecciones-poblacionales-cantonales/#:~{}:text=Segúnestosdatos%2CQuitoen,ElOrocon2.379habitantes (accessed on 15 May 2021).
81. Gigante, F.D.; Santos, J.P.V.; López-Bao, J.V.; Olea, P.P.; Verschuuren, B.; Mateo-Tomás, P. Farmers' Perceptions towards Scavengers Are Influenced by Implementation Deficits of EU Sanitary Policies. *Biol. Conserv.* **2021**, *259*, 109166. [CrossRef]
82. Bhuiyan, M.A.H.; Darda, M.A.; Hasan, M.R. Tourist Perception and Satisfaction on Safari Tourism at Bangabandhu Sheikh Mujib Safari Park in Bangladesh. *Int. J. Geoheritage Park.* **2021**, *9*, 430–440. [CrossRef]
83. Kiruthiga, K.; Thirumaran, K. Visual Perception on the Architectural Elements of the Built Heritage of a Historic Temple Town: A Case Study of Kumbakonam, India. *Front. Archit. Res.* **2017**, *6*, 96–107. [CrossRef]
84. Jaafar, M.; Noor, S.M.; Rasoolimanesh, S.M. Perception of Young Local Residents toward Sustainable Conservation Programmes: A Case Study of the Lenggong World Cultural Heritage Site. *Tour. Manag.* **2015**, *48*, 154–163. [CrossRef]
85. Zhao, S.; Timothy, D.J. Tourists' Consumption and Perceptions of Red Heritage. *Ann. Tour. Res.* **2017**, *63*, 97–111. [CrossRef]
86. Wang, J.; Wang, Z. Strengths, Weaknesses, Opportunities and Threats (SWOT) Analysis of China's Prevention and Control Strategy for the COVID-19 Epidemic. *Int. J. Environ. Res. Public Health* **2020**, *17*, 2235. [CrossRef]
87. Jiang, R.; Mao, C.; Hou, L.; Wu, C.; Tan, J. A SWOT Analysis for Promoting Off-Site Construction under the Backdrop of China's New Urbanisation. *J. Clean. Prod.* **2018**, *173*, 225–234. [CrossRef]
88. Sztando, A. Analiza Strategiczna Jednostek Samorządu Terytorialnego [Strategic Analysis of the Local Governments]. In *Metody Oceny Rozwoju Regionalnego*; Akademia Ekonomiczna we Wrocławiu: Wrocław, Poland, 2006.
89. Gierszewska, G.; Olszewska, B.; Skonieczny, J. *Zarządzanie Strategiczne Dla Inżynierów*; Polskie Wydawnictwo Ekonomiczne: Warszawa, Poland, 2013; ISBN 8320820510.
90. Nazarko, J.; Ejdys, J.; Halicka, K.; Magruk, A.; Nazarko, Ł.; Skorek, A. Application of Enhanced SWOT Analysis in the Future-Oriented Public Management of Technology. *Procedia Eng.* **2017**, *182*, 482–490. [CrossRef]
91. Municipio de Guayaquil. Guayaquil es mi Destino. Available online: https://www.guayaquilesmidestino.com/sites/default/files/gastronomica2.pdf (accessed on 21 September 2021).
92. GeoRaymi. Fundación de Guayaquil. Available online: https://www.goraymi.com/es-ec/guayas/guayaquil/fundaciones/fundacion-guayaquil-agg38ze7p (accessed on 24 July 2022).
93. Silva Guijarro, V.H. Los Hijos del Guayas en Busca de la Libertad. Análisis Historiográfico sobre el Protagonismo Socio-Étnico en la Independencia de Guayaquil según los Textos Escolares de Historia (1915–2015). *Hist. Const.* **2021**, *22*, 195–215. [CrossRef]
94. Laviana Cuetos, M.L. *Guayaquil en el Siglo XVIII: Recursos Naturales y Desarrollo Económico*; Editorial CSIC-CSIC Press: Madrid, Spain, 2002.
95. Calle, J.; Reinoso, J.; Michalon, R.; Aviles, J.C. Recopilacion de Informacion Base para el Mapa Geoambiental del Area Urbana del Gran Guayaquil y Mapas en Formato SIG. Bachelor's Thesis, Superior Politécnica del Litoral, Guayaquil, Ecuador, 2005; pp. 132–153.
96. Menendez-Carbo, S.; Ruano, M.A.; Zambrano-Monserrate, M.A. The Economic Value of Malecón 2000 in Guayaquil, Ecuador: An Application of the Travel Cost Method. *Tour. Manag. Perspect.* **2020**, *36*, 100727. [CrossRef]
97. Ortíz González, A. Procedencia de las Rocas Sedimentarias de los Cerros Santa Ana y del Carmen mediante Análisis Geoquímico, Guayaquil-Ecuador. Bachelor's Thesis, Universidad de Guayaquil, Guayaquil, Ecuador, 2019.
98. Guadalupe, D.; Caicedo, S.; Jorge, L.; Murillo, R. Diseño de un Plan Piloto Turístico Educativo sobre el Cementerio Patrimonial de Guayaquil Dirigido a los Estudiantes de la Escuela Dr. Alberto Avellán Vite. Ph.D. Thesis, Universidad de Guayaquil, Guayaquil, Ecuador, 2017.
99. Morante-Carballo, F.; Aguilar-Aguilar, M.; Ramírez, G.; Blanco, R.; Carrión-Mero, P.; Briones-Bitar, J.; Berrezueta, E. Evaluation of Slope Stability Considering the Preservation of the General Patrimonial Cemetery of Guayaquil, Ecuador. *Geosciences* **2019**, *9*, 103. [CrossRef]

100. Ministerio del Ambiente Ecuatoriano. Refugio de Vida Silvestre Manglares El Morro | Sistema Nacional de Áreas Protegidas del Ecuador. Available online: http://areasprotegidas.ambiente.gob.ec/es/areas-protegidas/refugio-de-vida-silvestre-manglares-el-morro (accessed on 21 March 2022).
101. Machiels, L.; Garcés, D.; Snellings, R.; Vilema, W.; Morante-Carballo, F.; Paredes, C.; Elsen, J. Zeolite Occurrence and Genesis in the Late-Cretaceous Cayo Arc of Coastal Ecuador: Evidence for Zeolite Formation in Cooling Marine Pyroclastic Flow Deposits. *Appl. Clay Sci.* **2014**, *87*, 108–119. [CrossRef]
102. Borde, A. Plan Maestro Bosque Protector Cerro Blanco. Guayaquil, Ecuador, 2021. *ARQ* **2021**, *108*, 90–99. [CrossRef]
103. Horstman, E. Establishing a Private Protected Area in Ecuador: Lessons Learned in the Management of Cerro Blanco Protected Forest in the City of Guayaquil. *Case Stud. Environ.* **2017**, *1*, 1–14. [CrossRef]
104. Cruz Cordovez, C.; Herrera, I.; Espinoza, F.; Rizzo, K. New Record of a Feral Population of Lithobates Catesbeianus Shaw, 1802 in a Protected Area (Santay Island) in the Ecuadorian Coast. *BioInvasions Rec.* **2020**, *9*, 421–433. [CrossRef]
105. Diaz-Christiansen, S.; López-Guzmán, T.; Pérez Gálvez, J.C.; Muñoz Fernández, G.A. Wetland Tourism in Natural Protected Areas: Santay Island (Ecuador). *Tour. Manag. Perspect.* **2016**, *20*, 47–54. [CrossRef]
106. Wu, T.-C.; Xie, P.F.; Tsai, M.-C. Perceptions of Attractiveness for Salt Heritage Tourism: A Tourist Perspective. *Tour. Manag.* **2015**, *51*, 201–209. [CrossRef]
107. Outeiro, L.; Villasante, S.; Oyarzo, H. The Interplay between Fish Farming and Nature Based Recreation-Tourism in Southern Chile: A Perception Approach. *Ecosyst. Serv.* **2018**, *32*, 90–100. [CrossRef]
108. Hair, J.F.; Ringle, C.M.; Sarstedt, M. PLS-SEM: Indeed a Silver Bullet. *J. Mark. Theory Pract.* **2011**, *19*, 139–152. [CrossRef]
109. Kubalíková, L.; Kirchner, K.; Bajer, A. Secondary Geodiversity and Its Potential for Urban Geotourism: A Case Study from Brno City, Czech Republic. *Quaest. Geogr.* **2017**, *36*, 63–73. [CrossRef]
110. Dos Polck, M.A.R.; de Medeiros, M.A.M.; de Araújo-Júnior, H.I. Geodiversity in Urban Cultural Spaces of Rio de Janeiro City: Revealing the Geoscientific Knowledge with Emphasis on the Fossil Content. *Geoheritage* **2020**, *12*, 47. [CrossRef]
111. Pereira, L.S.; da Silva Farias, T. Assessing the Cultural Values of the Geodiversity in a Brazilian City: The Historical Center of João Pessoa (Paraíba, NE Brazil), Mata Da Aldeia Chart. *Int. J. Geoheritage Park.* **2020**, *8*, 59–73. [CrossRef]
112. Cañizares, A.D.; Bourotte, C.L.M.; Garcia, M.d.G.M. Estudo Exploratório sobre a Percepção Da Geodiversidade e Das Geociências Pela População Da Região Metropolitana de São Paulo. *Anuário do Inst. Geociências* **2019**, *42*, 375–386. [CrossRef]
113. Russ, B.R.; Csek, M. Revealing Geodiversity Trough Environmental Education: Perceptions of Students about the Manga Do Céu Geosite. *Anuário do Inst. Geociências* **2012**, *35*, 271–280. [CrossRef]
114. Ervural, B.C.; Zaim, S.; Demirel, O.F.; Aydin, Z.; Delen, D. An ANP and Fuzzy TOPSIS-Based SWOT Analysis for Turkey's Energy Planning. *Renew. Sustain. Energy Rev.* **2018**, *82*, 1538–1550. [CrossRef]
115. Markovska, N.; Taseska, V.; Pop-Jordanov, J. SWOT Analyses of the National Energy Sector for Sustainable Energy Development. *Energy* **2009**, *34*, 752–756. [CrossRef]
116. Laviana Cuetos, M.L. Una Descripción Inédita de Guayaquil. *Temas Am.* **1982**, *1*, 73–84.
117. Valle Reyes, A.S. Diseño de una Ruta Turistica Colonial "Encantos del Cerrito Verde" como Medio de Interpretacion del Patrimonio Cultural en el Cerro Santa Ana de La Ciudad de Guayaquil Provincia del Guayas. Bachelor's Thesis, Universidad de Guayaquil, Guayaquil, Ecuador, 2014.
118. El Telégrafo. Los Manglares Restauran Suelos Hasta en el 89%. Available online: https://www.eltelegrafo.com.ec/noticias/guayaquil/1/manglares-suelos-guayaquil-estero-salado (accessed on 21 February 2022).
119. Mariscal-Santi, W.E.; Garcia-Larreta, F.S.; Mariscal-Garcia, R.S.; Cornejo-Ortega, A.V.; Ortega-Ramirez, P.M.; Montiel-Rivera, T.A.; Ponce-Solórzano, H.X.; De La Torre-Quinonez, E.F. Evaluación de la Contaminación Físico-Química de las Aguas Del Estero Salado, Sector Norte de la Ciudad de Guayaquil-Ecuador-2017. *Polo del Conoc.* **2018**, *3*, 133. [CrossRef]
120. Beigi, H.; Pakzad, P. Investigating Geotourism Capabilities of the Gavkhoni Wetland According to the SWOT Model. *WIT Trans. Ecol. Environ.* **2010**, *139*, 169–179.
121. Datta, K. Application of SWOT-TOWS Matrix and Analytical Hierarchy Process (AHP) in the Formulation of Geoconservation and Geotourism Development Strategies for Mama Bhagne Pahar: An Important Geomorphosite in West Bengal, India. *Geoheritage* **2020**, *12*, 45. [CrossRef]
122. Vardopoulos, I.; Tsilika, E.; Sarantakou, E.; Zorpas, A.; Salvati, L.; Tsartas, P. An Integrated SWOT-PESTLE-AHP Model Assessing Sustainability in Adaptive Reuse Projects. *Appl. Sci.* **2021**, *11*, 7134. [CrossRef]
123. Gkoltsiou, A.; Paraskevopoulou, A. Landscape Character Assessment, Perception Surveys of Stakeholders and SWOT Analysis: A Holistic Approach to Historical Public Park Management. *J. Outdoor Recreat. Tour.* **2021**, *35*, 100418. [CrossRef]
124. Savari, M.; Amghani, M.S. SWOT-FAHP-TOWS Analysis for Adaptation Strategies Development among Small-Scale Farmers in Drought Conditions. *Int. J. Disaster Risk Reduct.* **2022**, *67*, 102695. [CrossRef]
125. Jiskani, I.M.; Shah, S.A.A.; Qingxiang, C.; Zhou, W.; Lu, X. A Multi-Criteria Based SWOT Analysis of Sustainable Planning for Mining and Mineral Industry in Pakistan. *Arab. J. Geosci.* **2020**, *13*, 1108. [CrossRef]
126. Panigrahi, J.K.; Mohanty, P.K. Effectiveness of the Indian Coastal Regulation Zones Provisions for Coastal Zone Management and Its Evaluation Using SWOT Analysis. *Ocean Coast. Manag.* **2012**, *65*, 34–50. [CrossRef]
127. Navarro-Martínez, Z.M.; Crespo, C.M.; Hernández-Fernández, L.; Ferro-Azcona, H.; González-Díaz, S.P.; McLaughlin, R.J. Using SWOT Analysis to Support Biodiversity and Sustainable Tourism in Caguanes National Park, Cuba. *Ocean Coast. Manag.* **2020**, *193*, 105188. [CrossRef]

128. Jędrusik, M.; Lisowski, A.; Mouketou-Tarazewicz, D.; Ropivia, M.-L.; Zagajewski, B. Touristic Development of the La Lopé National Park (Gabon) in Light of the SWOT Analysis. *Misc. Geogr.* **2015**, *19*, 5–13. [CrossRef]
129. Rezagama, A.; Setyati, W.A.; Agustini, T.W.; Sunaryo; Devi, S.A.; Deswanto, E.; Budiati, I.M. Approaching SWOT Analysis to Develop Strategies of Marine-Ecotourism in Bedono Village, Sayung, Demak. *IOP Conf. Ser. Earth Environ. Sci.* **2021**, *750*, 012059. [CrossRef]
130. Carrión-Mero, P.; Borja-Bernal, C.; Herrera-Franco, G.; Morante-Carballo, F.; Jaya-Montalvo, M.; Maldonado-Zamora, A.; Paz-Salas, N.; Berrezueta, E. Geosites and Geotourism in the Local Development of Communities of the Andes Mountains. A Case Study. *Sustainability* **2021**, *13*, 4624. [CrossRef]
131. Liu, R.; Wang, Y.; Qian, Z. Hybrid SWOT-AHP Analysis of Strategic Decisions of Coastal Tourism: A Case Study of Shandong Peninsula Blue Economic Zone. *J. Coast. Res.* **2019**, *94*, 671. [CrossRef]
132. Gkoltsiou, A.; Mougiakou, E. The Use of Islandscape Character Assessment and Participatory Spatial SWOT Analysis to the Strategic Planning and Sustainable Development of Small Islands. The Case of Gavdos. *Land Use Policy* **2021**, *103*, 105277. [CrossRef]
133. Abbas, S.; Shirazi, S.A.; Qureshi, S. SWOT Analysis for Socio-Ecological Landscape Variation as a Precursor to the Management of the Mountainous Kanshi Watershed, Salt Range of Pakistan. *Int. J. Sustain. Dev. World Ecol.* **2018**, *25*, 351–361. [CrossRef]

 geosciences

Article

The Geosites of the Sacred Rock of Acropolis (UNESCO World Heritage, Athens, Greece): Cultural and Geological Heritage Integrated

Hara Drinia *, Fani Tripolitsiotou, Theodora Cheila and George Zafeiropoulos

Department of Geology and Geoenvironment, National and Kapodistrian University of Athens, Panepistimiopolis, 15784 Athens, Greece
* Correspondence: cntrinia@geol.uoa.gr

Abstract: Athens, the capital of Greece, is notable for its distinctive environment. Numerous archaeological and historical monuments contribute to the city's cultural wealth. These cultural monuments should include geological monuments, which are part of Athens' natural heritage. The Acropolis of Athens is one of the world's most recognizable and admired monuments, renowned for its archaeological, historical, and touristic significance. The Acropolis Rock is also a spectacular geological heritage monument. This article is about the Acropolis monuments, which are of great geological interest in addition to their cultural value. In recognizing each monument's unique geological features and quantitatively evaluating them, in terms of educational, touristic, cultural, and accessibility value, we document their special value in geoeducation and geotourism, not only for the public, but also for the scientific community. The potential for exploiting these geosites, in terms of geotourism, is very high and important not only for strengthening the local economy, but also for raising visitor awareness of environmental, geological, and geoconservation issues. However, the lack of understanding of geological heritage in relation to cultural heritage is underlined. The need for strategic educational planning and integration of geo-environmental education into school practice is evident.

Keywords: Athens; Acropolis; geosites; quantitative assessment; geotourism; geoeducation

Citation: Drinia, H.; Tripolitsiotou, F.; Cheila, T.; Zafeiropoulos, G. The Geosites of the Sacred Rock of Acropolis (UNESCO World Heritage, Athens, Greece): Cultural and Geological Heritage Integrated. *Geosciences* **2022**, *12*, 330. https://doi.org/10.3390/geosciences12090330

Academic Editors: Jesús F. Jordá Pardo and Jesus Martinez-Frias

Received: 10 August 2022
Accepted: 25 August 2022
Published: 30 August 2022

Publisher's Note: MDPI stays neutral with regard to jurisdictional claims in published maps and institutional affiliations.

1. Introduction

Greece is one of the few regions on earth where geology has been the most important factor in shaping its unique and beautiful natural environment, as well as its social, economic and historical development from ancient times to the present day.

Indeed, complex geological processes are responsible for the genesis and evolution of the unique island complexes, the many kilometres of coastline and the landscapes of unusual beauty, for the unique climate conditions, for the soil which supports a wide variety of flora and fauna, for the mineral raw materials and, of course, for its culture. As a result, there has been a strong scientific movement in Greece in recent years to evidence its geological heritage and to manage its geosites (e.g., [1–5]), with studies aimed at recording geotopes in specific areas that represent important moments in the history of the country and our planet.

Furthermore, the emergence of another, alternative form of tourism, geotourism, has gained its own momentum and is seeking its own share in the country's economy. Geotopes, Geoparks and Natural History Museums are the best fields for geotourism activities and, therefore, geotourism could be an important opportunity for local community development and sustainability [6–9].

Geotourism is a "new" challenge, not only because it can redistribute the country's tourism products in areas that have not been tourist destinations to date, but also because it can create a new quality of tourist stream in the country.

Although, the term "urban geotourism" is relatively new, the use of geology, geomorphology and other associated man-made activities and features (building stones or anthropogenic landforms) within urban areas for tourism and education is much older.

Del Lama [10] defines urban geotourism as tourism in places anywhere within the boundaries of a city (whether in the form of built heritage or geological formations) related to geological concepts and characteristics. As evidenced by numerous publications and books from various parts of the world with diverse cultural and social environments, the number of studies highlighting the use of geodiversity and geological heritage in urban areas is growing, yet mainly focuses on European cities. The rest of the world remains rather "unexplored" in this respect [11,12].

Urban geodiversity is defined by Palacio-Prieto [13] as "buildings and other man-made features of the city that developed under specific geological conditions or that have undergone specific geological processes over time". Such anthropogenic geodiversity elements can be beneficial to community cultural development, leading to the conclusion that cultural heritage associated with the abiotic, natural environment can be treated as a component of geocultural diversity and included in the urban geodiversity typology.

There are several classifications of urban geodiversity, but we focus on the one developed by Habibi et al. [11], who distinguished between in situ and ex situ geodiversity, which is consistent with other classifications [14] that include both in situ and displaced geodiversity features, such as those held in museum collections or used as building/decorative stones.

An "urban geotope" is defined as a place of geological or geomorphological interest within a city [15]. This location can be natural, originating from geological processes, or it can be the result of artificial constructions where characteristic rock types were used for their construction [13,16]. Urban geotopes highlight the relationship between geology-geomorphology and society. They are preserved not only because of their geological value, but also for aesthetic, cultural and economic reasons.

Urban geotopes usually occupy small territorial areas and their location depends on the size of the city, and its population, as well as its spatial structure. Therefore, only a few urban geoparks exist in the center of cities, while many of them are located on the outskirts of cities where there are more open spaces. As a result of their locations, they are more affected by anthropogenic activity, especially by urbanization. Some scholars refer to anthropogenic pressure to describe the urgency and necessity of their conservation [17].

In recent years, examples of geotourism with urban characteristics (urban geotourism) have appeared in many countries around the world, such as Brazil [18], which combines geological heritage with cultural heritage [19].

Athens, the capital of Greece, combines history and the past with modern reality. There are numerous archaeological and historical monuments which constitute the cultural wealth of the city. To these cultural monuments should also be added the geological monuments, which constitute the natural heritage of the country.

The city, although it gathers several important inherent advantages (such as archaeological sites, cultural heritage, and climatic conditions), fails to show itself as an attractive city break destination, mainly due to its functional disadvantages (it is not a tourist-friendly city, there is traffic congestion, high costs, an unattractive urban landscape, an absence of large conference spaces, etc.) and its lack of promotion. Whilst points of interest located in the area do exist, there is a decline in Attica's position in the tourism sector [20].

However, especially for Athens, the potential of dynamic tourism development remains extremely favorable. The promotion of the geo-cultural heritage of the city, which is directly related to the preservation and protection of the environment, can strengthen and diversify the development of tourism, identity, culture, and interests of the local population.

The well-known Acropolis is an ancient temple complex standing atop a rocky outcropping in the heart of Athens. The Acropolis of Athens is one of the most impressive and recognizable monuments in the world, world-renowned and of exceptional archaeological, historical, and touristic value. At the same time, however, the Acropolis Rock is also an admirable monument of geological heritage. Unfortunately, only a few people know that there are numerous and varied urban geosites within the sacred rock of Acropolis, reflecting its natural and cultural heritage, showing the connection between geology, geomorphology, and urban development, and which are interesting geotourist objects.

In this paper, we will deal with the monuments of the Acropolis, which, apart from their cultural value, are of high geological interest. The geotouristic potential of the city of Athens is highlighted through the identification and quantitative assessment of these geosites. The main goal of this assessment is to select the geosites that combine the best conditions for use in environmental education and that also have a high value for urban geotourism, due to their intrinsic value as representative geological sites, their connection with environmental issues (pollution, climate change, natural hazards, waste, and recycling, among others), and their accessibility conditions. The result of our study not only helps in scaling the regional tourism industry, adding value to tourism, but also reveals the geodiversity of the area, aiming at its geoconservation.

2. Material and Methods

The identification and assessment of geosites for geotourism was based on fieldwork and a detailed review of published literature and maps (both contemporary and historic). It not only focused on the traditionally accepted characteristics of geodiversity, but also covered aspects reflecting the interactions between geodiversity and culture.

According to today's holistic concept of geotourism, the inventory of geotourism resources should consider the following: (1) the natural features, geological, geomorphological, hydrological or paleontological, and ecological elements related to geodiversity and (2) the cultural aspects related to geodiversity and geographical heritage, e.g., churches and cemeteries, pavements and stone buildings or toponyms associated with geodiversity (e.g., [21]).

Particular attention was paid to anthropogenic landforms, which are of high interest, are very common in urban areas and have high potential for geotourism and education [22].

2.1. The City of Athens

Attica is one of the 13 regions of Greece and includes the Prefecture of Athens, which is the capital of the Attica region and of Greece (Figure 1).

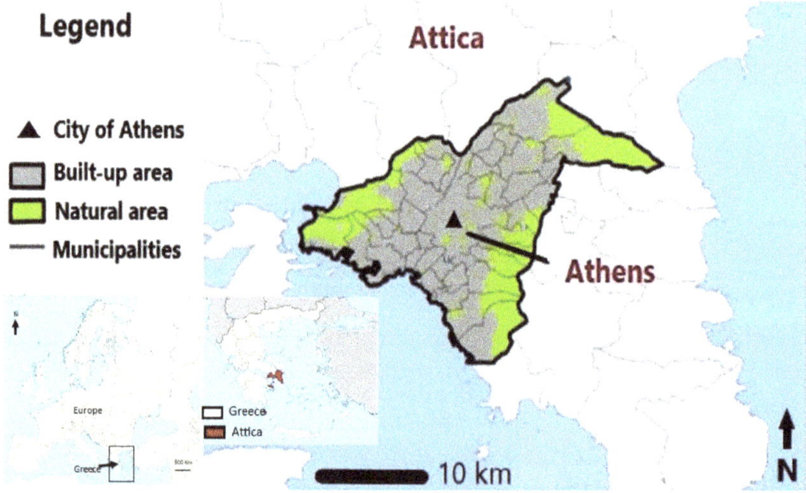

Figure 1. Map depicting the prefecture of Athens in Attica region. In inlets, the location of Greece in Europe and the location of Attica region in Greece.

Athens is a heavily urbanized area with severe traffic problems, both in the city center and in its harbour, Piraeus. The settlement of Athens covers a total area of 414.6 km^2, including the entire city of Athens and the harbour of Piraeus, as well as its suburbs. It has a subtropical Mediterranean climate with mild winters and hot, dry summers.

In the first post-war decades, a rapid concentration of the population took place in the central and western parts of the city. Since the mid-1970s, urbanization has shifted to the suburbs in the northeast and southeast. Unlike most European capitals, the urbanization of modern Athens is not related to the industrial revolution.

The city's population grew rapidly from 400,000 people in 1925 to about 1,000,000 by 1950. The population growth of modern Athens was due to the return of Greek refugees from Asia Minor in the 1920s after World War I and extensive internal migration after World War II.

Today, the urban areas of Athens and Piraeus have a population of approximately 3.2 million inhabitants in an area of 412 km^2. This number corresponds to approximately 1/3 of the Greek population. The Attica region is home to nearly half of Greece's population, more than 60% of the country's industrial production, as well as high-value real estate and infrastructure. The population density (people per km^2) is about 7500 and over 20,000 in some municipalities with a high frequency of residential, commercial, and business activities. There is no large-scale industry in Athens. Several industrial support services, including warehouses, commercial transport companies, and building materials yards, are located between Athens Center and Piraeus. Earlier industries in recent decades produced ceramics, textiles, footwear, and engaged in tanning and metallurgy.

Athens is mainly known for its ancient history and especially for the Golden Age of Pericles in 500 BC. Under his rule, Athens became the most powerful city-state in Greece. His main contribution was, of course, the establishment of democracy. Athens became the intellectual and artistic center of the ancient world. Among all other things, Pericles was also responsible for the construction of the Parthenon.

2.2. Geology–Geomorphology

The basin of Attica, also known as the basin of Athens, or the Athenian plain, is about 22 km long from NE to SW and 11 km wide across, and includes Athens, Piraeus and the municipalities of their surrounding suburbs. It is the most densely populated region of Greece. It is a large tectonic depression running NNE-SSW, bounded by Parnitha to the northwest, Pendeli to the northeast, Aegaleo and Poikilo to the west and Hymettus to the east, while the Saronikos Gulf opens up to the southwest (Figure 2).

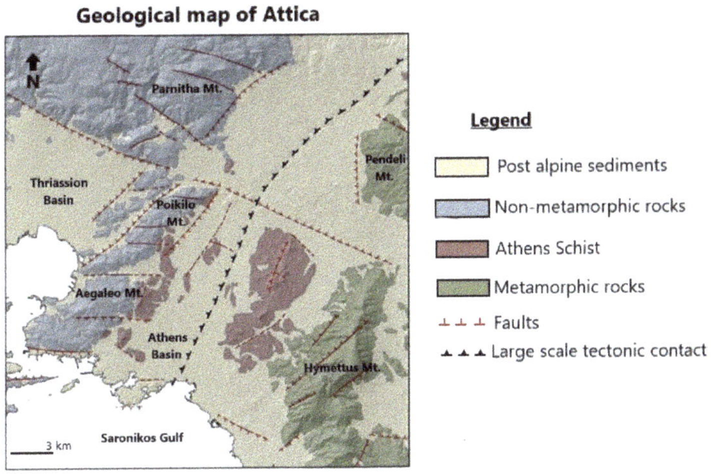

Figure 2. Simplified geological map of the basin of Attica [23], modified by us.

Inside the basin, a series of hills, the altitudes of which decreases towards the SW, has developed along the central axis in a NE-SW direction (Figure 3). The main ones, from north to south, are the Tourkovounia (339 m), Lykabettus (278 m), the Acropolis (156 m) and the Museum Hill or Filopappou (147 m), all made of a lowermost Upper Cretaceous (100 My) limestone, called locally the "Tourkovounia Formation" [24]. These hills have the same geological age and the same geological structure as the Acropolis hill.

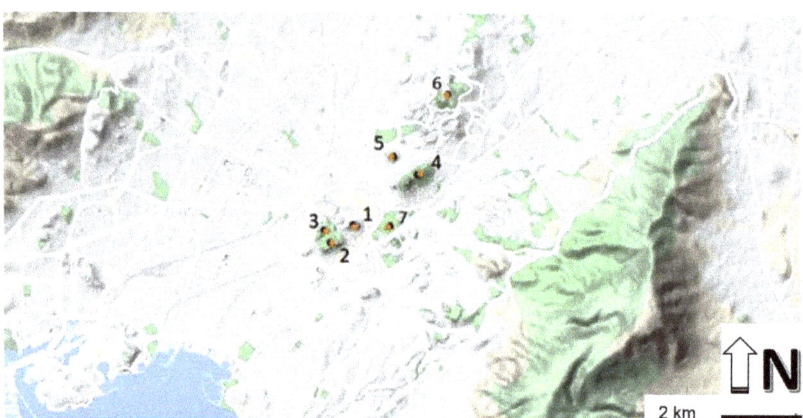

Figure 3. Satellite photo of Athens city indicating the location of the hills of the city. **1.** Akropolis, **2.** Filopappou **3.** Hill of the Nymphs **4.** Lycabettus **5.** Hill of Strefi **6.** Tourkovounia **7.** Ardittos Hill.

The four hills located on the surface of the basin are rocky remnants of a previous relief and are characterised as "inselberg hills," i.e., isolated hills [25]. There are also smaller ones, such as Ardittos (131 m), Strefi Hill, the Hill of the Nymphs, and Areios Pagos.

Pliocene marine deposits have been detected west of Lycabettus at a height of 120 m [26,27] and in the western part of Hymettus. In the area of Tourkovounia, clay deposits were found in the cracks and fractures of limestones, which chronologically belong to the Pleistocene [28].

The Attica basin is drained by two main hydrographic networks, the Kifissos and the Ilissos (Figure 4). Both hydrographic networks are characterised by the existence of mainly small and seasonal streams. Kifissos is the largest river in the region, 27 km long, originating from western Pendeli and southern Parnitha, flowing through the basin in a N-S direction into the bay of Faliro. Most of the river's route was formed by the mid-20th century into an open channel, which was later covered by the national highway A1, meaning that the Kifissos River has been largely modified and covered by the continuous development of the urban fabric [29]. Ilissos originates from Hymettus and also flows into Faliro bay (Figure 4).

Figure 4. Map showing the hydrographic networks of Attica [30], modified by us.

Since the mid-18th century, several researchers have studied the geological setting of the Athens area [31–38]. Despite this, the understanding of Athens' lithostratigraphic structure remains incomplete.

The Attica Basin was formed during the Alpine orogeny. It is located in the back-arc area of the Hellenic Arc and comprises Alpine basement rocks, both metamorphic and non-metamorphic, and post-Alpine sediments (Figure 2). The alpine formations today form the mountains that surround the basin, as well as the hilly area in its centre (Filopappou, Acropolis, Lycabettus, Tourkovounia, etc.), while the post-alpine sediments, i.e., the sediments that have been deposited in the Athens basin after the uplift of the mountains, come mainly from the erosion of the latter.

The main rocks that make up the uplifted alpine formations are as follows:

1. The limestones of the western mountains (Aegaleo, Poikilo Oros, Parnitha). These are rocks formed about 200 million years ago, from deposits in a marine environment, indicating that the area was once a large ocean.

2. The metamorphic rocks of the eastern mountains (Hymettus, Pendeli). These are rocks that were metamorphosed under conditions of high pressure and temperature. The main ones are marbles (an example is the well-known Pendelic marble) and schists.

3. The "Athens schists", as they have come to be called, albeit somewhat misleadingly, because, in reality, they are a complex system of different rocks (pelites, argillaceous shales, marls, sandstones), formed on the sea bed during the uplift of the mountain

ranges mentioned above. The Athens schists are overlain, in some places, by relatively young limestone, forming hilltops (e.g., Acropolis, Lycabettus, Tourkovounia).

2.3. The Geology and Geomorphology of the Acropolis Hill

The Acropolis hill rises 156 m above sea level, while the perimeter ground rises between 60 and 70 m. When viewed from above, the ridge of the hill resembles the shape of an ellipse. The largest axis of this ellipse, which runs east-west, is estimated to be about 250 m long, while the smallest axis, which runs across the width of the rock, is about 150 m long (Figure 5).

Figure 5. The hill of Acropolis which resembles the shape of an ellipse with its largest axis running east-west, of 250 m length (yellow line), and its smallest axis of about 150 m length (red line).

The lithostratigraphy of the Acropolis area is relatively simple. It is primarily made up of two lithostratigraphic units: the Athens schist and the Acropolis limestone [39].

The Athens schists are found in the basal zone of the area and at the base of the hillsides (Figure 6) and date back to the Cretaceous period (Maastrichtian to Eocene (?) in age) with the original sediments being deposited 72 million years ago, in a river delta. According to Marinos et al. [40] the Athens schist is a flyshoid phase of a delta-type deposit, which means that what we see today was once the talus of a huge river's delta. This formation has a distinct reddish color and is composed primarily of soft rock. The Athens schist is made up of a variety of low-grade metamorphic and relatively soft rocks. The stratigraphic formation was given its name because it covers a large portion of the city of Athens. The Athens Schist bedrock exhibits exceptional weathering, as well as intense folding, shearing, and extensional faulting, completing the structural "downgrading" of the rock mass.

The upper parts of the Acropolis hill (Figure 6) are made of limestone of the Cenomanian-Turonian age, about 30 million years older than the underlying Athens schist, that is no more than 40 m thick [39]. Compressional tectonic forces thrust the Acropolis limestone over the Athens schist, forming a nappe or overthrust sheet. Erosion of the limestone nappe caused the Acropolis to detach, resulting in the current feature. There are springs and karstic caves where the Athens schist and the limestone meet. The erosion of the same nappe that formed the Acropolis formed many of the hills in the Athens region. These include the Lykabettu, Areopagus, and Museum hills. The marble used to build the Acropolis came from the quarries of Mount Pendeli, a mountain to the northeast of Athens [39,41].

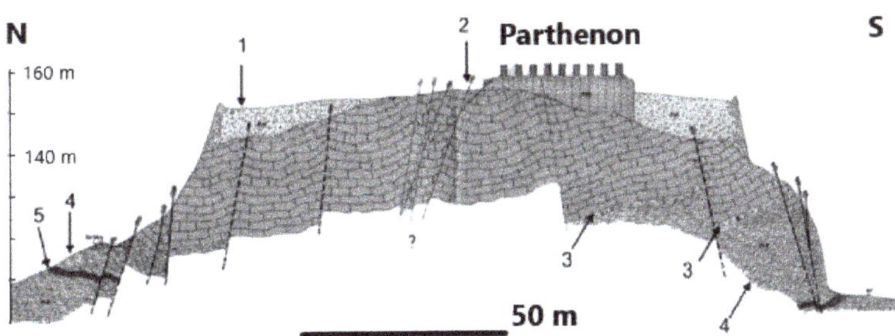

Figure 6. Geological section of the Acropolis hill, passing near the Klepsydra spring in the North, after Andronopoulos and Koukis [41], simplified; **1**: artificial earth fill, **2**: Upper Cretaceous limestones, **3**: conglomerate, **4**: the upper marly unit of the formation of the Athenian Schists composed of sandstones, limestones and marls, **5**: limestones interbedded in the marly unit, **?**: no data.

The geological composition and structure of the Acropolis Hill were crucial in its preservation. Perhaps it influenced human history as well, as we know that despite the major earthquakes that struck Attica during the historical years, this monument remained unaffected. Aside from the Parthenon, earlier and later buildings and constructions dating back over 2500 years have been preserved on the hill.

2.4. The Geosites of the Sacred Rock of Acropolis

The Sacred Rock of the Acropolis is an important urban geosite of archaeological, historical, ecological and geological interest [42], which has been designated a UNESCO World Heritage Site, since 1987. Acropolis and its monuments are global symbols of classical spirit and culture and constitute the largest architectural and artistic complex bequeathed by Greek antiquity to the world. Its monuments were developed by an extraordinary group of architects (such as Ictinus, Callicrates, and Mnisiklis) and sculptors (such as Phidias, Alcamenes, and Agoracritus), who transformed the rocky hill into a unique complex that heralded the emergence of classical Greek thought and art. The Acropolis hill, apart from its great archaeological and cultural value, also has very important geological value due to the numerous geosites it hosts.

In this study we will focus on the following geotopes (Figure 7):

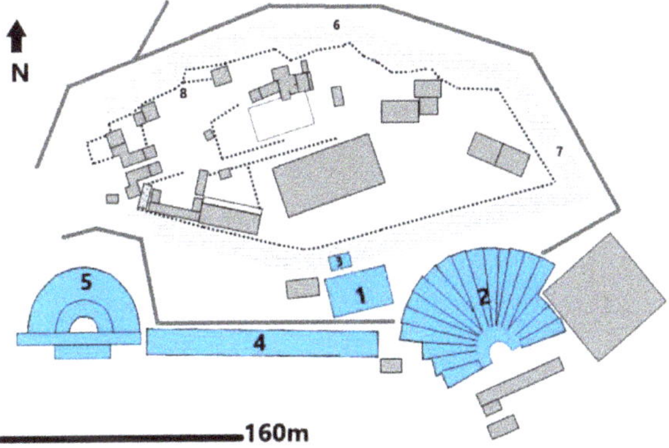

Figure 7. Plan of the Acropolis hill, showing the locations of the studied geotopes: **1**. GS1 Ascleipieion, **2**. GS2 Theater of Dionysus, **3**. GS3 The Choragic Monument of Thrasyllos, **4**. GS4. Stoa of Eumenes, **5**. GS5 Odeon of Herodes Atticus—Herodeion. **6**. GS6 Caves, **7**. GS7 Klepsydra, **8**. GS8 Mycenaean Fountain.

2.4.1. GS1: Asclepieion

The Asclepieion of Athens (Figure 7) was a sanctuary dedicated to the god Asclepius, one of the most important healing deities of Ancient Greece, and his daughter Hygeia. It is located on the south side of the Acropolis, west of the theater of Dionysus, and occupies the space between the ancient Promenade and the Acropolis rock. A spring inside a cave in this location still gushes water today. The action of rainwater causes chemical corrosion on limestone rocks. Water penetrates the soil and forms karst forms, such as hollows and caves, where there is no surface drainage. The limestone rocks on the Acropolis hill's slopes are intensely karstified, with caves and an aquifer at the contact of the permeable limestone with the underlying Athens schist, which is impenetrable, due to its clay composition. The ancient Greeks considered "clean water" to be essential for the purification and the healing of the sick. People worshiped the god while receiving medical treatments, surgical operations and advice in their sleep. The ritual of ascension was the most distinctive aspect of Asclepius' therapeutic practice. After being cleansed with spring water and sacrificing on the altar, the patients slept in order to have healing dreams or visions [43].

2.4.2. GS2: Theater of Dionysus

The Theater of Dionysus (Figures 7 and 8a,b) was a monumental stone theater, built in the decade of 330 or 320 BC. Its cavity had a capacity of between 17,000 and 19,000 seats. Due to its location and construction materials, it is an important geosite. By the end of the 6th century, a theater with wooden seats had already been built. Aktitis Lithos (Stone) eventually took their place. The coast of Piraeus, now known as Themistocles Coast, was famous for its quarries until a few decades ago and for the extraction of a white rock known as Aktitis Lithos (Figure 8c), which means stone from the coast. This is a marly limestone, a rock which was used to build many monuments in Attica [44]. The first row of seats, 67 in total, were made of Pendeli marble and were intended for prominent city people [45]. To emphasize the connection between the events and worship of Dionysus, an elevated and richly decorated marble throne of the priest Dionysus was placed in the center [46]. The beautiful orchestra floor, which was intended for the "Chorus," was made of Hymettus marble. The parapet around the stage was also built during the Roman era to safely fill the orchestra with water when boat races were held in the space [47]. The theater's location on the south side of the Acropolis meets the requirements of bioclimatic construction. As the theater was open, and the performances took place in early spring, the viewing conditions were heavily influenced by environmental factors, such as temperature, winds, sunshine, the orientation of the theater, the slope of the ground, and so on. Temperatures were higher on the south side of the hill than on the north during this period, as the Rock of Acropolis blocked the north winds. During the day, much of the orchestra was in the shade, but it gradually illuminated until noon. The clarity of the Athenian atmosphere, on the other hand, did not allow the shadow to significantly obscure the action space [48]. To remove rainwater, a sewer was carved into the rock or built around the orchestra. Its architectural design and implementation on sloping ground were archetypal in architectural history.

2.4.3. GS3: The Choragic Monument of Thrasyllus

The Choragic Monument of Thrasyllus (Figures 7 and 8d) is situated on the southeast side of the Acropolis, above the Theater of Dionysus. According to the epistle, Thrasyllu, an eminent Athenian, built the monument in the year 320/319 BC, during the reign of archon Neaichmos, to house the bronze tripod, his victory trophy, in the dramatic games held in the theater of Dionysus [49,50]. The monument is located at the entrance of a karst cave. The hidden parts are made of Piraeus Aktitis Lithos and Acropolis limestone. The exposed ones are made of Pendeli and Hymettus marble, while the architraves and the frieze are made of coarse-grained white marble from the Greek islands. The construction of the monument required the adaptation of the curved surface of the cave mouth to a large rectangular opening which presented many technical challenges [51]. The way the construction problems were solved demonstrates ancient Greek technology, and knowledge of cutting rocks, particularly marble, but also knowledge of geometry [52,53].

Figure 8. (**a,b**) GS2: Theater of Dionysus; (**c**) Aktitis Lithos; (**d**) GS3: The Choragic Monument of Thrasyllu and (**e**) GS4: Stoa of Eumenes; (**f**) Sanctuary of Aglauros; (**g**) GS7: Klepsydra; (**h**) GS8: Mycenaean Fountain; (photos by Fani Tripolitsiotou and Panos Karoutsos).

2.4.4. GS4: Stoa of Eumenes

The Stoa of Eumenes (Figures 7 and 8e) is located on the southern slope of the Acropolis, between the Theater of Dionysu and the Herodeion. It was granted to the city of Athens by the king of Pergamon, Eumenes II (197–159 BC), around 160 BC. Vitruvius (Roman writer, architect, and engineer), referring to the galleries built near theaters in general, mentioned that the Stoa served as a shelter for the spectators in inclement weather or as a storage area for the theater scenery. Its structural elements, such as triglyphs, metopes, cornices, columns, and architraves, were made of Pergamon marble which was shipped to Athens. Pergamon marble is not found in any building of Athens. The Stoa was built along the North Wall, which had been built earlier, during the reign of Lycurgus (330 BC), to hold back the embankments of the Promenade level to the north [54]. The North Wall was reinforced with buttresses joined by arches and was made of hard limestone with gray Hymettos marble uprights. It is regarded as one of the finest examples of extensive pre-construction in the ancient world. Indeed, its architectural members were numbered and marked with masonic letters, to make assembly easier [54]. The Stoa was also supported in an unusual way, on the Northern Wall of limestone with marble uprights. The Stoa is considered a great technical project of antiquity, the value of which was not fully appreciated until today [55].

2.4.5. GS5: Odeon of Herodes Atticus–Herodeion

The Odeon of Herodes Atticus (Figures 7 and 9a,b) is located at the western end of the Acropolis' southern slope. Tiberius Claudius Herodes of Atticus donated the land for its construction in the second century AD. The richness of the construction was impressive. White and black marbles were laid on the floor of the orchestra, while colorful marbles adorned the walls. More than 3000 trees, primarily cedars and cypresses, were required to house the hollow. It had the largest roof of any other ancient theater, and it appears to have burned down during the Heroulian raid in 267 AD. The roof covered an area of 3 acres and weighed between 750 and 800 tons. Adding the estimated weight of 180 tons of tiles, the total weight must have been 1000 tons. It was conical and built in one piece, with no intermediate posts, and had a span of 50 m. It was an admirable achievement at the time, and methods used for the construction of bridges were most likely used [56]. Furthermore, it was initially constructed in the Stoa of Eumenes, which had been converted into a construction site, where the construction's strength tests were carried out. Later on, with special scaffolding a height of 30 m was built, where the final installation took place. Until the "Hilton" hotel was built, the Odeon was the tallest building in the country. According to Korres [56], it is a construction marvel comparable to Trajan's bridge (105 AD), the work of the famous Trajan of Damascus. Its acoustics were suitable for musical performances when there was a ceiling, as opposed to its current outdoor form, where its acoustics are primarily suitable for speech reproduction, such as in the large outdoor ancient theaters used for ancient drama performances [57].

2.4.6. GS6: Caves

The landscape on the north slope of the Acropolis is steep, wild, and imposing, with karst caves and springs (Figure 7). It is the site of ancient myths and the worship of the earthy powers of vegetation and fertility [58]. There are temples of Zeus, Panas, and Apollo Hypoakraios. The Sanctuary of Aglauros (Figure 8f) is located on the eastern side of the Acropolis and is the city's largest cave (22 m east-west, mouth 14 m). According to Herodotu, the Persians invaded the Acropolis from here in 480 B.C. Aglauros, the daughter of the mythical king Kekrops, jumped from the Acropolis to save Athens from an extended siege. It was to her shrine that Athenian ephebes, once they had reached the age of 18, brought their military gear, swearing to protect the "sacred and holy" to the death, following the example of the Nymph.

2.4.7. GS7: Klepsydra

Klepsydra (Figures 7 and 8g) is the oldest water source on the Acropolis, dating back to before the city of Athens. It is situated on the northwest side of the rock, at the crossroads of the ancient Panathenaia and Promenade streets. The hidden spring was discovered by Neolithic people, who dug 22 wells 3–5 m deep to draw its water. The Athenians discovered water at the bottom of a deep fissure in the Late Bronze Age and built a stone and a wooden staircase to reach it [59]. It is thought to have formed as a result of an earthquake that caused a 35 m fault to cut through the rock formations on the north-east side of the Acropolis and reach the underlying red shale. This fault contributed to the formation of a cave and a karst spring within it. The spring was dedicated to the worship of Empedus, the water nymph. The Athenians named it "Klepsydra," which means "water thief," because its water rose and overflowed when the annual winds, the meltemia, blew, but vanished when they stopped [60]. This water supply alternation is typical of karst springs, where rainwater infiltrates through cracks in the limestone rocks. Today, the spring's ruins can be found deep underground, where a 69-step staircase leads to a 10 m deep marble-mouthed well. The ancient spring's water can be found at the bottom. During rainy seasons, there is so much water that it reaches the Ancient Agora.

Figure 9. (**a**) Panoramic view of GS5: Odeon of Herodes Atticus–Herodeion (**b**) Details of the building stones in GS5: Odeon of Herodes Atticus–Herodeion (Photo by Fani Tripolitsiotou).

2.4.8. GS8: Mycenaean Fountain

The Mycenaean fountain (Figures 7 and 8h) was built in the second half of the 13th century AD. It was discovered alongside the Mycenaean wall during the 1937 excavations [61]. It only lasted 30 years before being destroyed by an earthquake rendering its deepest part useless. Then a large part of the rock detached resulting in the formation of a fault. Its depth is 40 m while its width is 1 to 1.5 m. It was accessible via an invisible ladder, constructed with timber ties and built-in sections. Despite its destruction, it remained a secret passage throughout ancient times. The scale was divided into eight sections. The two upper ones had wooden steps attached to both sides of the rock that led to the northern slope's cavernous exit. The remaining sections were made of stone slabs, with a well carved in the background, 8 m deep and 2 m in diameter on the upper surface and 4 m on the bottom. The water was pumped with containers, while there was a cavity to collect sediments [58]. According to Pausanias' description, the scale is related to the festival of the "Arrifores", and it is believed that the Arrifores, who were four aristocratic girls aged seven to eleven, descended from here one night in early summer, bringing the "Arrita", the unspoken things, to the Sanctuary of Aphrodite [58].

2.5. Quantitative Assessment

Although there are numerous methods for quantitatively assessing geodiversity sites and elements, few of them were designed for urban areas [62], and urban geodiversity inventories are still uncommon [63].

In this study, the Suzuki and Takagi [64] method is used to evaluate geosites in urban environments, modified by Vegas [65], who evaluated the geosites of the Spanish city of Segovia.

This method consists of six main axes (fields): educational value (Ved), scientific value (Vsc), tourism value (Vtr), safety and accessibility assessment (Vsa), conservation-sustainability status (Vcs) and tourist information material (Vti).

Suzuki and Takagi [64] set three sub-criteria for each axis, while Vegas [65] set one extra sub-criterion for each axis, so that the method now has a total of 24 sub-criteria, four for each main axis. Each sub-criterion is scored from 1 to 4, with 1 indicating no score, 2 indicating low, 3 indicating medium, and 4 indicating high.

The six criteria and sub-criteria designed for this assessment are described in Tables 1–6.

Table 1. Value for environmental education and sub-criteria [64,65].

		Ved: Value for Environmental Education			
		Score			
ID	Sub-Criteria	1	2	3	4
Ved1	Ease of understanding of the geological history	Difficult to understand, even with the explanation from a geoguide	Understandable with an explanation from a geoguide	Can be understood with an explanation board or other information sources	Easy to understand without any additional explanations
Ved2	Representativeness	None, only for scientific use	Low, content is complicated or not very representative of what the geosite is intended to illustrate	Moderate, it is representative of the geosite, but it is not the best example in the city	High, it is very representative of the geological framework and is the best example in the city
Ved3	Ease of understanding information at the geosite (panels, brochures, website …)	There is no interpretational information	Content is complicated or is not adequate for environmental education	The interpretational material is simple and easy to understand, but without any content for environmental interpretation	The interpretational material is simple and easy to understand and contains aspects for environmental education
Ved4	Flexibility for educational use	Low, only by adults with secondary or higher education levels	Moderate, can only be understood by adults with basic studies	Can be understood by adults with basic studies and studentsof secondary education	High, can be understoodby adults and students of all educational levels

In the present study, it was considered necessary to redefine certain criteria due to the specificity of some geotopes of the Acropolis rock, which is located at a central point of the city of Athens. Specifically, in the third axis Vtr, in the last sub-criterion concerning the proximity to tourist information points, there was the following change in the scoring steps: score 1 corresponded to a time interval of more than half an hour; score 2 corresponded to a time interval of less than 30 min and more than 15 min; score 3 corresponded to a time interval of less than 15 min and more than 5 min; and finally, score 4 corresponded to a time interval of less than 5 min.

Table 2. Scientific value and sub-criteria [64,65].

		Vsc: Scientific Value			
		Score			
ID	Sub-criteria	1	2	3	4
Vsc1	Research significance	Low, there are no exclusive scientific publications on the geosite, nor is it the object of research	Moderate, there are scientific publications or it has been the object of study or visits by research teams with a national scope	High, there are over five scientific publications and it has been the object of study by international teams	Very high, it has been highlighted as a site of reference in multiple international publications or has aroused interest from the international scientific community
Vsc2	Clarity and non-obsolescence of the scientific story in the interpretational material (panels, guide books and web sites)	Low: there is no scientific material or it is obsolete	Partially explained: there is scientific information but not specific to the geosite	Explained: there is scientific information but it is incomplete or not as up-to-date as it could be	Clearly explained: the scientific information is clear, up-to-date and specific to the geosite
Vsc3	Rarity in the city of Athens and its municipal district	Not rare: there are many examples in the city	Moderate: there are at least five examples in the city	Rare: there are fewer than five examples in the whole municipal district	Very rare: there is only one example at the provincial or national level
Vsc4	Representativeness of the geosite within the city's geological frameworks	It is not representative of any of the geological frameworks	The geosite is only representative of a single geological framework	The geosite allows the representation of more than one geological framework	The geosite is the most representative example of its geological framework

Table 3. Tourism value and sub-criteria [64,65] (modified by us).

		Vtr: Tourism Value			
		Score			
ID	Sub-Criteria	1	2	3	4
Vtr1	Aesthetic/Emotional value, such as beauty or iconic site	Low: it has no scenic or aesthetic significance (small-scale reliefs, no chromatic variety, no watercourses …)	Moderate: it has morphological expression or some chromatic variety	High: it forms part of the relief in a sector in the city, with chromatic variety and the presence of water courses	Very high: it defines the structural landscape of the city or has emotional links to many sectors of the population
Vtr2	Other natural andcultural values	None, it has no more value than geological	Unimportant, there is some natural or cultural element, but it is not relevant	Important, there are natural or cultural elements	Very important, there are natural or cultural heritage elements and they are catalogued and recognised (property of cultural interest, singular trees, nesting sites …)
Vtr3	Other tourist attractions in the vicinity	None	Exist, but they are not in themselves important enough to attract tourism	Exist, and they are of interest to tourists, but are outside the Athens Tourist Board's main routes and guided visits	Famous attractions. They are among the most popular guided visits in the city
Vtr4	Proximity to the city's tourist offices and information centres (measuredby walking time)	More than half an hour	Between half an hour and fifteen minutes	Between fifteen and five minutes	Less than five minutes

Table 4. Safety and accessibility and sub-criteria [64,65] (modified by us).

		Vsa: Safety and Accessibility			
		Score			
ID	Sub-Criteria	1	2	3	4
Vsa1	Safety conditions of the geosite and the route leading to it	Relatively dangerous: it is on the verge of a street or road without a pavement or it is necessary to cross very busy roads, or near dangerous places (escarpments or sites with natural hazards)	Moderate risk of danger: there is a pavement or a small space for accommodating large groups, or the need to cross roads on zebra crossings	Not very dangerous: located in parks or on broad pavements, without nearby traffic, or on roads closed to traffic	Safe geosite: located in broad spaces without vehicle traffic, very little pedestrian transit, flat and with no natural hazards
Vsa2	Transit time walking between environmental interpretation centres in the city of Athens	More than one hour	More than 30 min	More than 15 min	Less than 15 min
Vsa3	Walking time from the closest bus stop to the geosite	More than 30 min	More than 15 min	More than 5 min	Right beside the bus stop
Vsa4	Accessibility for people with mobility problems	Not accessible for people with mobility problems	It is accessible for people with mobility problems, but assisted by carers	It is accessible to people with mobility problems, but not throughout the geotope	It is accessible to people with mobility problems and even throughout the geotope

Table 5. Conservation and site sustainability and sub-criteria [64,65].

		Vcs: Conservation and Site Sustainability			
		Score			
ID	Sub-Criteria	1	2	3	4
Vcs1	Current state of conservation	Low: the geological values of the geosite are not conserved	Partially conserved: there are some elements of the geosite that are not conserved	Moderately conserved: all the geological elements can be recognised	Well conserved: the geosite is entirely conserved
Vcs2	Legal protection	Not protected: it is not covered by any law or regulations	Existing plans for protection, but with a general scope for the geosite, not specifically referring to its geological values	Partially protected: it is recognised in the GUOP, but for its natural or cultural values rather than its geological value	Protected due to its geological value
Vcs3	Natural site sustainability	Difficult to preserve, subjected to natural processes (geomorphological processes or vegetation that conceals it) with high activity or frequency, or possible extreme catastrophic events	It may be damaged by medium- term natural processes (return periods of ten or more years)	It could be affected by some natural disaster or exceptional vegetation growth with return periods of over 100 years	With no incidence of natural processes that could affect its conservation
Vcs4	Anthropic sustainability	Difficult to preserve, subjected to constant, direct and damaging anthropic impacts on the geosite	It may be damaged by medium-term anthropic activities (over ten years)	It may suffer some anthropic impact but exceptionally, and with a frequency of over one century.	No incidence of anthropic activities that affect its conservation

Table 6. Value for environmental information for geotourism and sub-criteria [64,65].

| | | Vti: Value of Environmental Information for Geotourism | | | |
| | | Score | | | |
ID	Sub-Criteria	1	2	3	4
Vti1	Information panels of the approach to geosite	No information or panels that do not contain any environmental information	Limited information, but they are at high risk of vandalism or deterioration	There are panels but they only moderately facilitate environmental education	There are panels and they provide useful information for environmental education
Vti2	Geosite information on the Internet and in brochures and guide books	No information	There is information on websites or in printed matter	There is information both on the Internet and in printed matter	There is a wide range of information resources and also guides specialising in environmental interpretation
Vti3	International character of the environmental information (in different languages)	No information	One language only (Greek)	Two languages (Greek and English) or in Braille	More than two languages or the people who perform the environmental interpretation speak more than two languages (including sign language)
Vti4	Transmission of values and attitudes for the geoconservation	No value or attitude	Some vague reference to geoconservation	It contains references to recommendations and prohibitions related to geoconservation	The information includes clear and explicit references encouraging values and attitudes for geoconservation

Regarding the fourth axis, Vsa, the last sub-criterion which concerns the accessibility of geosites by people with physical or mental disabilities, was limited only to accessibility by people with mobility problems, including people who have limitations (temporary or permanent) who require accessibility to the environment: the elderly, children, pregnant women, families with pushchairs, people in plaster casts, etc. For this reason, there were the following changes in the scoring steps: score 1 indicated the non-accessibility of the geosite to people with mobility problems, score 2 indicated the accessibility of people with mobility problems but with the help of caregivers, score 3 indicated the accessibility of people with mobility problems but not throughout the entire geosite, and finally, score 4 indicated the accessibility of people with mobility problems throughout the entire geosite.

To make reading and interpreting the results easier, the six axes were grouped into pairs to represent a broader concept. The first axis of educational value, for example, in conjunction with the fourth axis of assessment for safety and accessibility, could indicate educational utility. This value encompassed all age groups, as well as all the characteristics that govern each visitor group. Similarly, the second axis of scientific value, in conjunction with the fifth axis, the assessment of conservation-sustainability status, could encompass all the elements that promote the enhancement of geological heritage. Finally, the third axis, which refers to tourism value, in conjunction with the sixth axis, which refers to the availability of information material on the area's environment and geology, could indicate the possibility of promoting geotourism.

3. Results

Table 7 shows the six axes with the four sub-criteria and their respective scores.

Table 7. Score of the sub-criteria for each assessment axis, for each geotope.

	Asklepeion (GS1)	Theater of Dionysus (GS2)	Thrasyllus Monument (GS3)	Stoa Eumenos (GS4)	Herodeion (GS5)	Caves (GS6)	Klepsydra (GS7)	Mycenaean Fountain (GS8)
Ved_1	3	2	3	3	3	3	3	2
Ved_2	3	4	4	4	4	3	4	2
Ved_3	3	3	3	3	3	3	3	3
Ved_4	4	4	4	4	4	4	4	4
Vsc_1	4	4	3	4	4	4	4	4
Vsc_2	4	2	2	3	2	3	2	2
Vsc_3	3	3	3	4	4	2	4	2
Vsc_4	3	3	3	3	4	3	4	3
Vtr_1	2	4	4	4	4	4	3	2
Vtr_2	4	4	4	4	4	4	4	3
Vtr_3	4	4	4	4	4	4	3	3
Vtr_4	3	3	3	3	3	2	2	2
Vsa_1	2	2	1	3	4	1	1	1
Vsa_2	4	4	3	4	4	2	3	2
Vsa_3	3	3	2	3	3	2	2	2
Vsa_4	2	3	1	2	4	1	1	1
Vcs_1	3	2	3	3	3	3	2	3
Vcs_2	3	3	3	3	3	3	3	2
Vcs_3	2	2	2	3	3	3	2	3
Vcs_4	2	2	3	3	3	3	3	2
Vti_1	1	3	1	1	1	1	1	1
Vti_2	2	2	2	2	2	2	2	2
Vti_3	3	3	3	3	4	3	3	2
Vti_4	2	1	1	1	1	1	1	1

Table 8 shows the mean values for each axis, obtained by summing the corresponding sub-criteria for each axis, dividing by 4.

Table 8. Mean value for each axis, for each geotope.

Average	Asklepeion (GS1)	Theater of Dionysus (GS2)	Thrasyllus Monument (GS3)	Stoa Eumenos (GS4)	Herodeion (GS5)	Caves (GS6)	Klepsydra (GS7)	Mycenaean Fountain (GS8)
Ved	3.25	3.25	3.5	3.5	3.5	3.25	3.5	2.75
Vsc	3.5	3	2.75	3.5	3.5	3	3.5	2.75
Vtr	3.25	3.75	3.75	3.75	3.75	3.5	3	2.5
Vsa	2.75	3	1.75	3	3.75	1.5	1.75	1.5
Vcs	2.5	2.25	2.75	3	3	3	2.5	2.5
Vti	2	2.25	1.75	1.75	2	1.75	1.75	1.5

At first glance, according to the results presented in the tables above, the Odeon of Herodes Atticus (Herodeion) exhibited the highest scores in each axis, except the last one, which concerns environmental information for geotouristic purposes. It was the most widely recognised geotope, with significant scientific and tourist value. Indeed, it presents interdisciplinarity as it combines architecture with physics (acoustic) and geology (building stones). It is relatively well-preserved, and it is also valuable for environmental education. However, it has limited geological information, which is required for geotourism exploitation.

On the other hand, the Mycenaean Fountain had the lowest scores on each axis, with safety and accessibility being extremely low, and environmental information for geotourism almost non-existent. However, the average score on the first and second axes for environmental education and scientific value was 2.75, indicating that it is interesting in terms of geological heritage.

Regarding the first axis (Value for environmental education—Ved) scores, all geotopes had high scores, ranging from 2.75 to 3.5, with the Mycenaean fountain having the lowest, indicating high value for Environmental Education. The Asclepeion, the caves and the Klepsydra are ideal geotopes for informing and educating the public about the phenomenon of karstification. The Theatre of Dionysus, the Monument of Thrassylus and the Stoa of Eumenes provide information about marly limestone (Aktitis Lithos), the marble of Pendeli and the marble of Hymettus, as well as their use as building materials. Moreover, the bioclimatic construction of the Dionysu Theatre can provide additional information on the climatic conditions prevailing at the time of its operation. The Klepsydra, apart from the phenomena of karstification, is directly linked to natural hazards, since its origin is due to seismic activity and the creation of a fault line. Consequently, because these geotopes are part of the world-famous Acropolis rock, their educational use is especially beneficial to students around the world. However, geo-environmental education has not taken advantage of this very important urban geotops. Students are more concerned with the cultural and historical significance of the Acropolis rock and its monuments than with its geological and scientific significance. They are perplexed by the close relationship between geological and cultural heritage [66].

As far as the second axis is concerned, which is about the scientific value, the studied geotopes scored highly, ranging from 2.75 to 3.5. Indeed, these geotopes are studied from an interdisciplinary perspective, since they serve as reference points in different areas of study and research (petrology, paleoclimatology, architecture, mechanics, physics, hydrology, natural hazards).

The scores on the third axis, which reflects the geotopes' tourist value, remained high, ranging from 2.5 to 3.75, indicating that the studied geotopes have great potential for tourist development and appeal to a diverse range of tourists.

In the fourth axis, which concerns safety and accessibility, four of the eight geotopes had very low accessibility, less than 2 (caves, Mycenaean fountain, Klepsydra, Thrasyllus monument), while the others appeared to be more accessible and safer for a tour. The Herodeion demonstrated the highest level of safety and accessibility.

In the fifth axis, which concerns conservation and sustainability, scores ranging from 2.25 to 3 were observed. In its long history, the geological and cultural heritage of the Acropolis rock have been severely damaged. As an example, in June 1641, many buildings in Athens were severely damaged by stones falling from the Acropolis. In November 1805, a strong earthquake caused significant damage to the Parthenon. However, if there is one thing that has been of great concern to scientists and engineers, it is the secret of the Acropolis rock's flawless seismic behaviour in a 25-century history of earthquake tremors. In general, the Acropolis and its monuments are genius structures in terms of their seismic risk behaviour. However, these unique monuments have suffered man-made and natural disasters over the course of history from various causes (wars, earthquakes, etc.), requiring a special study by experts to restore them, which is a particularly demanding and difficult task. Since the foundation of the new Greek state in 1830, restoration efforts have been made and projects have been launched to repair any damage caused by natural disasters. However, it is always imperative that man-made interventions protect and enhance these geological sites so that they are preserved as unchanged as possible over time, as they are monuments of historical and geological value of global interest.

Finally, the sixth axis had rather low scores, ranging from 1.5 to 2.25. While the historical and cultural information material is very rich, the same is not true for the geoenvironmental information. Therefore, the visitor's knowledge of the geological characteristics of the area is minimal, which means that geological information to promote geotourism is almost non-existent. Furthermore, there is no information leaflet on the geoheritage of the Acropolis Rock as one of the most intensively developed and visited areas, nor are there any signs or QR codes. There is, therefore, a lack of awareness of the fact that the Acropolis Rock has high geological value, and its promotion is expected to increase geotourism and public awareness of geoenvironmental and geoconservation issues.

The scores for each axis for each geotope were plotted with radar type diagrams, as shown in Figure 10.

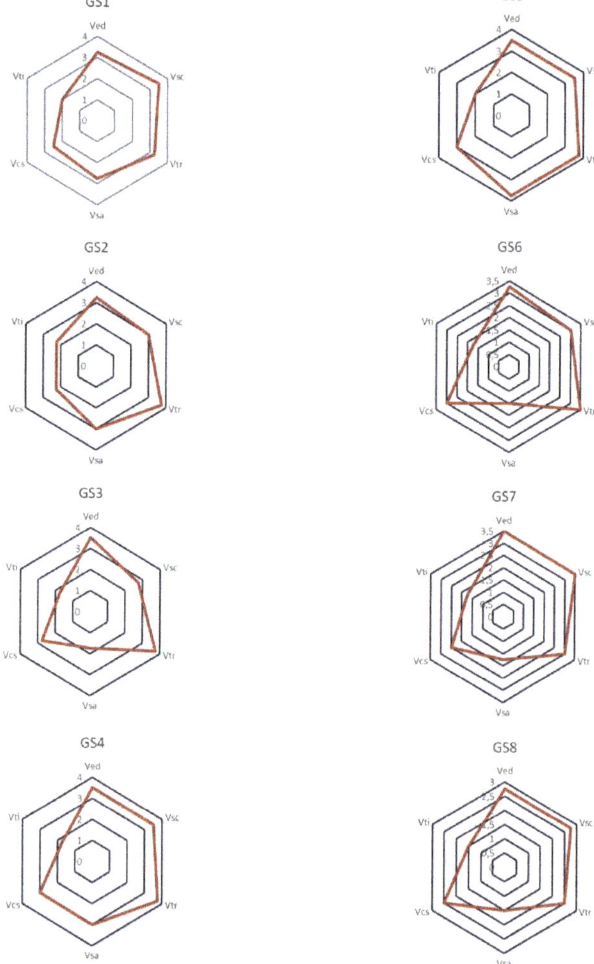

Figure 10. Radar diagrams which depict the axes scores for each geotope where Ved: educational value; Vsc: scientific value; Vtr: tourism value; Vsa: safety and accessibility assessment; Vcs: conservation-sustainability status and Vti: tourist information material.

According to the radar charts, all the geotopes studied had a high educational value (Ved) (score > 3), except for the Mycenaean Fountain (GS6), which was close to score 3. However, in addition to educational value, the majority of the geotopes had relatively high accessibility scores (Vsa). This is critical, because educational activities for all age groups could be developed without difficulty.

All the geotopes had a high scientific value (Vsc) (values/scores > 3) (except the Thrasyllus Monument-GS3 and the Mycenaean Fountain-GS6, which were close to score 3), while also having satisfactory conservation status values (Vcs).

As a result of the above combinations, there is the possibility of promoting and highlighting the geological heritage reflected in the study areas.

Finally, all the study sites had quite high scores in terms of the tourism value (Vtr) they project (six of the eight study sites had values greater than 3). However, they did not quite have satisfactory scores in relation to the information material (Vti) provided (all study sites had scores < 3). This highlights an issue that must be addressed immediately in terms of informing and promoting the values held by these sites. This will only highlight their geotourism value.

4. Discussion

Athens is a city with high geotourism potential due to the large concentration of unique, easily accessible geotourist sites in the area, well-developed tourist infrastructure and good information support. The geotourist sites of Athens are carriers of historical and cultural information, elements of the urban ecosystem.

The Acropolis of Athens, a monument with worldwide fame and recognition, is a combination of natural and cultural heritage. Its natural heritage cannot be separated from the building heritage, according to the UNESCO Convention for the Protection of the World Cultural and Natural Heritage, adopted by the General Conference of UNESCO on 16 November 1972. The study of building and decorative stones used throughout the history of Acropolis hill is, therefore, one of the most important approaches to combining geology and cultural heritage in an urban environment. It is worth noting that natural stones were used as building and/or decorative materials in historical cities. Many experts have focused on building materials that reflect geology at the local/regional level. Morra et al. [67], for example, investigated volcanic materials used in various construction projects in Naples (Italy) and discovered links between geological and architectural heritage. D'Atri et al. [68] and Borghi et al. [69] attempted something similar for Milan and Turin, respectively, in Italy. No doubt, this reflects an aspect of urban geoheritage that cannot be ignored. Del Lama et al. [18], in their publication on urban geotourism opportunities in the old centre of Sao Paulo (Brazil), highlighted the value of building stones in addition to the importance of local geomorphological features for the development of the city.

The importance of the cultural history of the Acropolis, which is linked to its archaeological and mythological history, is considered indisputable. Its geological history is evident in the field and one can easily perceive how the geological features have contributed to the preservation of the monument over the centuries. However, geo-environmental education has not utilized this very important urban site. Cheila [66], in her study, revealed that students perceive, to a high degree, the cultural and historical value of the Acropolis rock and its monuments, in contrast to its geological and scientific value. They struggle to understand the close interaction between geological and cultural heritage. The case of the Acropolis of Athens is an excellent opportunity for the implementation of geo-environmental education programmes that will help students to understand the close relationship between the two types of heritage [66]. Unfortunately, students lack the fundamental understanding of geosciences, which is important for the daily life of citizens [70]. Therefore, it is vital to include geoeducation in school curricula, which will take place in the field under the supervision of qualified personnel, using a multidisciplinary, holistic approach, with the goal of producing environmentally responsible citizens with well-developed geoethical values [66]. Geoeducation is called on to help students recognize the need for management of the natural environment, with a sense of responsibility and participation. Furthermore, geoeducation makes it clear that the preservation of geodiversity is a basic condition for the preservation of biodiversity [71].

Unfortunately, in the Greek educational system, this knowledge is provided in primary school through a series of course-thematic units of a few teaching hours, in the context of Geography course, taught by unskilled geology teaching staff [72], while in Lower High School Education, through the subject of Geology-Geography, where there is a lack of topics on geoheritage [73]. Moreover, the number of environmental groups in Greek schools that choose to develop an environmental program with a geoenvironmental theme is very limited [74], according to annual data. As a result, student education is characterised as very limited or non-existent, and it is regarded as incomplete to negligible in the fields of geosciences and geoenvironmental sciences [66].

In contrast to geological heritage, the orientation of curricula in cultural heritage understanding is evident through various disciplines (History, Religion, Literature, Visual Arts) and several cultural programs that are annually and voluntarily prepared in Greek schools. However, no research has been conducted on pupils' and students' geological understanding, geoethical awareness, or cultural heritage understanding. The work of Georgousis et al. [75], who investigated the younger generation's geocultural consciousness, focused on the geological dimension of the world-famous cultural heritage object, Meteora Geomorphes, in Greece, to establish the younger generation's understanding of the geological heritage concerning cultural heritage, and is the only exception. According to their results, pupils and students in Greece perceived, to a high degree, the aesthetic value and cultural value of geodiversity, with its historical, archaeological and religious aspects, compared to geological value, which was perceived only to a moderate degree. At a moderate grade, they also perceived the scientific value of geoheritage, but recognised the need for geoeducation.

Geotourism has several considerations as well. The high historical and cultural values of the studied geotopes have enormous geotourism potential, which could be developed by establishing a visitor center and organizing more frequent tours. The geoheritage of the Acropolis rock could be promoted through geotours that offer a special tourist (and educational) package and provide an overview of the Acropolis' geodiversity, and could be included in Athens' tourism strategy and promotion. Similar cases of geotours exist in London [76], São Paulo [18] and Rome [77].

Except for the caves, the geosites studied in this work are open to the public. This is critical because visitors can see the rocks and unique geological features of each monument. They can tell the difference between the various rocks from which they are formed, such as Pendeli marble, Aktitis Lithos, and so on. However, for the description, understanding, and importance of the geological value, a specialised person who knows geology is required to transmit this knowledge.

Some key considerations and future projects that should be adopted by local governments and incorporated into city planning processes, are proposed. This initiative could be developed into a geodiversity action plan for the city of Athens, which would be the first of its kind for a city in Greece. Such a plan is urgently required to avoid the degradation and destruction of the Acropolis rock's geological heritage.

The assessment of these geosites should serve as a catalyst for further research into the following: the recording of fauna and flora, in order to highlight biodiversity (biodiversity is listed as an additional value in several inventories, and its detailed assessment should be done separately by appropriate experts); cultural heritage relevance, inviting experts to record the possible associations of each site in this regard; safety and conservation by landscape engineers and architects, who can survey sites to find creative ways to ensure safety while preserving heritage and integrating it into the urban fabric sustainably.

Furthermore, our findings demonstrated the utility of applying the quantitative assessment method developed by Suzuki and Takagi [64] to urban geosites. This is not the first time this method has been used to assess the potential for geotourism and geoeducation in an urban setting. However, it should be emphasised that in order for the assessment results to be representative of the studied geosites and the target audience group (audience specificities), the peculiarities that may be concealed in each geotope must be taken into account. Modification of the sub-criteria (addition or removal of sub-cases) is required in this case.

5. Conclusions

The Acropolis Rock is one of the most admirable and recognisable monuments in the world, known worldwide for its archaeological, historical, and tourist value. At the same time, it is a remarkable monument of geological heritage, which is, however, unknown to the public. The study of building and decorative stones used throughout the history of the Acropolis hill is one of the most important approaches to combining geology and cultural heritage in an urban environment.

This study described and evaluated eight geosites on the Acropolis rock.

From the assessment data, it was concluded that the studied geosites presented high scientific and touristic value, as well as high value for environmental education. Through the discussion, the immediate need for the integration of geoeducation into the curricula of schools became apparent, so that, apart from the archaeological and cultural value of this world-famous monument, its connection with geological heritage can be made known. At the same time, however, in most geosites there is difficulty in accessibility for different reasons in each one, as they are located in different parts of the rock of the Acropolis, which presents special geological and topographical interest.

The possibility of exploiting these geosites in terms of geotourism is very high and important for strengthening the economy of the place, but mainly for raising the awareness of visitors on matters of the environment, geological heritage, and geoconservation. The sustainability and utilisation of geosites, based on the principles of sustainable development, provide multiple benefits for society and future generations.

Author Contributions: Conceptualization, H.D.; methodology, H.D. and G.Z.; investigation, H.D., F.T., G.Z. and T.C.; resources, F.T. and T.C.; writing—original draft preparation, F.T. and T.C.; writing— review and editing, H.D.; supervision, H.D. All authors have read and agreed to the published version of the manuscript.

Funding: This research received no external funding.

Institutional Review Board Statement: Not applicable.

Informed Consent Statement: Not applicable.

Data Availability Statement: The data presented in this study are available on request from the corresponding author.

Acknowledgments: The authors gratefully thank the journal editor and the three reviewers for their thorough consideration of this paper. Special thanks to Panos Karoutsos for his assistance in collecting photographic material and for entrusting us with his photographic archive.

Conflicts of Interest: The authors declare no conflict of interest.

References

1. Drinia, H.; Tsipra, T.; Panagiaris, G.; Patsoules, M.; Papantoniou, C.; Magganas, A. Geological Heritage of Syros Island, Cyclades Complex, Greece: An Assessment and Geotourism Perspectives. *Geosciences* **2021**, *11*, 138. [CrossRef]
2. Zafeiropoulos, G.; Drinia, H. Kalymnos Island, SE Aegean Sea: From Fishing Sponges and Rock Climbing to Geotourism Perspective. *Heritage* **2021**, *4*, 3126–3146. [CrossRef]
3. Evelpidou, N.; Karkani, A.; Tzouxanioti, M.; Spyrou, E.; Petropoulos, A.; Lakidi, L. Inventory and Assessment of the Geomorphosites in Central Cyclades, Greece: The Case of Paros and Naxos Islands. *Geosciences* **2021**, *11*, 512. Available online: https://www.mdpi.com/2076-3263/11/12/512 (accessed on 25 June 2022). [CrossRef]

4.   Spyrou, E.; Triantaphyllou, M.V.; Tsourou, T.; Vassilakis, E.; Asimakopoulos, C.; Konsolaki, A.; Markakis, D.; Marketou-Galari, D.; Skentos, A. Assessment of Geological Heritage Sites and Their Significance for Geotouristic Exploitation: The Case of Lefkas, Meganisi, Kefalonia and Ithaki Islands, Ionian Sea, Greece. *Geosciences* **2022**, *12*, 55. Available online: https://www.mdpi.com/2076-3263/12/2/55 (accessed on 25 June 2022). [CrossRef]
5.   Golfinopoulos, V.; Papadopoulou, P.; Koumoutsou, E.; Zouros, N.; Fassoulas, C.; Zelilidis, A.; Iliopoulos, G. Quantitative Assessment of the Geosites of Chelmos-Vouraikos UNESCO Global Geopark (Greece). *Geosciences* **2022**, *12*, 63. Available online: https://www.mdpi.com/2076-3263/12/2/63 (accessed on 25 June 2022). [CrossRef]
6.   Zouros, N.; Valiakos, I. Geoparks Management and Assessment. *Bull. Geol. Soc. Greece* **2017**, *43*, 965–977. [CrossRef]
7.   Karamani, P.; Drinia, H.; Panagiaris, G. Geotourism as a tool for the protection and promotion of the Cave of Galaxidi. In Proceedings of the 15th International Congress of the Geological Society of Greece Athens, Athens, Greece, 22–24 May 2019.
8.   Zafeiropoulos, G.; Drinia, H.; Antonarakou, A.; Zouros, N. From Geoheritage to Geoeducation, Geoethics and Geotourism: A Critical Evaluation of the Greek Region. *Geosciences* **2021**, *11*, 381. [CrossRef]
9.   Vlachopoulos, N.; Voudouris, P. Preservation of the Geoheritage and Mining Heritage of Serifos Island, Greece: Geotourism Perspectives in a Potential New Global Unesco Geopark. *Geosciences* **2022**, *12*, 127. [CrossRef]
10. Del Lama, E.A. Urban geotourism with an emphasis on the city of São Paulo, Brazil. In *Handbook of Geotourism*; Dowling, R.K., Newsome, D., Eds.; Edward Elgar Publishing: Cheltenham, UK, 2018; pp. 210–223.
11. Habibi, T.; Ponedelnik, A.A.; Yashalova, N.N.; Ruban, D.A. Urban geoheritage complexity: Evidence of a unique natural resource from Shiraz city in Iran. *Resour. Policy* **2018**, *59*, 85–94. [CrossRef]
12. Habibi, T.; Ruban, D.A. Geoheritage of the Neyriz ophiolite-related radiolarite sequence (Cretaceous; southwest Iran): First report and evaluation in regional and global contexts. *J. Afr. Earth Sci.* **2018**, *145*, 227–233. [CrossRef]
13. Palacio, J. Geoheritage Within Cities: Urban Geosites in Mexico City. *Geoheritage* **2014**, *7*, 365–373. [CrossRef]
14. Brilha, J. Inventory and quantitative assessment of geosites and geodiversity sites: A review. *Geoheritage* **2016**, *8*, 119–134. [CrossRef]
15. Reynard, E.; Pica, A.; Coratza, P. Urban geomorphological heritage. An overview. *Quaest. Geogr.* **2017**, *36*, 7–20. [CrossRef]
16. Petrović, M.D.; Lukić, D.M.; Radovanović, M.; Vujko, A.; Gajić, T.; Vuković, D. "Urban geosites" as an alternative geotourism destination - evidence from Belgrade. *Open Geosci.* **2017**, *9*, 442–456. [CrossRef]
17. Ng, C.Y.; McManus, P.; Dragovich, D. Using geoparks to conserve geological features in urban areas: A critical assessment of the potential and challenges of developing geotourism in Hong Kong. Abstract. In Proceedings of the UNESCO Global Geopark Conference, Geological Society of Northern Ireland, Belfast, Northern Ireland, 21 September 2006; p. 157.
18. Del Lama, E.A.; Bacci, D.D.L.C.; Martins, L.; Garcia, M.D.G.M.; Dehira, L.K. Urban Geotourism and the Old Centre of São Paulo City, Brazil. *Geoheritage* **2015**, *7*, 147–164. [CrossRef]
19. Riganti, A.; Johnston, J.F. Geotourism—A focus on the urban environment. In *Handbook of Geotourism*; Dowling, R.K., Newsome, D., Eds.; Edward Elgar Publishing: Cheltenham, UK, 2018; pp. 192–201.
20. Asprogerakas, E. City Competition and Urban Marketing: The Case of Tourism Industry in Athens. *Tour. Int. Multidiscip. Refereed J. Tour.* **2007**, *2*, 89–114.
21. Ólafsdóttir, R.; Tverijonaite, E. Geotourism: A Systematic Literature Review. *Geosciences* **2018**, *8*, 234. [CrossRef]
22. Brandolini, P.; Mandarino, A.; Paliaga, G.; Faccini, F. Anthropogenic landforms in an urbanized alluvial-coastal plain (Rapallo city, Italy). *J. Maps* **2021**, *17*, 86–97. [CrossRef]
23. Foumelis, M.; Fountoulis, I.; Papanikolaou, I.D.; Papanikolaou, D. Geodetic evidence for passive control of a major Miocene tectonic boundary on the contemporary deformation field of Athens (Greece). *Ann. Geophys.* **2013**, *56*, S0674. [CrossRef]
24. Karfakis, J.; Loupasakis, C. Geotechnical characteristics of the formation of "Tourkovounia" Limestones and their influence on urban construction-City of Athens, Greece. In *IAEG 2006, Paper number 794*; The Geological Society of London: London, UK, 2006.
25. Pavlopoulos, K.; Kotambasi, C.; Skentos, A. Geomorpgological evolution of the Athens basin. *Bull. Geol. Soc. Greece Thessalon.* **2005**, *38*, 1–13.
26. Gournelos, T.; Maroukian, H. Geomorphological observations concerning the evolution of the Basin of the Athens. *Geol. Balc. Sofia* **1990**, *20*, 15–24.
27. Papanikolaou, D.; Lozios, S.G.; Soukis, K.I.; Skourtos, E.N. The geological structure of the allochthonous "Athens schists". *Bull. Geol. Soc. Greece Thessalon.* **2004**, *36*, 1550–1559. [CrossRef]
28. Sindowski, F.K.H. Results and problems of heavy mineral analysis in Germany; a review of sedimentary-petrological papers, 1936–1948. *J. Sediment. Res.* **1949**, *19*, 3–25. [CrossRef]
29. Diakakis, M. An inventory of flood events in Athens, Greece, during the last 130 years. Seasonality and spatial distribution. *J. Flood Risk Manag.* **2013**, *7*, 332–343. [CrossRef]
30. Bathrellos, G.D.; Karymbalis, E.; Skilodimou, H.D.; Gaki-Papanastassiou, K.; Baltas, E.A. Urban flood hazard assessment in the basin of Athens Metropolitan city, Greece. *Environ. Earth Sci.* **2016**, *75*, 319. [CrossRef]
31. Lepsius, R. *Geologie von Attika. Ein Beitrag zur Lehre vom Metamorphismus der Gesteine*; D. Reimer: Berlin, Germany, 1893; p. 81.
32. Kober, L. *Der Hallstätter Salzberg*; Hölder-Pichler-Tempsky: Wien, Austria, 1929.
33. Marinos, G.; Petrascheck, W.E. Laurium. Geoligal and Geophysical Research. *Inst. Geol. Subsurf. Res.* **1956**, *5*, 266.
34. Tataris, A. Recent researches on the structure of Salamis Island and the opposite area of Perama. *Bull. Geol. Soc. Greece* **1967**, *7*, 36–51.

35. Niedermayer, J. Geological map and underground map of Athens, scale 1:10000. *Tech. Chamb. Greece Greek Geol. Soc.* **1971**.
36. Marinos, G.; Katsikatsos, G.; Georgiadou-Dikaioulia, E.; Mirkou, R. The Athens Shale system. Stratigraphy and tectonics. *Ann. Geol. Pays Hell* **1971**, *23*, 183–260.
37. Trikkalinos, J. The age of the metamorphic rock of Attica. *Ann. Géologiques Des Pays Helléniques* **1955**, *VI*, 193–198.
38. Paraskevaidis, H.; Chorianopoulos, P. An intersection through Mount Aegaleo. The schist of Athens. The hills of Athens. *Bull. Geol. Soc. Greece* **1979**, *13*, 116–141.
39. Regueiro, M.; Stamatakis, M.; Laskaridis, K. The geology of the Acropolis (Athens, Greece). *Eur. Geol.* **2014**, *38*, 45–52.
40. Marinos, G.; Katsikatsos, G.; Georgiadou-Dikeoulia, E.; Mikrou, R. The Athens schist formation, I. Stratigraphy and structure (in Greek). *Ann. Géologiques Des Pays Hell.* **1971**, *22*, 183–216.
41. Andronopoulos, B.; Koukis, G. *Engineering Geology Study in the Acropolis Area, Athens*; Engineering Geology Investigations 1976, Internal Reports, No. 1; IGME: Athens, Greece, 1976.
42. Aravantinos, A.; Kosmaki, P. *Outdoor Spaces in the City, Athens 1988, Greece*; Symeon Publications: Athens, Greece, 1988; pp. 150–152.
43. Panagiotidou, O. Illness and Healing in the Sanctuaries of Asclepius: A Cognitive Approach. Thessaloniki: Aristotle University of Thessaloniki. Acropolis Museum 2014. Available online: https://www.theacropolismuseum.gr/idrytiki-stili-toy-ieroy-toy-asklipioyamfiglyfo-toy-tilemahoy (accessed on 7 July 2022).
44. Milesis, S. Aktitis Piraeus Stone—The Quarries of Piraeus 2012. Available online: http://pireorama.blogspot.com/2012/10/blog-post_20.html (accessed on 8 July 2022).
45. Albini, U. *Nel Nome di Dioniso: Il Grande Teatro Classico Rivisitato Con Occhio Contamporaneo*; Prezzo Garzani Libri: Milano, Italy, 1999.
46. Moretti, J.C. *Theater and Society in Ancient Greece*; Translated by Dimitrakopoulos, E.; Patakis Publications: Athens, Greece, 2004.
47. Boletis, K.; Samara, A.; Aslanidis, K.; Papastamati Von Mook, H.; Parianou, E. *Restoration Work in the Theater of Dionysos*; Scientific Committee of the theater and sanctuary of the Asklepiion South Slope of the Acropolis; Ministry of Foreign Affairs-T.D.P.E.A.E.: Athens, Greece, 2009.
48. Blume, H.D. *Introduction to the Ancient Theater*; National Bank Educational Foundation: Athens, Greece, 1986.
49. Welter, G. Das choregische Denkmal des Thrasyllos. In *Archäologischer Anzeiger*; Walter de Gruyter: Berlin, Germany, 1938; pp. 36–68.
50. Greco, E. *Topografia di Atene, Sviluppo Urbano e Monumenti Dale Origini al III Secolo D.C.*; Scuola archeologica italiana: Athens, Greece, 2011; Volume 2.
51. Boletis, K. Sponsored Thrasyllus Monument, Athens 2012, Ministry of Culture and Sports. Available online: http://odysseus.culture.gr/h/2/gh251.jsp?obj_id=19820 (accessed on 7 July 2022).
52. Townsend, R. *Aspects of Athenian Architectural Activity in the Second Half of the Fourth Century B.C.*; University of North Carolina: Chapel Hill, NC, USA, 1982.
53. Coulton, J.J. *The Architectural Development of the Greek Stoa*; Oxford University Press: Oxford, UK, 1977.
54. Lefantzis, M.; Mpriana, M. *New Research During the Restoration of the Stoa of Eumenes the II*; Journal Anthemion; Centre of the Acropolis studies: Athens, Greece, 2005; Volume 13.
55. Giannikapani, E. In Efmenou. Ulysses Junction Athens 2012. Ministry of Culture and Sports. Available online: http://odysseus.culture.gr/h/2/gh251.jsp?obj_id=6623 (accessed on 9 July 2022).
56. Korres, M. *Herod's Roof and Other Giant Bridges*; Melissa Publications: Athens, Greece, 2014; pp. 150–160.
57. Vassilantonopoulos, S.L.; Mourjopoulos, J.N. The Acoustics of Roofed Ancient Odeia: The Case of Herodes Atticus Odeion. *Acta Acust. United Acust.* **2009**, *95*, 291–299. [CrossRef]
58. Tsakos, K. *Adorations and Loves on the Northern Slopes of the Acropolis*; Society of Friends of the National Archaeological Museum: Athens, Greece, 2016.
59. Hurwit, J.M.; Newton, A.D. *The Acropolis in the Age of Pericles*; Cambridge University Press: Cambridge, UK, 2004.
60. Parsons, A.W. Klepsydra and the Paved Court of the Pythion. *Hesperia: J. Am. Sch. Class. Stud. Athens* **1943**, *12*, 191–267. [CrossRef]
61. Broneer, O. A Mycenaean Fountain on the Athenian Acropolis. *Hesperia: J. Am. Sch. Class. Stud. Athens* **1939**, *8*, 317–433. [CrossRef]
62. Kubalíková, L.; Kirchner, K.; Kuda, F.; Bajer, A. Assessment of urban geotourism resources: An example of two geocultural sites in Brno, Czech Republic. *Geoheritage* **2020**, *12*, 7. [CrossRef]
63. Vegas, J.; Díez-Herrero, A. An assessment method for urban geoheritage as a model for environmental awareness and geotourism (Segovia, Spain). *Geoheritage* **2021**, *13*, 27. [CrossRef]
64. Suzuki, D.A.; Takagi, H. Evaluation of geosite for sustainable planning and management in geotourism. *Geoheritage* **2018**, *10*, 123–135. [CrossRef]
65. Vegas, J. El patrimonio geológico de la provincia de Segovia: Geodiversidad y geoconservación. *Colección Nat. Y Medio Ambiente Segovia* **2000**, *26*, 1–69.
66. Cheila, T. The importance of geoenvironmental education in redefining the relationship of students with the environment and society. The case of the sacred rock of the Acropolis. MsS Thesis, Post-graduate Program Environmental, Disaster & Crisis Management Strategies. Master's Thesis, National and Kapodistrian University of Athens, Zografou, Greece, 2022.

67. Morra, V.; Calcaterra, D.; Cappelletti, P.; Colella, A.; Fedele, L.; de' Gennaro, R.; Langella, A.; Mercurio, M.; de' Gennaro, M. Urban Geology: Relationships between Geological Setting and Architectural Heritage of the Neapolitan Area. *J. Virtual Explor.* **2010**, *36*, 27. [CrossRef]
68. D'Atri, A.; Borghi, A.; Martire, L.; Castelli, D.; Costa, E.; Dino, G.; Favero Longo, S.; Ferrando, S.; Gallo, L.M.; Giardino, M.; et al. Sedimentary rocks in the urban geological heritage of the Torino city. *J. Mediterr. Earth Sci.* **2013**, *5*, 51.
69. Borghi, A.; D'Atri, A.R.; Martire, L.; Castelli, D.; Costa, E.; Dino, G.A.; Longo, S.E.F.; Ferrando, S.; Gallo, L.M.; Giardino, M.; et al. Fragments of the Western Alpine Chain as Historic Ornamental Stones in Turin (Italy): Enhancement of Urban Geological Heritage through Geotourism. *Geoheritage* **2014**, *6*, 41–55. [CrossRef]
70. Trikolas, K.; Ladas, I. The necessity of teaching earth sciences in secondary education. In Proceedings of the 3rd International GEOschools Conference, Teaching Geosciences in Europe from Primary to Secondary School, Athens, Greece, 28–29 September 2013; pp. 28–29.
71. Fermeli, G.; Markopoulou-Diakantoni, A. Selecting Pedagogical Geotopes in Urban Environment. *Bull. Geol. Soc. Greece Thessalon.* **2004**, *36*, 649–658. Available online: https://ejournals.epublishing.ekt.gr/index.php/geosociety/article/view/1677 (accessed on 8 July 2022).
72. Rokka, A.C. Geology in Primary Education: Potential and Perspectives. *Bull. Geol. Soc. Greece* **2018**, *34*, 819–823. (In Greek) [CrossRef]
73. Meléndez, G.; Fermeli, G.; Koutsouveli, A. Analyzing Geology textbooks for secondary school curricula in Greece and Spain: Educational use of geological heritage. *Bull. Geol. Soc. Greece* **2007**, *40*, 1819–1832. [CrossRef]
74. Spartinou, M.; Zerlentis, I. The geological heritage of Cyclades and the Environmental Education. In Proceedings of the 6th Pan-Hellenic Geographical Conference of the Hellenic Geographical Society, Thessaloniki, Greece, 3–6 October 2002; Volume III. (In Greek)
75. Georgousis, E.; Savelides, S.; Mosios, S.; Holokolos, M.-V.; Drinia, H. The Need for Geoethical Awareness: The Importance of Geoenvironmental Education in Geoheritage Understanding in the Case of Meteora Geomorphes, Greece. *Sustainability* **2021**, *13*, 6626. [CrossRef]
76. Robinson, E. Geological walks around the city of London—Royal Exchange to Aldgate. *Proc. Geol. Assoc.* **1982**, *93*, 225–246. [CrossRef]
77. Pica, A.; Reynard, E.; Grangier, L.; Kaiser, C.; Ghiraldi, L.; Perotti, L.; Del Monte, M. GeoGuides, urban geotourism offer powered b GeoGuides, Urban Geotourism Offer Powered by Mobile Application Technology. *Geoheritage* **2018**, *10*, 311–326. [CrossRef]

 *geosciences*

Article

Large-Scale Accessibility as a New Perspective for Geoheritage Assessment

Yuri A. Fedorov [1], Anna V. Mikhailenko [1] and Dmitry A. Ruban [2,*]

1   Department of Physical Geography, Ecology, and Nature Protection, Institute of Earth Sciences, Southern
     Federal University, Zorge Street 40, Rostov-on-Don 344090, Russia
2   Department of Organization and Technologies of Service Activities, Higher School of Business, Southern
     Federal University, 23-ja Linija Street 43, Rostov-on-Don 344019, Russia
*   Correspondence: ruban-d@mail.ru

Abstract: The exploitation of geoheritage resources depends on their accessibility. The latter is usually established for geosites, whereas reaching the areas where geosites concentrate also deserves attention. Here, a novel, multi-criteria, score-based approach for assessing the large-scale accessibility of geoheritage-rich areas is proposed. The study takes into account various information about external and internal public transportation, road infrastructure, local services (including accommodation opportunities), and general settings. This approach is applied to the Russian South, where there are three geoheritage-rich areas, namely Lower Don, Abrau, and Mountainous Adygeya. Using new criteria, these areas differ by their large-scale accessibility, which is excellent in Lower Don and moderate in Abrau and Mountainous Adygeya. It is established that the co-occurrence of geoheritage-rich areas and popular tourist destinations does not guarantee excellent accessibility. The findings of the present study seem to be important for the development of optimal geoheritage resources policy, as well as for planning research and educational activities, such as the currently realized geochemical investigations and the regular field educational campaigns in the Russian South.

Keywords: geotourism; geosites; infrastructure; research projects; Russian South

Citation: Fedorov, Y.A.; Mikhailenko, A.V.; Ruban, D.A. Large-Scale Accessibility as a New Perspective for Geoheritage Assessment. *Geosciences* **2022**, *12*, 414. https://doi.org/ 10.3390/geosciences12110414

Academic Editors: Jesús F. Jordá Pardo and Jesus Martinez-Frias

Received: 28 September 2022
Accepted: 8 November 2022
Published: 10 November 2022

Publisher's Note: MDPI stays neutral with regard to jurisdictional claims in published maps and institutional affiliations.

1. Introduction

Current progress in geoheritage studies [1–10] is followed by the development of the concept of geoheritage resources [11–16]. The assessment of geoheritage sites (geosites) is an important procedure [7,17], but it needs significant reconsideration when applied to large areas in which geosites concentrate. Although some of these areas can be termed as geodiversity spots [18–20], the term "geodiversity" has already become so vague and indefinite that it can be left for theoretical needs. There is also a need to distinguish geosites sensu stricto and geodiversity sites due to their functional differences [21]. Therefore, the term "geoheritage-rich area" can be preferred for practical usage.

One of the most important properties of geoheritage is its accessibility, which determines the very opportunity to identify, describe, conserve, promote, and utilize unique geological features. Nonetheless, one should note that this property is only technical, and it is one of many other properties; thus, it does not determine the overall value of geoheritage. Many approaches have been proposed for geoheritage assessment, and almost all pay attention to the noted property (among the other properties). They have much in common, but differences and alternatives can also be found. The most popular approach has been proposed by Brilha [22], for whom accessibility is related to the educational and touristic values of geosites. Surprisingly, it is not related to scientific value, although scientists do not differ from students and tourists by their need to reach geosites (nonetheless, Brilha [22] noted use limitations). Accessibility is assessed by the distance between a given geosite and the road, the quality of the latter, and the availability of public transport (only buses are indicated). Such criteria matter in particular cultural and socio-economical contexts, but they are not universal. The other approach can be found in the work by Warowna et al. [23]. These specialists opposed the possibilities of reaching geosites by cars and public transport, and they also paid attention to the physical difficulties in reaching them. Analyzing the strengths and weaknesses of the previous proposals, Mikhailenko et al. [24] developed a multi-criteria approach for dealing with accessibility, which seems to be more or less independent on contexts and situations. Particularly, they emphasized the differences between outer and inner accessibility and paid attention to some other parameters, such as the need for permissions and entrance fees.

All the above-mentioned developments focus chiefly on geosites. Although the latter can be exploited for the purposes of research, education, and tourism taken alone, geoheritage-rich areas are more promising in this regard, and their accessibility needs special assessment. Although the importance of whole areas was already noted briefly by Brilha [22], the related approaches are still lacking. The accessibility-related developments for geoparks [25–29] either focused on internal infrastructural developments or openness to the community, which are significant, but there other aspects of the problem.

Road infrastructure is essential in geoheritage management due to its accessibility and connectivity functions [30–34]. A high-quality, paved road opens a given geosite to visitors. However, the presence of such a road means almost nothing if it is limited to an area connected to the other country by unpaved roads, or if such a road requires hours of driving without the possibility of stopping for dinner. In other words, it is important to realize that geoheritage is accessible not only locally, but also regionally and nationally. Assessing the related property for each particular geosite is unreasonable, except for the cases of single localities with global uniqueness isolated from other geosites. This means that accessibility can be assessed jointly for geoheritage-rich areas with multiple geosites. It can be termed as large-scale accessibility to be defined as the spectrum of opportunities to visit geoheritage-rich areas from other, more or less remote territories. Assessing this property is especially important in large countries such as Brazil, China, India, Russia, and Sudan, where geoheritage resources are distributed heterogeneously.

The objective of the present work is to introduce a novel approach for assessing the large-scale accessibility of geoheritage. It is tested for the territory of the Russian South, where three geoheritage-rich areas are known (Lower Don, Abrau, and Mountainous Adygeya). This development does not repeat what has already been proposed [22–24], although some previous experience is taken into account; regardless, this work focuses on a very novel perspective for the understanding of geoheritage resources.

2. Study Territory

This work deals with the territory known as the Russian South (Figure 1). This is a traditional label for the regions of the Southern and North Caucasian federal districts of the Russian Federation, which are situated in the very southwest of the country. This territory is known for its natural (mild climate and steppes); socio-economical (advanced agriculture, high entrepreneurial activity, and touristic importance); and cultural–historical (multiculturalism at the transition between Europe and Asia) peculiarities. Geographically, this huge territory encompasses grassy plains in the north and the center and forested mountains in the south (Figure 1). From the west, it is washed by the Azov and Black seas, the coasts of which form an almost continuous chain of famous resorts. Researchers have already examined the outstanding touristic and recreational potential of the Russian South [35–37].

Geoheritage resources are distributed highly heterogeneously within the Russian South. Presently, three geoheritage-rich areas are established there, namely Lower Don [38], Abrau [39], and Mountainous Adygeya [14] (Figure 1). To avoid repetition of the published information, the geoheritage characteristics of all three areas are summarized in Table 1, and some representative examples are shown on Figure 2.

Figure 1. Geographical location of the three considered geoheritage-rich areas.

Figure 2. Typical examples of geoheritage from the Russian South: Paleocene siliciclastic turbidites from Abrau (**a**), Late Jurassic carbonates from Mountainous Adygeya (**b**), and Miocene skeletal limestones from Lower Don (**c**).

Table 1. Basic characteristics of the considered geoheritage-rich areas (compiled from [14,38,39]).

Characteristics	Geoheritage-Rich Areas		
	Lower Don	Abrau	Mountainous Adygeya
Location (administrative affinity)	Rostov Region	Krasnodar Region	Republic of Adygeya and Krasnodar Region
Approximate size	>10,000 km^2	300 km^2	>2000 km^2
Geographical domain	Hilly plain, alluvial plain, seashore	Low mountains, seashore	Low and high mountains
Dominating landscape	Steppe (grassland)	Deciduous forests	Deciduous, mixed, coniferous forests
Number of geosites	>20 (inventory in progress)	2	16
Dominance of geosites	Low	Moderate	Moderate
Number of geoheritage types	14	7	14
Selected attractive features	Neogene outcrops with fossils, mud lakes, coal waste heaps	Cretaceous–Paleogene outcrops with trace fossils, lakes	Permian–Cretaceous outcrops with fossils, Paleozoic granitoids, waterfalls
Use in geoscience research	Low	Moderate	High
Use in geoscience education	Moderate	High	High
Use in geotourism	Low	Low	Moderate
Biodiversity	Low	Low	High
Human intervention	High (urbanization)	Moderate (touristic infrastructure)	Moderate (touristic infrastructure)
Landscape aesthetic attractiveness	High	High	High
Typical geosite visitors	University students and lecturers	Researchers, university students, and lecturers	Researchers, university students, and lecturers; geology amateurs and other geotourists

Lower Don is a vast area embracing the lower part of the Don River, its delta, the coasts and the near-coastal areas of the Taganrog Bay of the Azov Sea, and some adjacent plots (Figure 1). Geologically, it corresponds to the Rostov Dome of the Russian Platform, where Precambrian crystalline basement is overlain by Cretaceous and Cenozoic deposits; carboniferous sedimentary complexes and mid-Mesozoic igneous rocks are known from its northern periphery [40]. On the modern tectonic reconstructions, this area looks like the edge of the huge Precambrian block [41]. Although an inventory of the geoheritage resources of this area is in progress, the "core" knowledge about them was summarized by Nebabina and Ruban [38].

Abrau is a relatively small area near Novorossiysk, which stretches between the Black Sea coast in the south and the northern shore of the Abrau Lake in the north (Figure 1). Geologically, it represents the western edge of the Greater Caucasus orogen dominated by Late Cretaceous and Paleocene turbiditic deposits [42]. Tectonically, this is an Alpine orogenic domain formed in the late Cenozoic [41,43]. The geoheritage resources of this area were characterized comprehensively by Ruban [39].

Mountainous Adygeya is a rather large area and popular tourist destination embracing the Belaya River watershed southward of Maykop (Figure 1). Geologically, it is dominated by Mesozoic sedimentary complexes (siliciclastic turbidites and carbonates), although Paleozoic igneous rocks and thick red-bed sequences as well as Precambrian metamorphic rocks are also known there [44]. The area represents the long-term evolution of active marine basins, which existed there in the Mesozoic [45,46], after which the area (together with the entire Greater Caucasus) experienced orogenic uplift [41,43]. The geoheritage resources and their current exploitation have been described in detail by Ruban et al. [14].

These three geoheritage-rich areas provide unique opportunities to comprehend a broad spectrum of geological phenomena (Table 1), as well as to learn about the geological history of the Russian South. The geosites of Mountainous Adygeya represent the active tectonic development of the territory from the Precambrian to the mid-Cretaceous, and they shed light on the Variscan and Cimmerian deformations and the Mesozoic Caucasian Sea [14]. The geosites of Abrau inform about the regional geological evolution at the Mesozoic–Cenozoic transition [38]. The geosites of Lower Don reflect the "passive" tectonic development of the northern part of the Russian South in the late Cenozoic [39]. Importantly, all three areas also represent modern geological processes and the Anthropocene themes.

3. Material and Methods

The material for this work has been collected during field works in all three geoheritage-rich areas of the Russian South presented in this work. Experience with organizing major research projects, geo-ecological conferences, and field educational campaigns for students from the Southern Federal University (Rostov-on-Don, Russia) has been helpful. The information has been obtained by observations, map-based measurements, and a search for some information available online. A part of this material has been collected to plan and realize geochemical investigations within the framework of the Strategic Academic Leadership Program of SFedU Priority-2030.

As explained above, the existing methodology for assessing geoheritage accessibility focuses on single geosites [22–24], and it cannot be employed for entire geoheritage-rich areas (indeed, it can be used for assessing individual geosites in these areas). A new approach has to be proposed; it requires finding the proper criteria and establishing the scoring system. Several starting points for the development of such an approach can be outlined (these are only the premises—the finally used criteria are explained below).

First, it is reasonable to link large-scale accessibility to transport infrastructure. Indeed, geoheritage can be interesting to hikers, but only very rare, occasional hikers with extraordinary skills would decide to reach a geoheritage-rich area from their permanent locations due to distances measured by dozens and hundreds of kilometers. Hiking opportunities are reasonable to consider, but only in the case of single geosites [24]. Second, accessibility depends on public transportation because not all people can use cars, and geoheritage-rich areas can be too remote for many drivers. Accessibility depends on the number of options for transportation, i.e., if visitors can reach a given area by plane, bus, train, and boat. Like in the case of geosite accessibility [23,24], outer and inner accessibility should be distinguished. The latter depends on the number of stops of public transport within a given area. However, to take into account the timetable of public transport would be challenging because different visitors would judge it differently, and collecting the related information is not always possible. Third, road infrastructure allowing travel to the area by car and travel within it should be taken into account.

Fourth, special attention should be paid to local services. In the case of geosites, the presence of a restaurant or hotel located near them is unimportant to their accessibility. However, their absence creates significant difficulty for visiting geoheritage-rich areas because one would either need to organize a very long, one-day trip, or be specially prepared for staying without any comfort for more than one day; indeed, such difficulties would complicate or prohibit visiting some areas. Local services are not restricted to accommodation and meals, but also include transport rental. Fifth, there are various specific conditions that limit the accessibility of areas. Particularly, these are linked to settlement pattern and visiting restrictions, which are as follows. It is reasonable to pay attention to the biggest available settlements and not population density because the opportunity to find services (for instance, any technical support) is higher in towns and cities, even if these are fewer than in small villages, or if they are numerous. Brilha [22] preferred to focus on population density; this was reasonable, as he paid attention to the other aspects of geosite management. As for restrictions, the influence of struggling for visit permissions and paying entrance fees was explained by Mikhailenko et al. [24], and this seems to be equally important to geosites and geoheritage-rich areas.

The criteria are summarized in Table 2. It should be added that the proposed approach aims at assessing only the large-scale accessibility of geoheritage-rich areas, not their general value. Thus, the number of employed criteria should be limited to only those most related to large-scale accessibility.

Table 2. Criteria for assessment of large-scale accessibility of geoheritage-rich areas.

Criteria	Grades	Scores
	External public transportation	
Airport	Within area	10
	<50 km from area	7
	50–200 km from area	2
	>200 km from area	0
Railway station	Within area	10
	<50 km from area	7
	50–200 km from area	2
	>200 km from area	0
Bus station	Within area	10
	<50 km from area	5
	50–200 km from area	2
	>200 km from area	0
Port (river/lake/sea)	Within area	10
	<50 km from area	5
	50–200 km from area	2
	>200 km from area	0
	Internal public transportation	
Minor stops of trains, buses, boats in area	Numerous	20
	Few	10
	Absent	0
	Road infrastructure	
Best available road to area	Principal (paved)	20
	Secondary (paved)	15
	Unpaved	5
	Absent	0
Prevailed roads within area	Principal (paved)	20
	Secondary (paved)	15
	Unpaved	7
	Absent	0
	Local services	
Accommodation	Wide choice (numerous hotels, lodges, camps of different quality) within area	30
	Wide choice (numerous hotels, lodges, camps of different quality) <50 km from area	15
	Limited choice (few hotels, lodges, camps offering elementary services) within area	15
	Limited choice (few hotels, lodges, camps offering elementary services) <50 km from area	5
	Absent	0

Table 2. *Cont.*

Criteria	Grades	Scores
Car/boat rental	Within area	10
	<50 km or too limited	5
	>50 km or absent	0
Taxi	Within area	30
	<50 km	15
	>50 km or absent	0
Excursion bus rental	Within area	20
	<50 km or too limited	5
	>50 km or absent	0
Meals	Big choice (numerous restaurants and cafes) within area	30
	Limited choice (few restaurants and cafes) within area	10
	Absent	0
General setting		
Maximum rank of settlements within area	City (population >0.2 mln)	30
	Town (population <0.2 mln)	10
	Village (population <5000)	5
	absent	0
Seasonality in area	Unimportant for accessibility	50
	Somewhat important for accessibility	25
	Important for accessibility	5
Severe weather conditions in area	Rare (<1 event per year)	20
	Common (1–5 events per year)	10
	Frequent (>5 events per year)	0
Permissions for visiting area or its significant parts	Not required	15
	Required	0
Fees/tickets for visiting area or its significant parts	Not required	15
	Required	0
Grades of geoheritage-rich areas by their large-scale accessibility		
CATEGORY		TOTAL SCORES
Excellent		251–350
Moderate		151–250
Limited		0–150

Finding the proper criteria should be followed by development of the scoring system, which means establishing grades for all criteria and ascribing scores. The latter should be done to make the total scores (sum of all scores) meaningful characteristics, allowing judgment about the true accessibility of geoheritage-rich areas. Different criteria should have different grades and receive different scores depending on their relative importance. Indeed, some conditions increase and the others decrease the accessibility, and the total scores should reflect the balance between them. The proposed grades and scores (Table 2) reflect the opportunities to access areas from outside. Finally, the sum of all scores allows attributing a given area to one general category of large-scale accessibility, i.e., it is established whether it is characterized by excellent, moderate, or limited accessibility (Table 2).

4. Results

The application of the proposed approach to the considered geoheritage-rich areas of the Russian South indicates their differences (Table 3). First, one should note the differences by external public transport. Lower Don corresponds to the densely urbanized area in which a city with a population exceeding one million people is located, namely, Rostov-on-Don. One can reach this area by many types of transport, and there are also various transport opportunities to make trips within it. The situation differs in Abrau and Mountainous Adygeya, which are chiefly accessible by bus. Airports, railway stations, and even ports (in the case of Abrau) are located not so far from these areas, but trips from them to the area require using either public buses or taxis. Moreover, public transportation is absent within Abrau and limited in Mountainous Adygeya. The state of road infrastructure is so that one can easily use their own car or take a local taxi to reach the geosites, although rental opportunities are restricted in two areas (Table 3).

Table 3. Scoring large-scale accessibility of the considered geoheritage-rich areas.

Criteria	Geoheritage-Rich Areas		
	Lower Don	Abrau	Mountainous Adygeya
Airport	10	2	2
Railway station	10	7	2
Bus station	10	10	10
Port	10	5	0
Minor stops of trains, buses, boats in area	20	0	10
Best available road to area	20	15	20
Prevailed roads within area	15	7	15
Accommodation	30	30	30
Car/boat rental	10	5	0
Taxi	30	30	30
Excursion bus rental	20	5	5
Meals	30	30	10
Maximum rank of settlements within area	30	5	10
Seasonality in area	25	25	25
Severe weather conditions in area	10	10	10
Permissions (e.g., to natural reserves) to visit area or its significant parts	15	15	15
Fees/tickets for visiting area or its significant parts	15	15	0
TOTAL SCORES	310	216	194
Grade	Excellent	Moderate	Moderate

In all three areas, there are many options for accommodation (Table 3). In Lower Don, there are hundreds of hotels, lodges, and hostels (not only in Rostov-on-Don, but also in its vicinity and other settlements). In Abrau, the choice is more limited, but some can be accommodated at the Limanchik camp of the Southern Federal University or several hotels and lodges in Abrau-Dyurso. In Mountainous Adygeya, the hotel industry experiences significant growth, with dozens of hotels and lodges available, including relatively remote places and even directly within forests. However, one should also note that the visitors of Mountainous Adygeya may face a challenge with finding places to dine (restaurants and cafes are not numerous, and even some luxurious hotels do not have them), although the situation is gradually improving.

Special attention should be paid to settlements (Table 3). As mentioned above, Lower Don is an urban area with two big cites, namely Rostov-on-Don and Taganrog, and several smaller towns, namely Bataysk, Novocherkassk, and Shakhty (Figure 1). In contrast, Abrau only hosts Abrau-Dyurso village, and Mountainous Adygeya hosts Kamennomostsky town and a few villages. Visitors of the two latter areas may be faced with limited services (for instance, if there is an urgent need for serious car maintenance). Another specific feature of Mountainous Adygeya is the common use of entrance fees. In this area, some geosites are situated in the Caucasian State Natural Biosphere Reserve, for which visitors are required to pay a fee. Moreover, access to such important attractions as Rufabgo Waterfalls and Khadzhokh Klamm also requires a fee.

Generally, the total scores imply that the Lower Don geoheritage-rich area has excellent large-scale accessibility, whereas Abrau and Mountainous Adygeya have moderate large-scale accessibility (Table 3). The difference between the former and two latter is significant. In particular, it is strongly determined by the differences in external public transportation. One should also note that Lower Don receives lower scores than the two other areas by none criterion (Table 3).

5. Discussion and Conclusions

The results of the present study reveal spatial heterogeneity of the Russian South by the large-scale accessibility of its geoheritage-rich areas. This heterogeneity can be explained by the trends of territorial development in recent years (particularly, with regard to settlement pattern and transport infrastructure), which influence the properties of geoheritage resources. Although the established values (Table 3) do not argue against good reasons for the exploitation of these resources for the purposes of science, education, and tourism [14,38,39], more efforts are required to exploit them fully in Abrau and Mountainous Adygeya. Surprisingly, Abrau is a part of the large recreational zone along the Black Sea coast, and Mountainous Adygeya itself is an important tourist destination. Their touristic infrastructure is developed well. However, it appears that the latter is not enough to determine large-scale accessibility of the areas.

Assessing the large-scale accessibility of geoheritage-rich areas seems to be important, not only for "purely" scientific needs, but also for developing policies at the national, regional, and municipal levels concerning geoheritage resources. These policies are necessary because the exploitation of these resources (particularly in the form of geotourism) may produce socio-economic benefits [14,47–51]. Geoheritage management and exploitation are too innovative and complex, and they are difficult to develop without attention and support from administrative authorities. Two principal directions for geoheritage policy developments can be proposed in light of the findings of this study. First, successful exploitation of the available geoheritage resources requires improvements in their large-scale accessibility and, particularly, attention to those parameters of the areas, for which low or zero scores have been specified (Table 3). Particularly, the internal public transportation needs better development in Abrau, and especially Mountainous Adygeya (taking into account its size). Second, the policies should focus on justifying territorial development programs and initiatives (including those related to tourism) to the desirable improvements in large-scale accessibility. If a given area is rich in geoheritage, this means its exploitation can be beneficial; therefore, it is reasonable to consider the actions facilitating this exploitation. For geotourism, the large-scale accessibility of geoheritage-rich areas is vital. However, it is similarly important for research and education. For instance, an advanced research project is currently underway at the Southern Federal University. It focuses on the use of some geoheritage objects of the Russian South for the purposes of geochemical investigations (with an emphasis on heavy metals, particularly, mercury). Indeed, its success depends strongly on the large-scale accessibility of Lower Don and Abrau. In Mountainous Adygeya, the limited opportunities to rent excursion buses challenges the organization of field educational campaigns for university students.

Geoheritage exploitation should aim at socio-economic benefits, but it must also be sustainable [12,13,52–54]. Increasing the large-distance accessibility of geoheritage-rich areas requires expanding transport infrastructure and other human interventions in natural landscapes. Additionally, to direct environmental impacts and landscape reorganizations, the related activities trigger aesthetic modifications and result in some pollution. Although large-scale accessibility is chiefly linked to infrastructural objects outside geoheritage-rich areas, the related environmental stress on surrounding areas cannot be ignored. Although addressing this challenge requires state-of-the-art solutions, it is necessary to stress that the geoheritage policy should take the noted issues into account. This is an additional argument for geoheritage management within large-scale territorial planning initiatives.

The decades-long observations imply that the three considered geoheritage-rich areas are the most demanded by visitors from Rostov-on-Don and less Krasnodar (Figure 3). This is unsurprising because these cities are important research and educational centers of the Russian South, and the universities within them, such as the Southern Federal University, have strong geoscience programs. If these observations are correct, one can wonder whether the distance between these cities and the geo-heritage-rich areas is also a factor of large-scale accessibility. In fact, larger distances require more time and funds for travel. However, such remoteness is relational, and it may change with time. It should be distinguished from the large-scale accessibility, which is a basic property not depending on the direction of visitor flows (Figure 3). The latter determines the very opportunity to reach the area and its geoheritage from the outside (including remote places), whereas remoteness is linked to the actual mode of exploitation of geoheritage resources of this area.

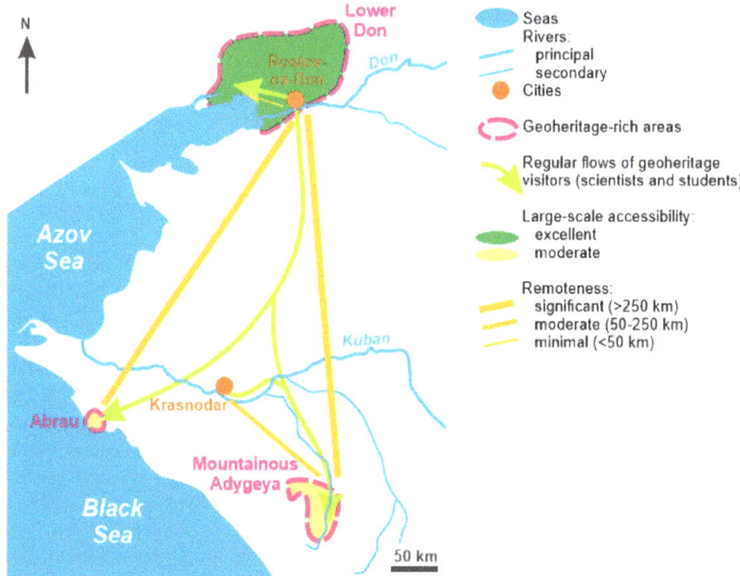

Figure 3. Relative remoteness of the three considered geoheritage-rich areas.

The criteria proposed in the present study for assessing the large-scale accessibility of geoheritage sites are objective and meaningful for all places irrespective of the socio-economical contexts. However, country-specific peculiarities should be noted and considered as possible limitations of the case studies. First, countries may differ by the number of car owners. Where this number is high, the need for public transportation is lower. Second, 50 km is a minor distance in Russia, and driving even 100 km to reach an airport or for accommodation is somewhat of a norm. However, the situation may differ significantly in smaller countries such as Hungary or Switzerland. Third, Russians prefer travelling by trains for large distances more readily than citizens of some other countries. These examples demonstrate that the proposed approach can be justified, taking into account the national contexts. Accessibility is a parameter that cannot be fully standardized because it strongly depends on people's experiences, feelings, and personal resources [55–57]. This challenge is less significant in the case of geosite accessibility, which more strongly depends on the local parameters [24]. However, the presence of the noted challenge does not mean that large-scale accessibility should not be investigated. It seems to be a particularly important property of geoheritage-rich areas, determining the success exploiting their resources. The present study, although anchored into the Russian reality, offers a general vision and criteria of the large-scale accessibility, which seem to be common for all contexts.

The proposed approach deals with absolute measures of the large-scale accessibility of geoheritage-rich areas. However, this important property may also have a relational aspect. One can hypothesize that willingness to travel and readiness to pay in order to reach a given area increase together with the overall value of geoheritage; thus, more valuable areas become more "proximal" to visitors with regard to their mode of thinking. Although various arguments supporting or disproving this hypothesis can be offered, only highly specialized research would permit judgments of this relational accessibility, which is outside the scope of the present work.

Conclusively, a novel approach is proposed to assess the large-scale accessibility of three geoheritage-rich areas of the Russian South. It is established that one of them (Lower Don) is highly accessible, and two others (Abrau and Mountainous Adygeya) are moderately accessible. The principal opportunity for further research is linked to the development of methodological frameworks, allowing adaptation of the proposed approach to the different country-specific contexts. The other opportunity is developing the approach to be applicable to submarine and non-populated domains such as Antarctica. Finally, it will be reasonable to address the relative nature of accessibility with psychological experiments.

Author Contributions: Conceptualization, Y.A.F., A.V.M. and D.A.R.; methodology, D.A.R.; investigation, A.V.M. and D.A.R.; writing—original draft preparation, A.V.M. and D.A.R.; writing—review and editing, Y.A.F., A.V.M. and D.A.R.; project administration, Y.A.F. and A.V.M. All authors have read and agreed to the published version of the manuscript.

Funding: The research was carried out within the framework of the Strategic Academic Leadership Program of SFedU Priority-2030; project no. SP-12-22-5 (to Y.A.F. and A.V.M.). The contribution of D.A.R. was not funded.

Institutional Review Board Statement: Not applicable.

Informed Consent Statement: Not applicable.

Data Availability Statement: Not applicable.

Conflicts of Interest: The authors declare no conflict of interest. The funder had no role in the design of the study; in the collection, analyses, or interpretation of data; in the writing of the manuscript; or in the decision to publish the results.

References

1. Brocx, M.; Semeniuk, V. Geoheritage and geoconservation—History, definition, scope and scale. *J. R. Soc. West. Aust.* **2007**, *90*, 53–87.
2. Henriques, M.H.; dos Reis, R.P.; Brilha, J.; Mota, T. Geoconservation as an emerging geosciences. *Geoheritage* **2011**, *3*, 117–128. [CrossRef]

3.   Herrera-Franco, G.; Carrión-Mero, P.; Montalván-Burbano, N.; Caicedo-Potosí, J.; Berrezueta, E. Geoheritage and Geosites: A Bibliometric Analysis and Literature Review. *Geosciences* **2022**, *12*, 169. [CrossRef]

4.   Kaur, G. Geodiversity, Geoheritage and Geoconservation: A Global Perspective. *J. Geol. Soc. India* **2022**, *98*, 1221–1228. [CrossRef]

5.   Neto, K.; Henriques, M.H. Geoconservation in Africa: State of the art and future challenges. *Gondwana Res.* **2022**, *110*, 107–113. [CrossRef]

6.   Pijet-Migoń, E.; Migoń, P. Geoheritage and Cultural Heritage—A Review of Recurrent and Interlinked Themes. *Geosciences* **2022**, *12*, 98. [CrossRef]

7.   Reynard, E.; Brilha, J. (Eds.) *Geoheritage: Assessment, Protection, and Management*; Elsevier: Amsterdam, The Netherlands, 2018.

8.   Urban, J.; Radwanek-Bąk, B.; Margielewski, W. Geoheritage Concept in a Context of Abiotic Ecosystem Services (Geosystem Services)—How to Argue the Geoconservation Better? *Geoheritage* **2022**, *14*, 54. [CrossRef]

9.   Williams, M.A.; McHenry, M.T.; Boothroyd, A. Geoconservation and Geotourism: Challenges and Unifying Themes. *Geoheritage* **2020**, *12*, 63. [CrossRef]

10.   Xu, K.; Wu, W. Geoparks and Geotourism in China: A Sustainable Approach to Geoheritage Conservation and Local Development—A Review. *Land* **2022**, *11*, 1493. [CrossRef]

11.   Bétard, F.; Hobléa, F.; Portal, C. Geoheritage as new territorial resource for local development. *Ann. Geogr.* **2017**, *717*, 523–543. [CrossRef]

12.   Garcia, M.D.G.; Queiroz, D.S.; Mucivuna, V.C. Geological diversity fostering actions in geoconservation: An overview of Brazil. *Int. J. Geoheritage Park.* **2022**, *10*, 507–522. [CrossRef]

13.   Henriques, M.H.; Castro, A.R.S.F.; Félix, Y.R.; Carvalho, I.S. Promoting sustainability in a low density territory through geoheritage: Casa da Pedra case-study (Araripe Geopark, NE Brazil). *Resour. Policy* **2020**, *67*, 101684. [CrossRef]

14.   Ruban, D.A.; Mikhailenko, A.V.; Yashalova, N.N. Valuable geoheritage resources: Potential versus exploitation. *Resour. Policy* **2022**, *77*, 102665. [CrossRef]

15.   Santangelo, N.; Valente, E. Geoheritage and Geotourism resources. *Resources* **2020**, *9*, 80. [CrossRef]

16.   Unjah, T.; Yusry, M.; Simon, N. Identification and characterization of geoheritage resources at Hulu Langat, Selangor. *Bull. Geol. Soc. Malays.* **2021**, *72*, 191–204. [CrossRef]

17.   Štrba, L.; Rybar, P.; Balaz, B.; Molokac, M.; Hvizdak, L.; Krsak, B.; Lukac, M.; Muchova, L.; Tometzova, D.; Ferencikova, J. Geosite assessments: Comparison of methods and results. *Curr. Issues Tour.* **2015**, *18*, 496–510. [CrossRef]

18.   Bétard, F.; Peulvast, J.-P. Geodiversity Hotspots: Concept, Method and Cartographic Application for Geoconservation Purposes at a Regional Scale. *Environ. Manag.* **2019**, *63*, 822–834. [CrossRef]

19.   da Silva, J.C.; dos Santos, D.S.; da Rocha, T.B. Identifying geomorphological diversity hotspots for conservation purposes: Application to a coastal protected area in Rio de Janeiro State, Brazil. *Appl. Geogr.* **2022**, *142*, 102689. [CrossRef]

20.   Gray, M. Geodiversity: Developing the paradigm. *Proc. Geol. Assoc.* **2008**, *119*, 287–298. [CrossRef]

21.   Frassi, C.; Amorfini, A.; Bartelletti, A.; Ottria, G. Popularizing Structural Geology: Exemplary Structural Geosites from the Apuan Alps UNESCO Global Geopark (Northern Apennines, Italy). *Land* **2022**, *11*, 1282. [CrossRef]

22.   Brilha, J. Inventory and quantitative assessment of geosites and geodiversity sites: A review. *Geoheritage* **2016**, *8*, 119–134. [CrossRef]

23.   Warowna, J.; Zgłobicki, W.; Kołodyńska-Gawrysiak, R.; Gajek, G.; Gawrysiak, L.; Telecka, M. Geotourist values of loess geoheritage within the planned Geopark Małopolska Vistula River Gap, E Poland. *Quat. Int.* **2016**, *399*, 46–57. [CrossRef]

24.   Mikhailenko, A.V.; Ruban, D.A.; Ermolaev, V.A. Accessibility of Geoheritage Sites—A Methodological Proposal. *Heritage* **2021**, *4*, 1080–1091. [CrossRef]

25.   Cheablam, O.; Tansakul, P.; Nantakat, B.; Pantaruk, S. Assessment of the Geotourism Resource Potential of the Satun UNESCO Global Geopark, Thailand. *Geoheritage* **2021**, *13*, 87. [CrossRef]

26.   Deng, L.; Zou, F. Orogenic belt landforms of Huanggang Dabieshan UNESCO Global Geopark (China) from geoheritage, geoconservation, geotourism, and sustainable development perspectives. *Environ. Earth Sci.* **2021**, *80*, 662. [CrossRef]

27.   Henriques, M.H.; Canales, M.L.; García-Frank, A.; Gomez-Heras, M. Accessible Geoparks in Iberia: A Challenge to Promote Geotourism and Education for Sustainable Development. *Geoheritage* **2019**, *11*, 471–484. [CrossRef]

28.   Rais, J.; Barakat, A.; Louz, E.; Ait Barka, A. Geological heritage in the M'Goun geopark: A proposal of geo-itineraries around the Bine El Ouidane dam (Central High Atlas, Morocco). *Int. J. Geoheritage Park.* **2021**, *9*, 242–263. [CrossRef]

29.   Wang, J.; Zouros, N. Educational Activities in Fangshan UNESCO Global Geopark and Lesvos Island UNESCO Global Geopark. *Geoheritage* **2021**, *13*, 51. [CrossRef]

30.   Bruno, D.E.; Perrotta, P. A geotouristic proposal for Amendolara territory (northern ionic sector of Calabria, Italy). *Geoheritage* **2012**, *4*, 139–151. [CrossRef]

31.   Camino, M.A.; Halpern, K.; Bó, M.J.; Meroi Arcerito, F.R. Sierra Bachicha: Proposal for a new site of geological interest in the Balcarce district, province of Buenos Aires. *Ser. Correl. Geol.* **2018**, *34*, 5–14.

32.   Lirer, L.; Petrosino, P.; Armiero, V. A proposal of some geosites in the framework of a new geological map of Campi Flegrei. *Alp. Mediterr. Quat.* **2008**, *21*, 39–46.

33.   Ranjbaran, M.; Sotohian, F. Development of Haraz Road geotourism as a key to increasing tourism industry and promoting geoconservation. *Geopersia* **2021**, *11*, 61–79.

34.   Štrba, L.; Baláž, B.; Lukác, M. Roadside geotourism—An alternative approach to geotourism. *E-Rev. Tour. Res.* **2016**, *13*, 598–609.

35.   Andreyanova, S.; Ivolga, A. The tourism potential of the North Caucasus: The formation, characteristics and development prospects. *Geoj. Tour. Geosites* **2018**, *22*, 347–358.

36. Ivlieva, O.V.; Shmytkova, A.V.; Sukhov, R.I.; Kushnir, K.V.; Grigorenko, T.N. Assessing the tourist and recreational potential in the South of Russia. *E3S Web Conf.* **2020**, *208*, 05013. [CrossRef]

37. Oborin, M.S.; Kozhushkina, I.; Gvarliani, T.; Ivanov, N. Socio-economic preconditions of resort agglomerations development in the south of Russia. *Worldw. Hosp. Tour.* **2018**, *10*, 467–477. [CrossRef]

38. Nebabina, E.I.; Ruban, D.A. *Geological Heritage Sites in the Southwest Rostov Region*; RGU: Rostov-na-Donu, Russia, 2006. (In Russian)

39. Ruban, D.A. On the Duality of Marine Geoheritage: Evidence from the Abrau Area of the Russian Black Sea Coast. *J. Mar. Sci. Eng.* **2021**, *9*, 921. [CrossRef]

40. Ivanitskaya, V.B.; Pogrebnov, N.I. *Geological Structure of the Lower Don and Lower Volga*; RGU: Rostov-na-Donu, Russia, 1962. (In Russian)

41. Hasterok, D.; Halpin, J.A.; Collins, A.S.; Hand, M.; Kreemer, C.; Gard, M.G.; Glorie, S. New Maps of Global Geological Provinces and Tectonic Plates. *Earth-Sci. Rev.* **2022**, *231*, 104069. [CrossRef]

42. Baraboshkin, E.Y.; Bondarenko, N.A.; Lyubimova, T.V. *Unique Geological Objects of the North-Western Caucasus*; KubGU: Krasnodar, Russia, 2012. (In Russian)

43. Van Hinsbergen, D.J.J.; Torsvik, T.H.; Schmid, S.M.; Matenco, L.C.; Maffione, M.; Vissers, R.L.M.; Gürer, D.; Spakman, W. Orogenic architecture of the Mediterranean region and kinematic reconstruction of its tectonic evolution since the Triassic. *Gondwana Res.* **2020**, *81*, 79–229. [CrossRef]

44. Nazarenko, O.V.; Mikhailenko, A.V.; Smagina, T.A.; Kutilin, V.S. *Natural Conditions of Mountainous Adygeya*; SFU: Rostov-on-Don, Russia, 2020. (In Russian)

45. Adamia, S.; Alania, V.; Chabukiani, A.; Kutelia, Z.; Sadradze, N. Great Caucasus (Cavcasioni): A long-lived north-tethyan back-arc basin. *Turk. J. Earth Sci.* **2011**, *20*, 611–628. [CrossRef]

46. Lordkipanidze, M.B.; Adamia, S.A.; Asanidze, B.Z. Evolution of the active margins of the ocean Tethys (by example of the Caucasus). In *Oceanology: Reports. 27 International Geological Congress*, 3rd ed.; Lisitsin, A.P., Ed.; Nauka: Moscow, Russia, 1984; pp. 72–83. (In Russian)

47. AbdelMaksoud, K.M.; Emam, M.; Al Metwaly, W.; Sayed, F.; Berry, J. Can innovative tourism benefit the local community: The analysis about establishing a geopark in Abu Roash area, Cairo, Egypt. *Int. J. Geoheritage Park.* **2021**, *9*, 509–525. [CrossRef]

48. Brilha, J.; Gray, M.; Pereira, D.I.; Pereira, P. Geodiversity: An integrative review as a contribution to the sustainable management of the whole of nature. *Environ. Sci. Policy* **2018**, *86*, 19–28. [CrossRef]

49. Kubalíková, L. Cultural ecosystem services of geodiversity: A case study from Stranska skala (Brno, Czech Republic). *Land* **2020**, *9*, 105. [CrossRef]

50. Pereira Balaguer, L.; da Glória Motta Garcia, M.; de Almeida Leite Ribeiro, L.M. Combined Assessment of Geodiversity As a Tool to Territorial Management: Application to Southeastern Coast of State of São Paulo, Brazil. *Geoheritage* **2020**, *14*, 60. [CrossRef]

51. Freire-Lista, D.M.; Becerra Becerra, J.E.; Simões de Abreu, M. The historical quarry of Pena (Vila Real, north of Portugal): Associated cultural heritage and reuse as a geotourism resource. *Resour. Policy* **2022**, *75*, 102528. [CrossRef]

52. Bentivenga, M.; Cavalcante, F.; Mastronuzzi, G.; Palladino, G.; Prosser, G. Geoheritage: The Foundation for Sustainable Geotourism. *Geoheritage* **2019**, *11*, 1367–1369. [CrossRef]

53. Kubalíková, L.; Bajer, A.; Balková, M.; Kirchner, K.; Machar, I. Geodiversity Action Plans as a Tool for Developing Sustainable Tourism and Environmental Education. *Sustainability* **2022**, *14*, 6043. [CrossRef]

54. Somma, R. The Inventory and Quantitative Assessment of Geodiversity as Strategic Tools for Promoting Sustainable Geoconservation and Geo-Education in the Peloritani Mountains (Italy). *Educ. Sci.* **2022**, *12*, 580. [CrossRef]

55. Diaz-Soria, I. Being a tourist as a chosen experience in a proximity destination. *Tour. Geogr.* **2017**, *19*, 96–117. [CrossRef]

56. Larsen, J. Tourism mobilities and the travel glance: Experiences of being on the move. *Scand. J. Hosp. Tour.* **2001**, *1*, 80–98. [CrossRef]

57. Tjørve, E.; Flognfeldt, T.; Tjørve, K.M.C. The Effects of Distance and Belonging on Second-Home Markets. *Tour. Geogr.* **2013**, *15*, 268–291. [CrossRef]

geosciences

Article

Frasassi Caves and Surroundings: A Special Vehicle for the Geoeducation and Dissemination of the Geological Heritage in Italy

Piero Farabollini and Fabrizio Bendia *

School of Sciences and Technologies, Section of Geology, University of Camerino, Via Gentile III da Varano, 62032 Camerino, Italy
* Correspondence: fabrizio.bendia@unicam.it

Abstract: One of the most fascinating aspects of the work of geologist lies in knowing how to read the physical landscape as an expression of the geological and geomorphological phenomena that shaped the Earth's relief over time. The necessity to disseminate these vast areas of knowledges and skills starts from here, with the aim to enhance the concept of geodiversity and to raise awareness for its protection and promotion. This work aims to share some examples of projects realized following the subscription of agreement between different public authorities in the Apennine territory of Marche Region in Italy, such as the Geology Section of the University of Camerino (MC), "Consorzio Frasassi" (which manages the underground karst complex of Frasassi caves) and municipality of Genga (AN), where Frasassi caves are located. Thanks to this partnership, our research group realized didactic geological notebooks for school groups visiting the caves and interactive museum laboratories in 3D, showing the geological evolution of the area. This sharing of knowledge will contribute to educate communities about the importance of the geological heritage. At the same time, this project can serve as motivation to establishing the same type of collaborations in those territories where similar projects can be replicated.

Keywords: geoeducation; geoheritage; Frasassi; Italy

Citation: Farabollini, P.; Bendia, F. Frasassi Caves and Surroundings: A Special Vehicle for the Geoeducation and Dissemination of the Geological Heritage in Italy. *Geosciences* **2022**, *12*, 418. https://doi.org/10.3390/geosciences12110418

Academic Editors: Hara Drinia, Panagiotis Voudouris, Assimina Antonarakou and Jesus Martinez-Frias

Received: 27 September 2022
Accepted: 8 November 2022
Published: 12 November 2022

Publisher's Note: MDPI stays neutral with regard to jurisdictional claims in published maps and institutional affiliations.

1. Introduction

The geologist is the only professional figure who does not focus only on the external appearance of a physical landscape. He asks why, where, when, and how that landscape formed and which geological and geomorphological processes created it. He doesn't stop on the mere observation of the outer portion of the Earth's crust, but rather penetrate it in depth, understanding why there may be mountains or plains on the surface. A geologist can reconstruct how the arrangement and the nature of rocks vary underground, identifying the presence of folds, faults, fossils, minerals, and aquifers, and thus determining the presence of water, applying their own knowledge towards something essential for the sustainability of life.

Therefore, there is vital importance in sharing this knowledge and explaining even the most fascinating and complicated processes in the best possible way in order to involve and arouse interest in students, scientists, and researchers, as well as tourists and simple passionate people.

The study area is contained within the regional natural park named "Parco Naturale Regionale della Gola della Rossa e di Frasassi", a protected area of 10,026 hectares instituted in 1997 and considered an extraordinary example of geological, biological, historical, and cultural heritage [1,2]. This is also the location of the hypogeum complex of Frasassi caves, in the Apennine territory of Marche Region, under the municipality of Genga (AN), within the Mount Valmontagnana, in the southern side of Frasassi gorge. Frasassi gorge is a fluvial incision with a W–E direction cut by Sentino River during the Quaternary period. Frasassi caves began to form in this period, when this portion of Apennine emerged above sea level, and continental processes (fluvial erosion and karst phenomena) began to incise the gorge and create the hypogeum complex. The underground cave network is composed of tunnels and pits and develops within the rocky walls in Calcare Massiccio Formation constituting the gorge, in an altitude range between 200 and 500 m. a.s.l. The karst tunnels are arranged on at least four main levels, the evolution of which is linked to standstill and deepening phases of the surface hydrographic network. The lower and most recent levels, between 200 and 300 m. a.s.l., probably originated in the middle Pleistocene and late Pleistocene periods, as indicated by the correlations with the terraced alluvial deposits [3] and by the dating of speleothems, with ages reaching 200,000 years old [4].

The incision of Frasassi gorge also acted by cutting the local sedimentary succession, representing about 120 million years; that is, from the lower Jurassic of the "Calcare Massiccio" Formation to the upper Cretaceous of the "Scaglia Rossa" Fm., outcropping on the top of Monte Vallemontagnana (southern side of the gorge). This interval of time is represented in about 600–700 m of the straight walls of the gorge, along which many geosites are recognizable, collected (with stars symbolism) in a map showing their distribution as well as other places of historical, religious, and cultural interest (Figure 1).

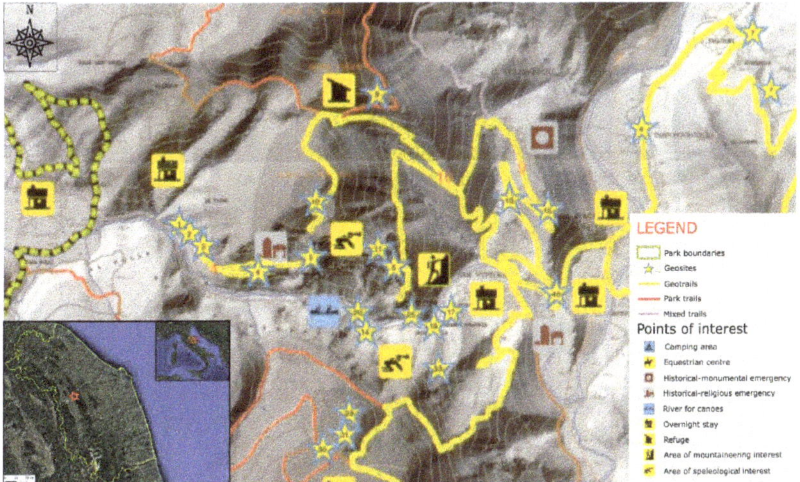

Figure 1. An excerpt of the map attached in "I Quaderni del Parco, N.8—Geositi e Geoescursionismo nel Parco Naturale della Gola della Rossa e di Frasassi" [1], with the geosites and historical/cultural places indicated along the Frasassi gorge (the red star in the overview map indicates its location). The coordinate grid and the scale bar are shown in the original map.

The locality of Frasassi is not only famous for its touristic caves, but also for its territory, where nature has greatly shaped the geomorphological landscape. In this area, in fact, there are many other tourist attractions (epigean caves, rocky slopes, fossil sites, rivers, waterfalls, sulfur springs, and thermal waters) that provide suitable conditions to appreciate the cultural heritage (many abbeys and hermitages are set in the caves and built with local stones), climbing (each year, the Frasassi climbing festival is organized), trekking and bike riding (in Frasassi caves and along the countless mountain trails), running (each year, the Frasassi sky race is organized), fishing (with carp, trout, chub, and perch), bird watching (eagles nest along the vertical walls of the gorge), and relaxing facilities (river baths and thermal structures). All of these attractions are guided by a common thread by which they have been originated and shaped: geology. Further, geology strictly conditioned the subsequent anthropization of this territory, from prehistoric times to today.

The awareness of the large geodiversity of the Frasassi area laid the groundwork for the subscription of an agreement between three authorities: "Consorzio Frasassi", "Geology

Section of University of Camerino", and "Municipality of Genga", with the aim to preserve, disseminate, and promote the enormous geological heritage of the area. On the one hand, this would have allowed "Consorzio Frasassi" and "Municipality of Genga" to deepen the geological knowledge of the area, counting on academic skills; on the other hand, it would have allowed the Geology Section of University of Camerino to facilitate the dissemination of geological sciences to a wider audience, allowing greater visibility (thanks to the around 250,000–300,000 annual visitors, between tourists and school groups).

The two case studies realized and illustrated in this work aim to explain the extraordinary geodiversity of Frasassi caves and surroundings to school groups and tourists. For the dissemination of the geological, geomorphological, and karst evolution of the territory of Frasassi and its geological heritage, together with the other signatory partners of the agreement, it was decided to set up two geo-educational projects:

-   Field laboratory with the use of didactic geological notebooks (for school groups visiting the caves);
-   3D interactive museum with virtual reality geological laboratories.

These two realized projects involve geo-products as a tool for geo-education [5], with a large didactic potential [6]. However, at the time of drawing up this article, the bibliography does not provide many similar projects in the international context, in terms of both activities with didactic geological notebooks and 3D interactive museum geological laboratories.

Through these two different projects, school groups and students can be brought closer to the understanding of the phenomena that generated the geological wonders of the territory, facilitated by "field teaching" and by original tools capable of effectively involving them (in both paper and digital form). The aim of this work is to recognize, review, and evaluate the usefulness of geo-cultural features in in terms of didactics.

2. Materials and Methods

Among the most important objectives of "Consorzio Frasassi", the institution that opened the cave to the public in 1973 and manages it, there is undoubtedly the desire of deepen the scientific aspects of the underworld, being aware that the enhancement, protection, and promotion of the geological heritage can be pursued only through a deep knowledge of the phenomena and laws that regulate the planet Earth. School groups are a very important target for "Consorzio Frasassi" because, each year, almost 70,000–80,000 students visit Frasassi caves, out of a total of approximately 250,000–300,000 visitors per year (data found on the official website of "Consorzio Frasassi"). In this regard, "Consorzio Frasassi" is in constant contact with schools and tour operators to attract school groups and tourists to its underground complex. On the other side, the University of Camerino is always looking for opportunities to disseminate the importance of its courses of study (in this case, Earth Sciences).

Thanks to the sharing of knowledge with the academic environment of the Geology Section of the University of Camerino, it was possible to deepen the studies and enhance the informative and educational aspects, which is very useful for students but also interesting for tourists.

Frasassi gorge shows a large variety of geosites concentrated in a rather small area. For this reason, during some characterizations of the area [2], it is interpreted as an "areal geosite", composed of more than one punctual geosite. In this work, it has not been necessary to provide a characterization of every single geosite but, in order to characterize the geosite from a qualitative point of view (Table 1), we decided to rely on international examples known in the bibliography. For this reason, we chose the geosite of Frasassi caves to represent the area. The selection of the case considered one criterion; that is, whether they are or have been in the process of patrimonialization (in this case an UNESCO candidacy). We modified the table provided by the authors of [7] for our case study.

Table 1. A characterization of the geosite of Frasassi caves.

Site	Presence in Catalogues	Origin	Extension	Visitors per Year	Accessibility	Semiological-Didactic Value
Frasassi caves	The Italian Geosites Inventory (ISPRA)	Karst catalyzed by the presence of sulfidic groundwater	About 25 km of caves and conduits	250,000–300,000	Very easily accessible with equipped and safe walkways. Only partial use for disabled	The geosite has a high value because the observation of its elements in the field allows people to understand the natural process that produced it; therefore, the geosite has great geo-educational potential

For this reason, it was decided to implement the didactic aspects by offering immersive experiences, an experiential approach conducted both directly in the cave (through didactic geological notebooks) and via digital means through mobile apps and 3D viewers (3D interactive museum geological laboratories).

These two approaches can be considered, in a certain way, as opposite but complementary. This is because the first aspect, the activity with didactic geological notebooks, represents the practical experience on the field, where students will literally dive into an underground environment; that is, they will enter in the cave, get dirty and wet, use paper, write with a pen and draw on the didactic notebook. The second aspect, the 3D interactive museum geological laboratories, represents the actual, digital period, where students will have an immersive but still detached experience through virtual reality. If, on the one hand, they will not be practically involved, on the other hand, they will fly through the landscape, having the possibility to observe it from another prospective. This experience is equally formative because, during the 3D travel at 360°, it will be possible to interact with items scattered along the surroundings of Frasassi, deepening the encountered topics. This activity also includes a final multiple-choice quiz, useful to fix the acquired knowledge.

These two realized solutions represent an effective approach to explain the complex phenomena that regulate geology, trying to transmit knowledge in the best possible way. This is possible because, in this territory, we are lucky to be able to closely observe and address a myriad of exemplary geological features and situations, so as to be able to understand the processes responsible for their origin. Thanks to this unique possibility offered to us by nature, the idea of enhancing this potential was born, leveraging immersive, experiential tourism.

The didactic geological notebooks were conceived, designed, and realized for lower secondary schools (in Italy called "scuola secondaria di primo grado", once known as "scuola media"). They were drawn up in Italian language, because Frasassi caves work almost entirely with Italian schools.

Didactic geological notebooks aim to represent a practical opportunity to deepen the topics covered in the classroom and during the experience inside the Frasassi caves. Thanks to the highly interactive mould attributed to them, didactic geological notebooks can be considered a real open-air laboratory (in this case, an underground laboratory). They were set up following a trail allowing to closely admire and understand rocks, structures, and landforms originated and modelled thanks to eternal and tireless geological processes.

An extremely informative, colloquial, and novelized approach was chosen for the drafting of the text, as the target is represented by students aged between 11 and 14. Therefore, we simulated that the students could play the role of a real explorer, a speleologist who will have to orient themselves in the cave with the support of a map, to be completed with the recognized elements (stalactites, stalagmites, columns, landslides, and so on) along the path. Yes, because, to become a true speleologist, students must first be able to read the landscape that surrounds them and know how to orient themselves even in environments that are very different from those with which they are familiar.

The active involvement of students in this educational trail represents a strong element of the proposed planning, convinced (empirically) that the consolidation of knowledge and understanding of the phenomena are facilitated by direct observation and experiences gained in the field.

The second project is under realization at the museum "Art, history, and territory" of the village of Genga. "Consorzio Frasassi" asked the University of Camerino to offer its scientific support to create an interactive, three-dimensional, immersive instrument at 360° with augmented reality. The Geology Section of University of Camerino provided a document called "Frasassi: water and geology", where all geological aspects were treated with to the leitmotif of water. Water, in fact, is responsible for the creation of rocks, for the modeling of the territory, for karst and speleogenesis, and for anthropic settlements. It is also a risk and a resource for each territory. All of these topics were treated in the document, which constitutes the backbone of the project, the loom on which to shape and develop the interactive 3D laboratories with augmented reality.

This activity foresees the 3D interactive scenario and is suitable for a much wider audience of recipients, not only school groups but also university students, tourists, and passionate common people.

The project consists of a 3D and 360° panorama of Frasassi gorge, acquired by a special drone and viewable by the user from above (Figure 2). The scenario is interactive. It is based on a 3D virtual reality with immersion of 360°, and users can view the panorama of the Frasassi surroundings while flying, zooming, and interacting where they want. Images were acquired through a special drone, an octocopter with eight propellers for the support of Titan cameras, with an overall resolution of 11 k.

Figure 2. Panoramic view of the Frasassi gorge at 360°, captured by the drone used for 3D virtual reality.

This digital content allows for online visiting of caves and the area around the caves by people who, because of health reasons, cannot do so in reality. Therefore, it will be an instrument to show the landscape from above that is useful for people with disabilities.

3. Results

As mentioned previously, the great intuition for the two projects under consideration lies in the possibility of providing to school groups or tourists a wide range of offers, depending on the type of experience they need; that is, the activities through didactic geological notebooks are an "in the field" experience, where students go in the cave, see the phenomenon with their eyes, and can direct enjoy the experience. Contrariwise, the 3D interactive museum geological laboratories represent, in a certain sense, the necessary digital innovation desired today by the users.

These are two completely different types of didactic approaches, but we believe that they are both experiences that should be engaged in, as one represents the completion of the other. Both activities will be illustrated in the following pages, although one should refer to the Supplementary Materials for specific details.

3.1. Didactic Geological Notebooks for School Groups Visiting the Caves

The didactic geological notebooks were attached as Supplementary Materials to the present work.

Two different types of didactic notebooks were realized: one for teachers and guides of the Frasassi caves and the other one for students, accompanied by the map of the explorable cave in plan view (Figure 3). The first one ("Guida al quaderno didattico") is designed to instruct teachers and guides on how to hold the didactic lesson in the cave, as they will then be the ones to guide and question the students. The second one ("Il taccuino del giovane speleologo") is considered the real didactic notebook for students and is composed of three sections: an initial didactic, a second related to the experience, and a third in which we will try to fix some key concepts of the experience in a lighter manner.

Figure 3. The two notebooks (one for teachers and guides and the other for students) with the map of the cave and the real scenario for the activities.

The first part of "Il taccuino del giovane speleologo" concerns the theory. It must be dealt with in the classroom as an extra didactic activity or during the visit to the Frasassi cave. Therefore, in the text, the following is explained: what a cave is, how the water is able to dissolve the limestone rock, how extensive the Frasassi caves are, if there is water in them, what structure the caves have, when they were formed, why in the cave we find gypsum deposits, why concretions are formed and how much they grow every year, and which organisms live in the cave. The topics that can be treated in the cave are practically a myriad: geology, sedimentology, stratigraphy, speleology, topography, cartography, hydrogeology, karst, structural geology, geomorphology, palaeontology, and speleobiology. As sciences, these disciplines are obviously based on the laws of chemistry, physics, and mathematics. The added value of educational notebooks in the cave thus lies in the fact that they offer an experiential visit with a high multidisciplinary content.

The second part of the didactic geological notebook represents the real activity, with practical exercises. It was conceived as an exploratory adventure and, for this reason, the cover has been given the appearance of an explorer's notebook, in A5 format, with the cartography map as a central insert in A3. This map is the "beating heart" of the experience, because all of the activities are based on it and it is also a useful tool to encourage students to take their first steps in orienteering.

We stated that the activity in the cave would be a real exploration, so we decided to reproduce the cartography of the cave in plan view, as a sort of ancient map, similar to a treasure map. Here, the pathway is traced and the stops at which the various activities are planned are indicated with crosses. There are five exercises. Each one requires at least 5–10 min of time for illustration and resolution, and the total foreseen time for a standard visit to the caves will add up to almost 2 h.

The first exercise will take place at stop 2 of the tour and will encourage students to become familiar with the map. They will orient the cartography with the north arrow and they will trace in it the rockfall deposit in the "Abisso Ancona" (the largest underground environment of the Europe caves) and some gypsum deposits (trying to respect the scale and dimensions).

A further method of representing the environment is the sectional one. In the second exercise (stop 3 of the tour), it will be possible to observe a north–south section of the "Abisso Ancona" (Figure 4). Based on some known points shown in the image, students must draw some surrounding elements, such as geological strata, rockfall deposits, principal stalactites, and the stalagmites known as "I Giganti". Older students may be further asked to be careful to maintain the correct proportions of scale.

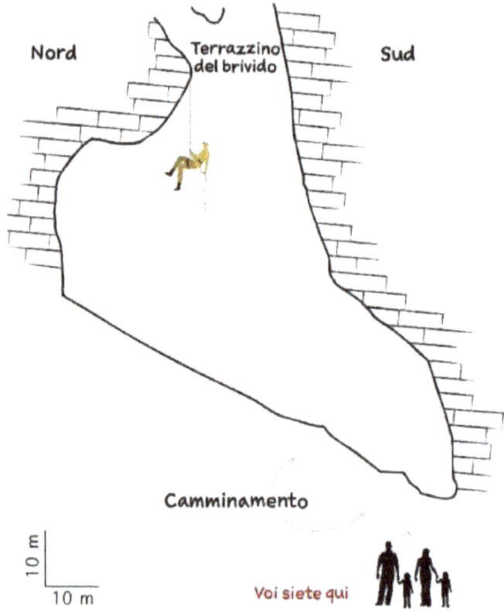

Figure 4. The second activity with a representation of the cave "Abisso Ancona" in the section.

The third exercise (Figure 5) refers to stop 4 of the tour. In this situation, students must connect, using an arrow, the exact terms with the elements they see in the photo of Figure 4. To help themselves, they can observe the image and make direct comparisons with what they see in front of their eyes.

This stop also has another exercise; in this case of geometry, the "Abisso Ancona" in plan view is similar to a square. Based on the scale bar on the map, students will have to calculate the approximate perimeter and area. In this way, they will understand that geology also involves mathematics and geometry.

The fourth experience was designed for adult students and is located at stop 7; this point is one of the very rare occasions in which it is possible to observe a fault directly from inside the mountain. By putting the point on paper and joining it with a straight line with another point where the fault is visible, the students will learn to draw a fault line on the map.

The last activity is designed for younger school children. It will be found at stop 9, the last portion of the cave. School children will have to keep their eyes open for what is considered the cave's real treasure: the eccentric stalactites (Figure 6). These are very peculiar concretions—they often form curls that can even develop upwards, defying gravity.

Figure 5. The third activity, where the elements visible in the cave must be connected with their names.

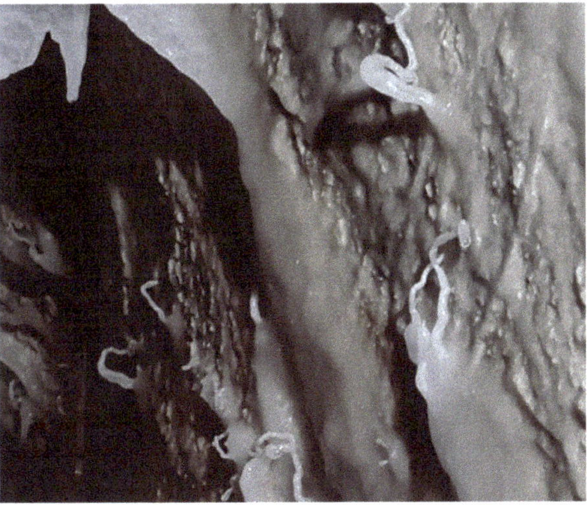

Figure 6. The eccentric stalactites in "Sala dell'Infinito". As it is possible to see, they are growing upwards.

The last, relaxing exercises aim to fix the key concepts through a crossword puzzle and a cryptogram (Figure 7). These may be considered simple exercises, but they are formative and fixative of the concepts learned in the cave.

All of these exercises are designed to promote geology, bringing students closer to Earth Sciences. The map in plan view will stimulate young students to orient themselves in the space, with the difference that, unlike the external environment, the reference points and objects used to orient oneself will be completely different, as well as the perception of distances. Based on the morphology of the cave (the shape of the walls or vaults), on a particular speleothem (stalactites, stalagmites, columns, curtains, draperies, and so on), in the presence of particular sediments (such as gypsum, a rockfall deposit, or a fossil-rich sedimentary level), rather than in the presence of groundwater, the students will observe in detail the cave, with a very keen eye on those elements useful for their orienteering.

Figure 7. The last exercises of the didactic geological notebooks.

3.2. 3D Interactive Museum Geological Laboratories

At this moment, it is not possible to provide a video representation of 3D interactive museum geological laboratories, because the work has been completed, but has not yet been set up in the museum.

In this second realized project, the 3D interactive museum geological laboratories, the activity aims to provide an immersive visual experience, which is currently being assembled at the museum "Art, history, and territory" of Genga. The acquired 360° landscape can be viewed on a monitor or through the use of a mobile app, and the 3D visualization is possible thanks to Oculus Quest viewers, provided by the museum. This typology of geo-education represents an example of STEM education, a teaching approach that combines Science, Technology, Engineering, and Math [8].

In addition to showing the whole landscape from above and allowing detailed zooms, the activity can be considered as a journey that accompanies the user through the geological history of the Frasassi area. Through accurate geological sections and 3D modeling, it has been possible to reproduce the paleoenvironments in which rock formed and evolved (Figures 8 and 9). This aspect is fundamental, because, through paleoenvironmental animations (Figure 10), it was possible to simulate the geological processes that accompanied rocks from the Jurassic period to the current day and that, following the emergence above sea level, have led to the geomorphological shaping of the relief and to the formation of underground cavities (even if sometimes, owing to technical limitations, it was not possible to replicate the precise evolution).

By modeling these environments, maximum attention was paid to details; therefore, it was sometimes necessary to model the marine organisms that populated the sea in the Jurassic and, today, we find fossils, such as ammonites, cuttlefishes, squids, and ichthyosaurs (Figure 11). Yes, ichthyosaurs, because the well preserved three-meter holotype of Jurassic Gengasaurus Nicosiai, found a few hundred meters from the caves in the 1970s, is kept in the speleo-paleontological museum of San Vittore.

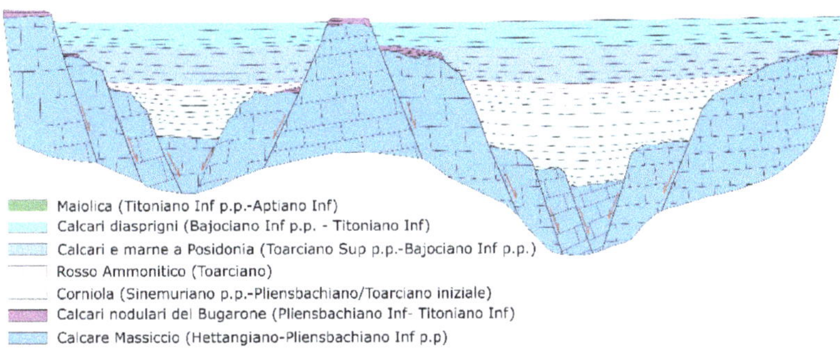

Figure 8. A schematic geological section used for animated modeling of the Jurassic paleoenvironment. During the animation, the activation of normal faults and the prosecution of sedimentation are shown.

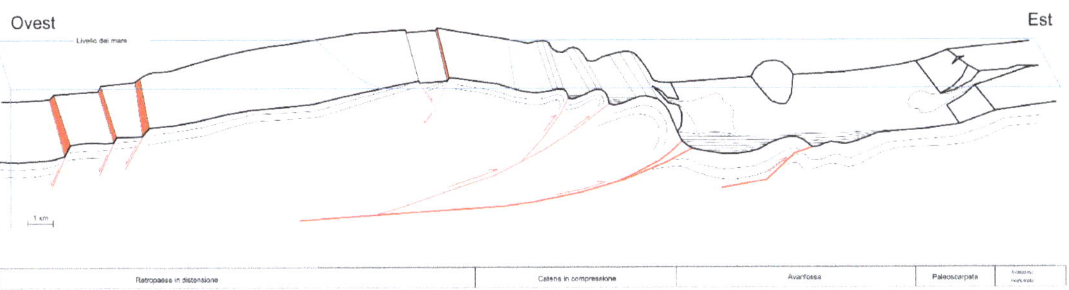

Figure 9. A schematic geological section used for animated modeling of the Umbria-Marche Miocene paleoenvironment. During the animation, the activation of normal faults is shown.

Figure 10. A frame from the animation of the Jurassic paleoenvironment modeled.

Figure 11. Some of the organisms modeled for the Jurassic marine paleoenvironment reconstruction.

4. Discussion

The dissemination of scientific heritage, using well-known and appreciated topics, could represent one of the new goals of Earth Sciences [9]. The examination of some of the dramatic events that have occurred in Italy related to its geo-environmental setting and the effects of interaction with anthropogenic pressure reveals the need to provide the general public with correct and clear information about the complex scenario characterizing this country [10].

The landscape plays a key role in communication and knowledge processes; it is the aspect of the environment that people perceive and with which they interact. The project's strategy is to involve the various social components, relying on unconventional and intergenerational communication, and to encourage active, extensive, effective, and lasting participation in risk prevention and sustainable resource enhancement activities. The tactics involve a range of methods that make scientific issues first accessible and then intriguing, until they become engaging. Communication is the medium and message [11]. Departure, path, and arrival are considered. In this regard, the use of topics more appealing to the general public as vectors of scientific concepts is a successful strategic solution when trying to share complex, or simply unpopular, concepts, information, or regulations. An effective tactic is to provide "unexpected" information through unconventional channels and to invite prominent figures to be actively involved as testimonials [12].

The extraordinary geo-environmental and cultural landscape that characterizes Italy is a powerful means of communicating territorial issues to all of society. Landscapes, if recognized and understood, become part of everyone's cultural heritage and reveal the dynamics that characterize their natural history, on the basis of the aforementioned "risk/resource" combination, providing the observer with knowledge indispensable to understanding complex environmental realities [13].

It is in this sense that the very perception of the landscape is the basis of a cognitive process that can trigger a virtuous circle, which revitalizes the roots that bind man and the environment, fostering more creative participation of society in balanced land management and sustainable development.

These projects were partially realized in an experimental manner, reaching an encouraging involvement of the local communities. A desirable outcome would be that the project will achieve further institutional support in order to be implemented.

The didactic geological notebooks were tested during their drafting through constant discussions with the administrations involved (Geology Section of the University of Camerino, Consorzio Frasassi, Municipality of Genga). In April 2022, a specific educational meeting was also organized with about 60 teachers from all over central Italy and the most experienced guides of the cave. On this occasion, the whole project was presented, because guides and teachers need to know the correct way to disclose it to students. Each aspect of the didactic notebooks was illustrated directly in the cave, in order to become familiar with the project, to know its facets, and to understand the best way to apply it by stimulating children during the experience in the cave. The reviews collected have been almost totally positive, although making people from other sectors understand the mechanisms that regulate Earth Sciences is not easy.

The 3D interactive museum with virtual reality geological laboratories is under construction, although we can count on positive reviews from insiders (geologists, graphic designers, local administrators, and marketing staff of the caves). The final realization will presumably take place in November 2022.

The use of innovative digital software in the framework of education has many barriers. Among them, the high cost of equipment for organizing the educational space, the need for additional competencies of the teaching staff [14], and the use of students' personal devices to ensure successful group interaction [15]. During this activity, special Oculus Quest viewers will be provided to the students in order to implement the 3D vision and the immersion in the environments reproduced at 360°.

5. Conclusions

The territorial proximity between the premises of Geology Department of Camerino and the Frasassi caves stimulated collaboration between the public bodies, united by geological excellence. This represents an element of strength on which to set up projects capable of disseminating the importance of Earth Sciences and the figure of the geologist. The environmental context facilitates teaching in the field and geo-education, above all thanks to the welcoming structures offered by the district. Here, in fact, starting from the 1970s, the tourist reception has undergone great development, and now the territory is able to accommodate large numbers of school groups (around 80,000–90,000 per year).

Several characteristics provide the mountain environment areas with great potential for geoheritage, geoconservation, and geotourism studies [16], creating the best conditions for geo-education. The underground karst complexes, in fact, have always attracted the curiosity of people, arousing in them sublime feelings of attraction and repulsion. In Frasassi, the attractive effect of the hypogeum complex is amplified because the fascinating caves are located inside a breathtaking gorge, characterized by steep slopes that are hundreds of meters high, hosting the lively Sentino River, embedded at their feet. These features are fundamental to enrich the fascinating landscape in which it is possible to recognize the impressive geomorphological elements and to observe the processes (active and well understood) that condition their development.

For all of these reasons, geomorphosites (especially mountain ones) seem to be particularly interesting sites for developing educational activities on environmental issues [17]. All of these elements inspire attendance by school children, stimulating outdoor experiential teaching that facilitates the understanding of phenomena. The outdoor didactic is seen as a moment of leisure, an unusual method of teaching that is original and able to capture the attention of scholars.

The hope is that, thanks to the projects carried out and illustrated in this work, other virtuous behaviors can be established in other territories, thanks to their replicability in other contexts (although the conditions are not present in all of them).

Supplementary Materials: The following supporting information can be downloaded at https://www. mdpi.com/article/10.3390/geosciences12110418/s1. Quaderno didattico_Guida docenti: Guida al quaderno didattico; Quaderno didattico_Studenti: Il taccuino del giovane speleologo; Quaderno didattico_Mappa.

Author Contributions: Writing—original draft preparation, F.B.; writing—review and editing, P.F.; supervision, P.F. All authors have read and agreed to the published version of the manuscript.

Funding: This research received no external funding. The two projects of the didactic geological notebooks for school groups and the 3D interactive museum with virtual reality geological laboratories were founded by "Consorzio Frasassi". The authors and the Geology Section of the University of Camerino provided the project idea and the scientific support necessary for its realization, which was edited under a graphic and technological (modelling) point of view by "Brugiatelli Design", Serra de' Conti (AN), Italy and "L.D. Multimedia s.r.l.", Moncalieri (TO, Italy).

Data Availability Statement: Not applicable, no data were used for this article.

Acknowledgments: It is our duty to thank professor Emanuele Tondi for having allowed the realization of the two projects providing, as usual, very useful original ideas. The paleontologist Marco Peter Ferretti for the support offered in the modeling of Jurassic organisms; Sandro Galdenzi for the precise indications relating to the didactic notebooks; the archeologist Gaia Pignocchi for the archeological consultancy; and the entire working group that contributed to the creation of the 3D interactive museum with virtual reality geological laboratories, which, in addition to the authors of the present article, is composed of Claudio Di Celma, Marco Peter Ferretti, Sandro Galdenzi, Maria Chiara Invernizzi, Danica Jablonska, Marco Materazzi, Stefano Mazzoli, Pietro Paolo Pierantoni, Emanuele Tondi, and Tiziano Volatili. Finally, we would like to thank Marco Filipponi and David Bruffa of the Municipality of Genga; and Lorenzo Burzacca, Maurizio Tosoroni and Sara Bonacucina of the "Consorzio Frasassi" for allowing us to apply our knowledges and our projects to their caves and their territory, making them available for school groups, citizens and tourists.

Conflicts of Interest: The authors declare no conflict of interest.

References

1. Farabollini, P.; Spurio, E.; Aringoli, D.; Chiarini, C.; Kardos, E. *I Quaderni del Parco, N.8—Geositi e Geoescursionismo nel Parco Naturale della Gola della Rossa e di Frasassi*; Comunità Montana dell'Esino-Frasassi, Parco Naturale della Gola della Rossa e di Frasassi: Serra San Quirico, Italy, 2007.
2. Alessandroni, G.; Bendia, F.; Farabollini, P.; Gennari, E.; Nesci, O.; Tatali, B.; Valentini, L. I Geositi delle Marche—Un patrimonio di geodiversità di eccezionale valore. Presentazione risultati Gruppo di Lavoro SIGEA-APS Marche, Università di Urbino, Università di Camerino, Ordine Geologi Marche. In Proceedings of the Workshop I Geositi delle Marche, Ancona, Italy, 26 May 2022.
3. Bocchini, A.; Coltorti, M. Il complesso carsico Grotta del Fiume Grotta Grande del Vento e l'evoluzione geomorfologica della Gola di Frasassi. *Mem. Ist. It. Speleol.* **1990**, *4*, 155–180.
4. Taddeucci, A.; Tuccimei, P.; Voltaggio, M. Studio geocronologico del complesso carsico "Grotta del Fiume-Grotta Grande del Vento" (Gola di Frasassi, AN) e indicazioni paleoambientali. *Il Quat.* **1992**, *2*, 213–222.
5. Dryglas, D.; Miśkiewicz, K. Construction of the geotourism product structure on the example of Poland. In Proceedings of the 14th International Multidisciplinary Scientific GeoConference SGEM 2014, Albena, Bulgaria, 17–26 June 2014; Volume 3, pp. 155–162.
6. Miśkiewicz, K. Promoting geoheritage in geoparks as an element of educational tourism. In *Geotourism. Organization of the Tourism and Education in the Geoparks in the Middle-Europe Mountains*; Szponar, A., Toczek-Werner, S., Eds.; University of Business in Wrocław: Wrocław, Poland, 2016; pp. 37–48.
7. Hueso-Kortekaas, K.; Iranzo-García, E. Salinas and "Saltscape" as a Geological Heritage with a Strong Potential for Tourism and Geoeducation. *Geosciences* **2022**, *12*, 141. [CrossRef]
8. Shakirova, N.; Said, N.; Konyushenko, S. The use of virtual reality in geo-education. *Int. J. Emerg. Technol. Learn. (iJET)* **2020**, *15*, 59–70. [CrossRef]
9. Farabollini, P.; Lugeri, F.R.; Amadio, V.; Aldighieri, B. The role of Earth Sciences and Landscape Approach in the Ethic Geology: Communication and Divulgation for the Prevention and Reduction of Geological Hazards. In *Engineering Geology for Society and Territory*; Lollino, G., Arattano, M., Giardino, M., Oliveira, R., Peppoloni, S., Eds.; Springer: Berlin/Heidelberg, Germany, 2014; Volume 7, pp. 115–120.
10. Lugeri, F.R.; Farabollini, P.; Amadio, V.; Greco, R. Unconventional Approach for Prevention of Environmental and Related Social Risks: A Geoethic Mission. *Geosciences* **2018**, *8*, 54. [CrossRef]
11. McLuhan, M.; Fiore, Q. *Il Medium è il Massaggio*; Editore Corraini; Penguin Books: London, UK, 2011; p. 160. ISBN 8875702861.
12. Lugeri, F.R.; Farabollini, P.; De Pascale, F.; Lugeri, N. PPGIS applied to environmental communication and hazards for a community-based approach: A dualism in the Southern Italy 'calanchi' landscape. *AIMS Geosci.* **2021**, *7*, 490–506. [CrossRef]
13. Panizza, M.; Piacente, S. Geomorphosites: A bridge between scientific research, cultural integration and artistic suggestion Geomorphological sites and geodiversity. *Il Quat.* **2005**, *18*, 3–10.
14. Zeer, E.F.; Tretyakova, V.S.; Miroshnichenko, V.I. Strategic Directions of Pedagogical Personnel Training for the System of Continuing Vocational Education. *Educ. Sci. J.* **2019**, *21*, 93–121. (In Russian) [CrossRef]

15. Ezrokh, Y.S. HR Perspectives of Russian Universities: Who Will Teach in the Near Future? *Educ. Sci. J.* **2019**, *21*, 9–40. (In Russian) [CrossRef]
16. Giusti, C.; Reynard, E.; Bollati, I.; Cayla, N.; Coratza, P.; Hoblea, F.; Ilies, D.; Martin, S.; Megerle, H.; Pelfini, M.; et al. A new network on mountain geomorphosites. *Geophys. Res. Abstr.* **2013**, *15*, EGU2013-6706.
17. Reynard, E.; Coratza, P. The importance of mountain geomorphosites for environmental education: Examples from the Italian Dolomites and the Swiss Alps. *Acta Geogr. Slov.* **2016**, *56*, 291–303. [CrossRef]

Article

Virtual Fossils for Widening Geoeducation Approaches: A Case Study Based on the Cretaceous Sites of Figueira da Foz (Portugal) and Tamajón (Spain)

Senay Ozkaya de Juanas [1,2,*], Fernando Barroso-Barcenilla [2,3], Mélani Berrocal-Casero [2,3] and Pedro Miguel Callapez [1]

[1] Departamento de Ciências da Terra (Centro de Investigação da Terra e do Espaço), Universidade de Coimbra, 3030-790 Coimbra, Portugal
[2] Departamento de Geología, Geografía y Medio Ambiente (Grupo de Investigación PaleoIbérica), Universidad de Alcalá, 28805 Alcalá de Henares, Spain
[3] Departamento de Geodinámica, Estratigrafía y Paleontología (Grupo de Investigación Procesos Bióticos Mesozoicos), Universidad Complutense de Madrid, 28040 Madrid, Spain
[*] Correspondence: sjuanas@student.dct.uc.pt

Abstract: Accessible palaeontological sites conform highly adequate out-of-school environments for meaningful learning experiences regarding formal and non-formal teaching of geosciences. With a perspective of international cooperation, two correlative Cenomanian–Turonian (Upper Cretaceous) outcrops from the Iberian Peninsula have been chosen as the focus of this project—the sections of Figueira da Foz (Portugal) and Tamajón (Spain)—along with the Palaeontological and Archaeological Interpretation Centre of Tamajón (CIPAT). Virtualization of fossil samples and sites has been undertaken by means of phase-shift scanning, photogrammetry, and small object scanning by structured light and laser triangulation, resulting in three-dimensional virtual models of the main fossil tracks and invertebrate fossil samples. These virtual fossils have allowed the development of transdisciplinary didactic activities for different educational levels and the general public, which have been presented as file cards where the age of participants, objectives, multiple intelligences, European Union key competences, needed resources, development, and further observations are specified. This work aims to contribute to improving the design and development of didactic sequences for out-of-school education at these sites, organizing effective transdisciplinary teaching tools, and developing awareness, values, and responsibility towards geoheritage.

Keywords: didactics; geoeducation; geotourism; geoconservation; natural heritage; virtual palaeontology

Citation: Ozkaya de Juanas, S.; Barroso-Barcenilla, F.; Berrocal-Casero, M.; Callapez, P.M. Virtual Fossils for Widening Geoeducation Approaches: A Case Study Based on the Cretaceous Sites of Figueira da Foz (Portugal) and Tamajón (Spain). *Geosciences* **2023**, *13*, 16. https://doi.org/10.3390/geosciences13010016

Academic Editors: Hara Drinia, Panagiotis Voudouris, Assimina Antonarakou and Jesus Martinez-Frias

Received: 2 December 2022
Revised: 22 December 2022
Accepted: 27 December 2022
Published: 1 January 2023

1. Introduction

Palaeontology is a scientific discipline which studies the origin and evolution of life on the dynamic Earth, diving into the depths of time's arrow [1] through a Huttonian perspective of the sedimentary and fossil record. As it allies the extraordinary diversity and singularity of fossils with a rich conceptual and historical framework that allows palaeontologists to rediscover the worlds before Adam [2], it has been successfully explored by teachers as an attractive didactic resource for both formal (institutionalized, academic) and non-formal (outside the formal educational curriculum) educational contexts for geosciences, e.g., [3]. Content related to Earth sciences comes across as a valuable tool for the teaching–learning process to easily transmit the importance of science and its method, the role of researchers in society, how our planet evolved during millions of years, raise awareness about the relevance of protecting natural heritage, and as a whole, to address geoeducation. This term refers to an environmental learning, which also promotes geoethics, geoheritage, and geoconservation [4], to encourage people to understand and learn more about geosciences in general, aiming at organizing effective teaching tools [5] and applying modern means, such as three-dimensional (3D) modelling and its application to information and communications technologies (ICT).

Field trips implemented as part of out-of-school learning sequences allow for the development of important concepts and skills related to geosciences in general and Earth's history in particular, e.g., [6,7]. Therefore, accessible palaeontological sites conform a highly adequate out-of-school environment for meaningful learning experiences [8]. Specifically, Cretaceous geological formations and their commonest lithologies and facies can be easily found and accessed around the world, allowing for the replication of scientific education and outreach strategies in many regions [9], such as the "Cretaceous Viewpoint" in the Orígens Global Geopark (Spain, Mirador Cretaci: [10]) or the Cretaceous fossil sites in South Korea, which comprise a world-class geotourism resource [11].

Aiming to explore the potential of out-of-school learning through a perspective of international cooperation, two correlative palaeontological sites from the Iberian Peninsula have been chosen as the focus of the development of this project: the Cenomanian–Turonian (Upper Cretaceous) sections of Figueira da Foz (Portugal) and Tamajón (Spain) (Figure 1), where the virtualization of fossil samples and sites has been undertaken to develop a series of transdisciplinary didactic activities for different educational levels and the general public.

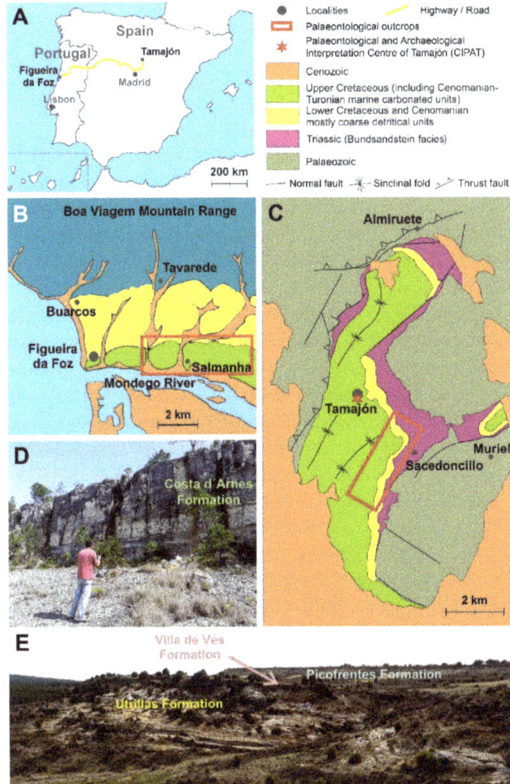

Figure 1. Geographical and geological context of the palaeontological sites of Figueira da Foz (Portugal) and Tamajón (Spain). (**A**) Map of the Iberian Peninsula locating the palaeontological sites of Figueira da Foz (Portugal) and Tamajón (Spain). (**B**) Geological sketch map of Figueira da Foz (Portugal). (**C**) Geological sketch map of Tamajón (Spain). (**D**) Photograph of the palaeontological site of Figueira da Foz showing the Costa d'Arnes Formation. (**E**) Photograph of the outcrops of Tamajón showing the Utrillas, Villa de Vés, and Picofrentes formations. Modified from [9].

2. Geological and Palaeontological Context

The fossil-rich palaeontological sites are exceptional examples of the diversity of marine fossil assemblages that were widespread in the warm and shallow carbonate platform environments of the Tethyan Realm during the Cretaceous Period, when sea-levels and global average temperatures were usually much higher than nowadays and where extraordinary flora and fauna evolved during millions of years (ca. 145–66 M.a.), e.g., [12,13].

The Cenomanian–Turonian section of Figueira da Foz (Portugal) is located near its namesake coastal town of west central Portugal, north of the Mondego River, near the estuary (Figure 1B) and very close to the entrances of the A14 and A17 highways. Local accessibility is excellent, including direct access to Coimbra, Aveiro, and the A25 highway to Spain. This site has been studied since 1849, and repositories of its fossil assemblages are housed at the Geological Museum (Museu Geológico, MG-LNEG, Lisbon) and the University of Coimbra (Universidade de Coimbra, UC, Coimbra), among other Portuguese institutions, demonstrating its scientific and educational value [14,15]. The main available exposures are found in two old quarries (Salmanha I and Salmanha II), where a set of mid-Cenomanian to lower-Turonian (Upper Cretaceous) marine beds of the West Portuguese Carbonate Platform are recorded by the Costa d'Arnes Formation. The 65 m thick of the stratigraphic succession holds several fossiliferous units with diverse ammonite content [16,17], such as the remarkable *Neolobites vibrayeanus* and *Vascoceras gamai* cephalopod assemblages, along with other abundant benthic invertebrates with Tethyan affinities, including many species of bivalves, gastropods, and echinoids (e.g., *Neithea hispanica, Tylostoma ovatum, Mecaster scutiger*) (Figure 1B,D). In addition, various sections of these sites are quite accessible and adequate for the development of outreach activities.

The Upper Cretaceous section of Tamajón (Spain) is located in the province of Guadalajara, approximately 1 h away from the northeast of Madrid, and combines high scientific, educational, and outreach values [9,18,19]. The sedimentary record of its Cenomanian interval reaches 35 m thick. It is included into the Utrillas Formation [20], where a coastal vertebrate tracksite has been described [21], and the Villa de Vés [22] and Picofrentes [23] formations, which yield a high diversity and abundance of marine invertebrates along with plant remains, bioturbations, and vertebrate fossil samples [18]. The latter has been biostratigraphically studied since the past century, having described a high diversity of invertebrate and vertebrate marine species, such as bivalves, gastropods, cephalopods, and echinoids (e.g., *Granocardium (Granocardium) productum, Tylostoma torrubiae, Vascoceras harttii, Hemiaster* spp.). Furthermore, the Tamajón vertebrate tracksite yields a high density of ichnotaxa, among which, to date, several tracks and trackways of crocodyliforms ("Galloping crocs": [24]), theropod dinosaur footprints, and fish fin traces have been described (Figure 1C,E). Furthermore, many representatives of these faunas are also housed in Portuguese and Spanish institutions, including the universities of Coímbra, Alcalá and Complutense de Madrid, and they are also partially available in public exhibitions, such as at the Palaeontological and Archaeological Interpretation Centre of Tamajón in Guadalajara, Spain (CIPAT for its acronym in Spanish: *Centro de Interpretación Paleontológica y Arqueológica de Tamajón*: [25]).

To date, the data and samples recollected from these sites have been studied with traditional palaeontological methods and tools. However, among other support resources for its research and didactics, virtual palaeontology (the study of fossils throughout 3D visualizations or virtual fossils: [26]) has become a popular non-destructive and non-invasive technique, as these new methodologies, along with traditional ones, offer a variety of advantages for the scientific and educational aspects of palaeontology [27], as well as for geotourism purposes.

Furthermore, the scientific and educational value of these palaeontological sites, along with the possibilities that virtual fossils can offer dissemination, outreach, and didactics, allow for addressing the crucial issue of educating society in preserving natural heritage through positive geoconservation actions. An effective strategy for this objective is to prompt geotourism in the localities and proximities of the outcrops. Geotourism is a relatively new form of alternative tourism, with significant European and global development potential [4], defined also by the Arouca Declaration [28]. Finding various definitions of the term, geotourism is viewed [29] as geological tourism that has a focus on geoheritage and with the main goal of attending geoconservation by education, being an essential and flexible tool to raise awareness about scientific findings and their role in society. Moreover, these initiatives also include geoethics, which widens the cultural horizon of geosciences knowledge and contributes to orienting scientists and society in the choices for responsible behaviour towards the future of humankind on planet Earth [30–32].

In the scientific literature, the terms geoheritage, geoethics, geoeducation, geoconservation, and geotourism are closely related to the development of geoparks and other geological points legally defined. However, the authors would like to state in this work the possibility of developing the mentioned terms in scientifically interesting geological and palaeontological areas, which may still not have been legally labelled as protected localities and yet could yield highly important scientific and didactic values where field trips, didactic activities, and even visitor centres could be developed. These are the cases of the natural environments of the palaeontological sites of Figueira da Foz and Tamajón, and the Palaeontological and Archaeological Interpretation Centre of Tamajón [25], which is a key component for the development of local geotourism based in the understanding of their identity or character and their natural attributes. Therefore, it should aim to provide education, especially for the benefit of young people and students, to encourage a wide range of the public to understand and learn about geology and the environment, in general, by using modern means.

3. Materials and Methods

Field work has been carried out in both Upper Cretaceous sites, Figueira da Foz (Portugal) and Tamajón (Spain), paying special attention to the exploration of didactic possibilities for on-site and virtual activities, along with recollection of fossil samples in each stratigraphic level, complemented by facies, biostratigraphic, and palaeoecological data for scientific and educational purposes. Furthermore, as the vertebrate tracksite of Tamajón must be virtually recorded in situ, a different approach was undertaken. To carry out an onsite preparation and the digitalization of the track surface by using laser scanning and photogrammetry techniques, an intensive field campaign was carried out during the summer of 2018. Firstly, the coverage of the surface was removed to prepare and clean the track surface (Figure 2A) for its posterior in situ study (Figure 2B). Then, researchers proceeded to the digitalization of the surface by means of phase-shift laser scanning (Figure 2C) and the systematic taking of photographs, essential to applying photogrammetry techniques. Lastly, the tracksite was covered again with geotextiles and local sediments as a way to take away the risk of subaerial deterioration and pillaging possibilities.

At the laboratory, the point cloud (combination of vertexes in a three-dimensional coordinate system) and over 600 photographs of the ichnite surface were obtained, respectively, by using laser scanning and photogrammetry techniques, and they were processed using different software. Moreover, small fossil scanning was carried out (Figure 2D) and followed by posterior treatment of the virtual results of the marine invertebrate specimens from both outcrops.

The sampled fossils were subsequently housed at the Department of Earth Sciences of the University of Coimbra (Portugal) and the Museum of Palaeontology of Castilla-La Mancha (Spain) and ceded temporarily to the Town Council of Tamajón to exhibit them at the CIPAT (Palaeontological and Archaeological Interpretation Centre of Tamajón [25]) (Figure 3).

Figure 2. Onsite and virtual palaeontology methodologies. (**A**) Cleaning of the track surface by means of traditional tools to enable correct taking of photographs and scanning of the tracksite. (**B**) Systematic on-site study of the main tracks, applying traditional methods such as drawings. (**C**) Phase-shift scanning of the track surface with the "Laser Scanner Focus 3D". (**D**) Laser triangulation scanning by using a small object scanner "NextEngine 3D Laser Scan" for the virtualization of invertebrate fossil samples.

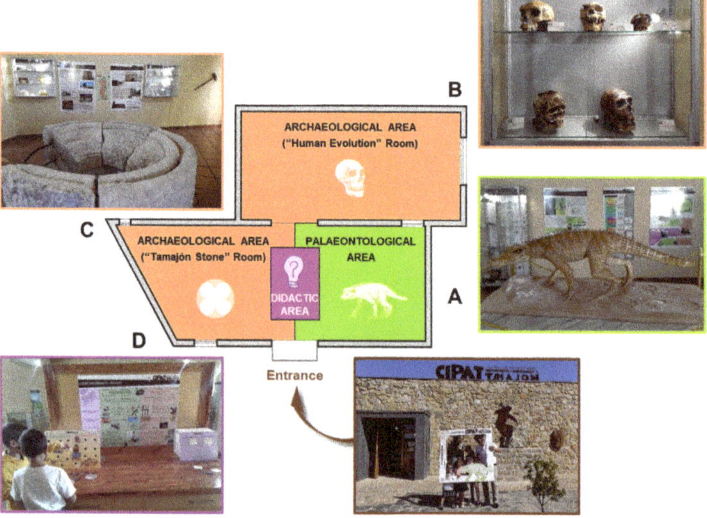

Figure 3. The Palaeontological and Archaeological Interpretation Centre of Tamajón (CIPAT). The map of the CIPAT in the middle, the Palaeontological Area (green, (**A**)), the Archaeological Area (orange, with the rooms of "Human Evolution" (**B**) and "Tamajón Stone" (**C**), and the Didactic Area (purple, (**D**)), and photographs of the front of the centre (entrance) and of each of the mentioned areas, following the anticlockwise itinerary.

Once the virtual palaeontology results were obtained, these made possible the design of several didactic activities, which are being carried out at the Didactic Area of the CIPAT [15] (Figure 3D), supporting the explanations of the infographic panels, real samples, and reconstructions held at the Palaeontological Area (Figure 3A).

3.1. Phase-Shift Laser Scanning

The device used is the "Laser Scanner Focus 3D" (Figure 2C), which uses LiDAR technology (Light Detection and Ranging), one of the leading choices for digital recording of tracksites [33,34]. Five measurements of 360° were completed with this technology, posteriorly obtaining a georeferenced point cloud.

Once on-site data were taken, the authors proceeded to process the information by using several software applications. The first set of data was extracted from "Trimble" (Geosptial software). To obtain the three-dimensional reconstruction, files were analysed and addressed with the "Autodesk ReCap" program.

3.2. Photogrammetry

The whole tracksite was photographed by ground-based techniques, using an "Olympus EM5 Mark 2" camera, with a resolution of 4608 × 3456 pixels. Considering the high density of different vertebrate ichnites, the tracksite surface was divided in a 1 × 1 m square metre grid. To process the photographs in the laboratory, the program "Autodesk ReCap Photo" was used, aiming to obtain 3D models of the main vertebrate crocodyliform tracks (CTA-1, CTA-2), the ichnite of a small theropod (DT-1), and the swimming track of a fish, assigned to *Undichna unisulca* (FST-1) [21].

3.3. Small Object Scanning

A small object scanner "CR-Scan01" from the Earth and Space Research Centre of the University of Coimbra (CITEUC, Portugal) and a "NextEngine 3D Laser Scan" from the Area of Palaeontology of the Faculty of Geological Sciences of the University Complutense of Madrid were used. The first scanner uses transmitted light, while the second one works with laser triangulation (MultiStripe Laser Triangulation; MLT). These devices are equipped with their corresponding software "CR Studio" and "Scan Studio", respectively, which were used to obtain in the laboratory a digital 3D model of the chosen samples, in this case marine invertebrate specimens from the sites of Figueira da Foz (Portugal) and Tamajón (Spain) (Figure 2D).

3.4. Geoscience Education

Didactics, understood as the way of teaching meaningfully to others, is essential for formal and non-formal educational contexts of geological sciences, as well as for outreach purposes, regarding this discipline to elaborate an adequate teaching methodology. Authors have considered the Earth Systems approach [9], with special emphasis on the development of thinking skills, critical analysis, and considering the objective of reaching environmental insight (*sensu* [35]).

The multiple intelligence theory [36] has been chosen as one of the main methodologies in which to base the didactic activities, e.g., [37]. Even though, since its first approach in the late past century, the multiple intelligences theory has been updated, only the first eight pillars have been considered for this work: the visual–spatial, linguistic–verbal, logical–mathematical, musical, bodily–kinaesthetic, interpersonal, intrapersonal, and naturalistic intelligences. Moreover, other approaches have also been considered, such as situated cognition [38,39] based on experiential learning [40], where learning experiences should involve similar types of activities as those which experts confront on a daily basis (e.g., taxonomical classification of invertebrate fossils), or the framework for museum practice (FMP: [41]), where visits to interpretation centres or to natural settings aim to develop the intrinsic motivation of pupils [42]. All together, they will enable reaching meaningful learning (*sensu* [43]), developing essential abilities and concepts such as high-order thinking skills (e.g., applying Bloom's taxonomy: [44]), critical thinking, or the affective domain towards others and the natural world.

This sequence considers as a main objective the role of education for citizenship and scientific literacy for society. Furthermore, taking into account that content regarding palaeontology is not usually considered or only briefly mentioned in the Portuguese and Spanish primary and secondary education curricula [7,37], the designed activities include the development of the European Union key competences [45] along with an insight to some of The Global Goals of the Agenda 2030 for Sustainable Development [46], such as goals number 4. Quality Education, 5. Gender Equality, 8. Decent work and economic growth, and 15. Life on Land.

The proposed activities have been mainly designed for their on-site implementation at the Figueira da Foz and Tamajón outcrops, as well as for the CIPAT (Palaeontological and Archaeological Interpretation Centre of Tamajón, Guadalajara, Spain: Figure 3). Here, the Didactic Area was projected between the Palaeontological and Archaeological sections to work as a nexus between them and as a complement to the permanent expositions where infographics, real fossil samples, and reconstructions are shown [15,25,47]. In this case, by giving supporting material and developing didactic activities for young and adult visitors, helping to understand different palaeontological aspects of the Cenomanian–Turonian of Tamajón (fossilization process, basic taxonomical classification, biostratigraphy, palaeoecology, palaeogeography, etc.), as well as more general content regarding geosciences, geoheritage, and geoethics, is facilitated.

These activities have been carried out in a non-formal education museum context, which implies that during their application participants can go back to the permanent exposition and look up information. Furthermore, these activities have been designed to be completed autonomously and in one single session of variable duration. Each activity covers a series of contents and objectives, not needing to follow a concrete sequence, constructing in this way different key ideas about science, palaeontology, and also specific aspects, such as regarding the Cretaceous life or the fossiliferous sites of Figueira da Foz and Tamajón. Nevertheless, ideally, a teaching agent (teachers, museum guides, counsellors, etc.) guides participants throughout the activities, facilitating tools which allow them to construct scientific knowledge, reaching meaningful learning experiences. Therefore, scaffolding techniques [48] are highly relevant for the correct implementation of the activities.

Feedback recollection has also been considered, as this information is essential to check the effectiveness of the proposed activities at the Didactic Area, compiling information regarding interests and the level of impact in relation to the set objectives and previous knowledge [49]. For the recollection of data, a variety of qualitative methods has been developed, mainly by means of didactic workshops and surveys. As part of an activity at the CIPAT, a feedback panel can be found at the Didactic Area, where participants can write something they have learnt during the visit and what they have enjoyed the most. This qualitative feedback has already been put into practice by the authors on several occasions throughout different outreach events [15,37], allowing the adaptation to educational interests and needs, especially for primary education levels; however, the results will not be discussed in this work.

4. Results

By means of phase-shift laser scanning, a three-dimensional model of the tracksite has been obtained (Figure 4). Applying photogrammetric techniques, 3D reconstructions of the selected tracks and ichnites [21] (Figure 5) have also been possible. Furthermore, by scanning different fossil specimens with the small object scanners, different 3D models have been obtained (Figure 6). Based on the digital reconstruction of the palaeontological material, a didactic proposal for geoscience education has been elaborated to support the integration of these results in the Palaeontological and Archaeological Interpretation Centre of Tamajón (CIPAT), as detailed below.

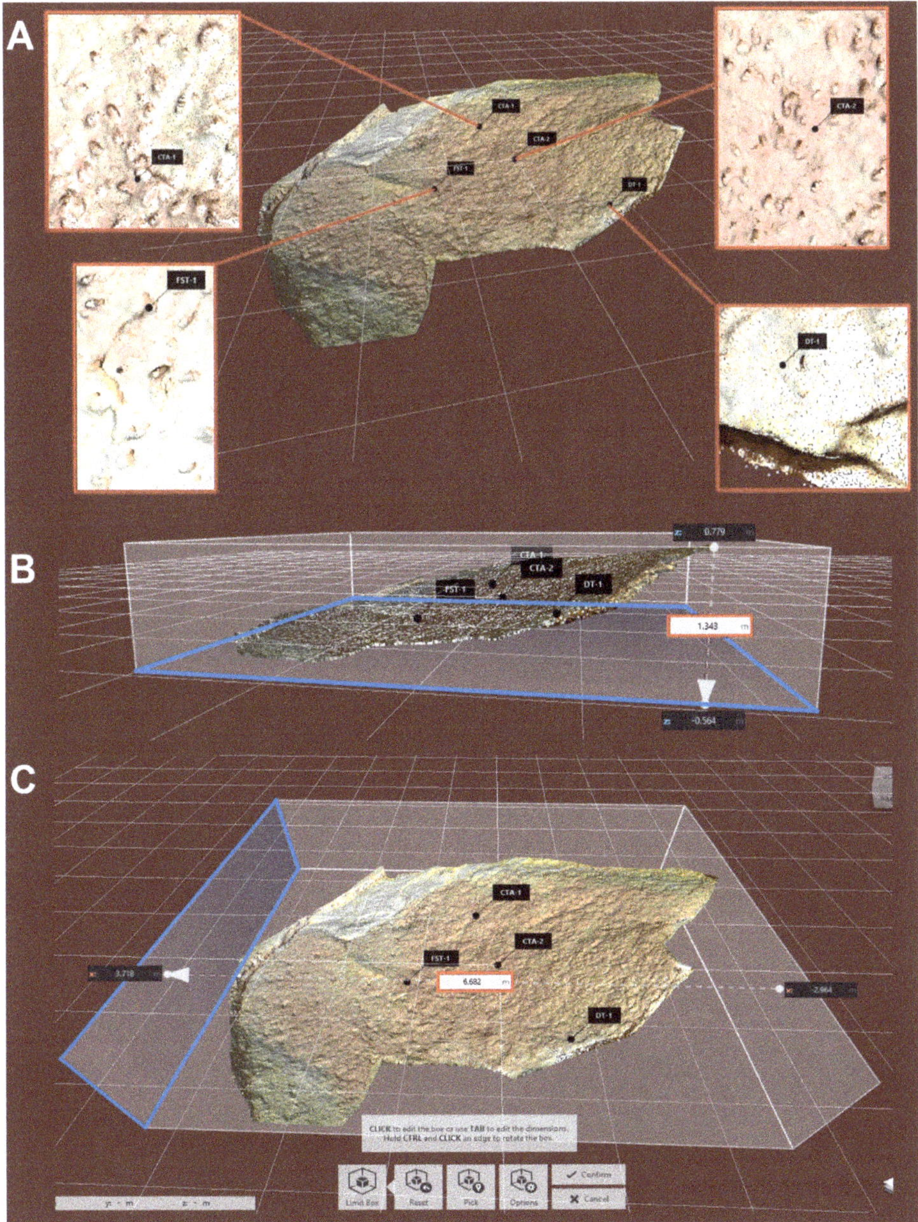

Figure 4. Three-dimensional model of the Upper Cretaceous track surface of Tamajón scanned with a phase-shift scanner. (**A**) Track surface and augmented view of the main crocodyliform tracks (top right and top left), small theropod print (bottom right), and fish fin trace *Undichna unisulca* (bottom left). (**B**) Lateral view of the tracksite. Total height (1.343 m) of the exposed channel surface. (**C**) Total length (6.662 m) of the exposed channel surface. Modified from [19].

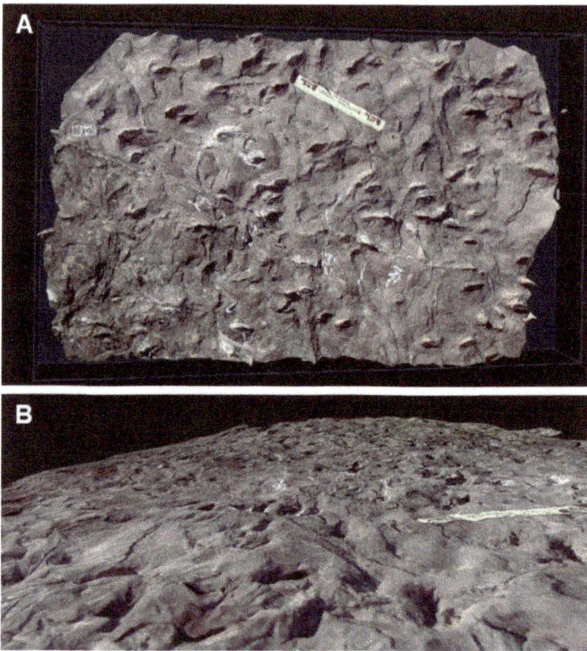

Figure 5. Photogrammetric three-dimensional modelling of the crocodyliform track CTA-2. (**A**) Top view of the crocodyliform track. (**B**) Detailed view of the negative relief of the crocodyliform track.

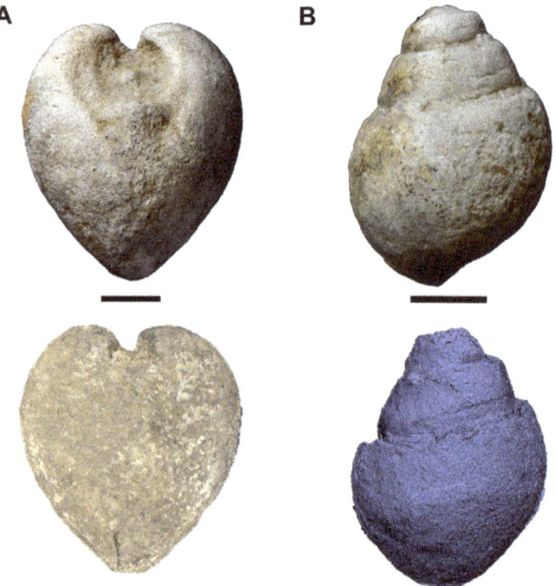

Figure 6. Three-dimensional models of the fossil casts of invertebrate specimens scanned with a small object scanner. (**A**) *Granocardium* (*Granocardium*) *productum* bivalve photograph (top view) and 3D model (bottom view). (**B**) *Tylostoma torrubiae* gastropod photograph (top view), and 3D model with "solid object" option (bottom view). Graphic scales: 1 cm.

4.1. Phase-Shift Laser Scanning

The processing and treatment of the on-site recollected data by means of a phase-shift scanner have resulted in a virtual 3D model that allows for permanent storage of the current state of the tracksite (Figure 4). The reconstruction shows the real dimensions of the exposed channel surface (length, width, and depth) along with the coordinates of each ichnite (Figure 4A). It also allows for automatically measuring the distance between any chosen points of the surface. Among the results, it is possible to clearly observe both planes of the track surface, as the program used allows 360° movement around the scanned area (Figure 4B,C). By observing both planes of the surface, diverse options to examine and study the ichnites are found from above (concave epireliefs) or from below (convex hyporeliefs). The latter is the virtual cast generated by the scanner, making it possible to measure the depth and shape of each of the ichnites. The utilized program also integrates the possibility to mark different points over the site, allowing the addition of notes, images, and photographs, creating a database and easing the exchange of the registered information with other users.

This digital reconstruction is currently being used to study in detail the ichnites and other non-biogenic sedimentary structures, to measure length, interdigital angles, and depth of prints, helping to identify new morphotypes that have not yet been described in this tracksite.

4.2. Photogrammetry

Regarding the resulting photogrammetric 3D models, various reconstructions have been obtained (e.g., Figure 5), where the main crocodyliform tracks (CTA-1, CTA-2) are represented, along with the ichnite of a small theropod (DT-1) and the swimming track of a fish, assigned to *Undichna unisulca* (FST-1). The program "ReCap Photo" allows one to view the surface of the prints in a full 360° view, making it possible to measure depth and angles, in addition to establishing scales, colouring functions, distortion, and mesh.

4.3. Small Object Scanning

The utilization of a small object scanner has allowed for the digitalization of the selected invertebrate fossil samples, recollected from the Upper Cretaceous marine sites of Figueira da Foz (Portugal) and Tamajón (Spain). The resulting 3D models show the real colouring as well as the complete and detailed external morphology that can be observed in a 360° view (Figure 6). Furthermore, applying the different visualization options, once the model is completed and finished, allows for in-depth study of the morphological details of the fossils, including their skeletal parts and inner or outer casts, thus providing useful additional information to the study of these samples.

The three-dimensional models will be used for a virtual 3D online repository as well as allowing 3D printing of these invertebrate samples, useful for both scientific and learning purposes and avoiding a potential deterioration of the originals due to repeated hands-on procedures.

4.4. Geoscience Education

The didactic activities designed for this project are presented as a set of file cards for a practical, dynamic use and understanding for different educational contexts, organized by complexity due to the abstraction level needed in each case. This can be observed by the recommended age of the participants in each activity. Before starting the sequence, the activity "Palaeontology vs. Archaeology" should be developed, by using the Spanish utility model "Interactive Didactic Panel—Diferencia2" (registration number ES1268764) [37], which helps to clarify, by considering previous knowledge, the difference between both disciplines and their main object of study.

The proposed activities have been titled "The Fossil Hunt" (Appendix A), "Cretaceous beach" (Appendix B), "Following Tami's tracks" (Appendix C), and "Discovering the Tracksite of Tamajón" (Appendix D). Each file card includes the title of the activity, the recommended age of participants, the objectives to be achieved, the multiple intelligences, and the European Union (EU) key competences covered, along with the needed resources, the sequence of the activity, and some further observations (Figures 7 and 8).

Figure 7. Sample of didactic activities. (**A**) Photograph of an *in situ* ammonite (red dashed circle) at the Figueira da Foz Site (Portugal) for the activity "The Fossil Hunt" (Appendix A). (**B**) Photograph of an *in situ* coral fragment (red dashed circle) at the Figueira da Foz site (Portugal) for the same activity. (**C**) Image of the online/virtual activity "Cretaceous beach" (Appendix B), showing two levels of difficulty regarding the name of the specimen.

Figure 8. Examples of some of the proposed didactic activities. (**A**) Track surface mat for the activity "Following Tami´s tracks". (**B**) Ichnological report for the activity "Following Tami´s tracks" (Appendix C). (**C**) Palaeobiological sample of the activity "Discovering the Tracksite of Tamajón" (Appendix D) where different questions are aroused to strengthen scientific thinking and reasoning.

5. Discussion

Palaeontological sites can be used as highly adequate spaces for the out-of-school learning of geosciences in general, and Earth history in particular. In this way, the Cenomanian–Turonian (Upper Cretaceous) outcrops of Figueira da Foz (Portugal) and Tamajón (Spain) demonstrate that the combination of different virtual palaeontology techniques is an asset to the scientific study of the palaeontological sites, allowing the implementation of transdisciplinary didactic activities for different educational levels and the general public. Furthermore, 3D techniques, such as phase-shift scanning, photogrammetry, and small object scanning, have been successful for the digital recording of the vertebrate tracksite of Tamajón and invertebrate fossil specimens of both Iberian sites, contributing to the deepening of scientific studies beyond the traditional methods.

Three-dimensional reconstructions facilitate the study and comparison of fossils without a direct or excessive manipulation of samples, facilitating long-lasting conservation of the fossil specimens at their corresponding Portuguese or Spanish institution. It is also an effective way to replace the traditional use of replicas or physical models for morphological studies, e.g., [50]. This aspect also enhances collaborative research throughout the globe [51], facilitating, in this case, Iberian cooperation strategies between two neighbouring countries with many common geological and historical traits. Regarding traditional methods for ichnite and track surface studies (detailed drawings, in situ casts, etc.), these can lead to the loss of scientific information, which hinders ichnotaxa comparison [33] and is usually technically complex and expensive. Therefore, digital 3D recording by means of laser scanning and photogrammetry techniques is an adequate alternative for these detailed ichnite studies, facilitating the systematic classification of the high abundancy of prints and tracks of the track surface of Tamajón, allowing the coverage of the site and avoiding its deterioration.

Digital recording also allows for the protection of didactic and outreach values of this important palaeontological heritage and facilitates its transposition to different social groups (e.g., scientific, educational, or geotouristic). For this matter, these digital reconstructions will be incorporated in a virtual repository of the Palaeontological and Archaeological Interpretation Centre of Tamajón (CIPAT), expanding inclusion possibilities and attending to diversity and the possibilities of different sectors of society and institutions. These data are also essential for the 3D printing of precise replicas of the ichnites, tracks, and even the tracksite at various scales, as well as of the marine fossil samples of the studied Iberian Upper Cretaceous sites.

Furthermore, 3D modelling has also allowed for the design and development of several didactic activities for the Didactic Area of the CIPAT, aimed at different educational levels and the general public. To base these activities on the multiple intelligence theory proved to be beneficial for the teaching–learning process, as it stimulates meaningful learning (*sensu* [43]) as well as prompting self-esteem, which is directly related to an increment on motivation in learning.

These didactic activities have been designed to help participants understand the relevance of palaeontology as a scientific discipline of geosciences with a strong interdisciplinary scope, showing in a diversity of ways how scientific reasoning is integrated within these extraordinary findings and how the Earth's systems have evolved during millions of years. Among the main objectives, the understanding of the geological dimension of time (geological time) is essential; therefore, guiding participants to adequately arrange these abstract concepts is crucial to allow the cognitive construction of correct scientific knowledge throughout long life learning, which is essential for the understanding of the world we live in. The work carried out here also aims to contribute to improving the design and development of didactic sequences for out-of-school education at these sites [8], organizing effective transdisciplinary teaching tools, developing awareness, values, and responsibility towards the natural heritage of these territories.

Therefore, by the research covered in this project, it is suggested that geoeducation can be managed at rich geological points and interpretation centres, which may not be necessarily part of geoparks, such as the studied examples of the palaeontological sites of Figueira da Foz and Tamajón and the CIPAT. These sites are adequate in situ geoeducation points which harbour interesting local, regional, and even international natural elements useful for learning natural science through palaeontological content and its related lithologies and sedimentary structures. Thus, it becomes even clearer that geoeducation constitutes the main tool to transmit knowledge and, at the same time, to emphasize the importance of geoheritage and geoconservation [4].

As a whole, this work aims to contribute to the emerging topic of geoeducation through out-of-school meaningful learning experiences, demonstrating the benefits of presenting science as a transdisciplinary discipline and allowing it to be more accessible among society.

6. Conclusions

On the bases of the above, it is highly important to adopt more strategies that make geoeducation widely available, not only for students of different educational levels but also for the general public. This includes the learning of palaeontology and the history of the Earth through several formal and non-formal activities explored in fieldwork contexts and interpretation centres, such as those exemplified in this work. Both topics and specifically designed activities also should be better integrated into special curricula programs at various primary and secondary school levels. In this way, there will be a major opportunity for future citizens to be informed about issues that raise geological and cultural interest [4], some of them belonging to the geological heritage of their living places and localities and municipalities. As a whole, these activities aim to cover geoeducation through the benefits of a multicultural perspective of international cooperation by making science understandable, achievable, and pursuable.

Author Contributions: Conceptualization, S.O.d.J., F.B.-B., M.B.-C. and P.M.C.; data curation, S.O.d.J. and M.B.-C.; funding acquisition, S.O.d.J., F.B.-B., M.B.-C. and P.M.C.; investigation, S.O.d.J., F.B.-B., M.B.-C. and P.M.C.; methodology, S.O.d.J., F.B.-B., M.B.-C. and P.M.C.; resources, F.B.-B.; software, S.O.d.J. and M.B.-C.; supervision, S.O.d.J., F.B.-B., M.B.-C. and P.M.C.; writing—original draft, S.O.d.J., F.B.-B. and M.B.-C.; writing—review and editing, S.O.d.J., F.B.-B., M.B.-C. and P.M.C. All authors have read and agreed to the published version of the manuscript.

Funding: Contracts UI/BD/150971/2021 (S.O.d.J.) through CITEUC of the Foundation for Science and Technology (Portugal) and Margarita Salas UCM CT31/21 (M.B.-C.) of the Complutense University of Madrid (Spain) and research project SBPLY/21/180501/000242 of the Junta de Castilla-La Mancha and the University of Alcalá (Spain).

Acknowledgments: To D. Eugenio Esteban de la Morena, remaining members of the Town Council, and residents of Tamajón (Spain) for the support given throughout this research. The authors would also like to thank their colleagues Vanda Faria dos Santos and Manuel Segura, along with the rest of members of PaleoIbérica, for their valuable contributions regarding these palaeontological sites. In addition, the authors thank the work and effort of the reviewers of this manuscript, who have improved its original content.

Conflicts of Interest: The authors declare no conflict of interest.

Appendix A

File card of "The Fossil Hunt". Multiple intelligences represented as symbols: 👄 linguistic–verbal; 👁 visual–spatial; 🤸 bodily–kinaesthetic; 👥 interpersonal; 🌿 naturalistic. European Union (EU) competences represented as coloured keys: 🔑 mathematical competence and competence in science, technology, and engineering; 🔑 literacy competence; 🔑 personal, social, and learning to learn competence; 🔑 citizenship competence; 🔑 cultural awareness and expression competence.

<table>
<tr><td colspan="2" align="center">The Fossil Hunt</td></tr>
<tr><td>Recommended age</td><td>> 3 years old.</td></tr>
<tr><td>Objectives</td><td>To search for geological samples. To understand the study of geological processes. To be aware of the importance of Geoconservation.</td></tr>
<tr><td>Multiple Intelligences</td><td></td></tr>
<tr><td>EU Key competences</td><td></td></tr>
<tr><td>Resources</td><td>"Fossil Hunt guide". Support material for explanations (optional).</td></tr>
<tr><td>Development</td><td>Participants will be handed out the Fossil Hunt guide to start with the search of in situ fossil samples. Once one has been found, they must complete the Fossil Hunt guide by answering the different questions and take a picture or make a drawing of their findings. Special attention should be given to the importance of protecting Natural Heritage and Geoconservation, as the fossil sample must not be recollected. The activity can be done individually, in pairs or groups.</td></tr>
<tr><td>Further observations</td><td>The Fossil Hunt guide has been designed to be used both, digitally and printed. The Fossil Hunt guide is based on Bloom's Taxonomy, allowing to adapt to different educational levels and low and high order thinking skills. Several versions of the Fossil Hunt guide can be designed by choosing preferential activities for set groups.</td></tr>
</table>

Appendix B

File card of "Cretaceous beach". Multiple intelligences represented as symbols: linguistic–verbal; visual–spatial; intrapersonal. European Union (EU) competences represented as coloured keys: mathematical competence and competence in science, technology, and engineering; personal, social, and learning to learn competence; digital competence.

Cretaceous beach	
Recommended age	> 5 years old.
Objetives	To use critical thinking to infer in which type of sedimentary environment could each palaeobiological entity have lived in.
Multiple intelligences	
EU Key competences	
Resources	Interactive online game. Support material for explanations (optional).
Development	The aim of the game is to drag the Upper Cretaceous fossils, to their corresponding sedimentary environment. If it is correct, the fossil will turn into its palaeobiological entity.
Further observations	To adapt it to different levels, there are three versions of the game. The first version only differentiates general environments (*e.g.*, open ocean) for younger ages. The level of complexity of the other two differs in the specific name of the fossils, scientific name (*e.g.*, *Hemiaster bufo*) or common name (*e.g.*, sea urchin). In this way, the advanced version will have the specific names, adequate for university levels.

Appendix C

File card of "Following Tami's tracks". Multiple intelligences represented as symbols: visual–spatial; logical–mathematical; interpersonal. European Union (EU) competences represented as coloured keys: mathematical competence and competence in science, technology, and engineering; literacy competence; personal, social, and learning to learn competence.

Following Tami's tracks	
Recommended age	> 7 years old.
Objectives	To know about ichnologists work methodology. To be aware of the variety of information fossilized prints can provide. Extract palaeontological knowledge from a simplified reconstruction of the Tamajón tracksite.
Multiple intelligences	
EU Key competences	
Resources	Tracksurface mat (Fig. 8A), "Ichnological report". Measuring instruments (ruler or measuring tape).
Development	The tracksurface mat must be placed on the floor. Then a teaching agent broadly explains the importance of ichnology, as tracks can give different information than other fossil samples. By means of scaffolding techniques, teaching agents should guide participants to complete the Ichnological report by discussing in group, encouraging critical thinking. For some questions, participants will have to measure several parameters to correctly answer their questions, contrasting scientifically their inferences.
Further observations	The Ichnological report has been designed to be used both, digitally and printed. To extend the activity, a previous explanation can take place, regarding the fossilization process of ichnites and different types of impressions (print, subimpression, cast) by using 3D printed models of the digitalized ichnites, as well as explaining the relevance of these "Galloping crocs".

Appendix D

File card of "Discovering the Tracksite of Tamajón". CIPAT: Palaeontological and Archaeological Interpretation Centre of Tamajón (Spain). Multiple intelligences represented as symbols: linguistic–verbal; visual–spatial; interpersonal. European Union (EU) competences represented as coloured keys: mathematical competence and competence in science, technology, and engineering; personal, social, and learning to learn competence; citizenship competence; cultural awareness and expression competence; digital competence.

<table>
<tr><td colspan="2" align="center">Discovering the Tracksite of Tamajón</td></tr>
<tr><td>Recommended age</td><td>> 3 years old.</td></tr>
<tr><td>Objectives</td><td>To obtain a joint and integrative vision regarding the tracksite exposition at the Palaeontological Area of the CIPAT. To understand how a same area has changed over millions of years of Geological influence.</td></tr>
<tr><td>Multiple Intelligences</td><td></td></tr>
<tr><td>EU Key competences</td><td></td></tr>
<tr><td>Resources</td><td>Video of the 3D reconstruction of the tracksite of Tamajón.</td></tr>
<tr><td>Development</td><td>The video allows to observe the totality of the digital 3D model obtained by the scanning of the tracksurface of Tamajón. Allowing everyone to have digital access to the site, as it is currently covered to avoid spoliation, deterioration, and to enable future studies onsite as well as its preservation. In addition, different questions will be popping up throughout the video, to encourage critical thinking and scientific literacy. Visitors can answer and debate with others about their thoughts on the topic.</td></tr>
<tr><td>Further observations</td><td>It can also be used as visual support for palaeontological explanations regarding the CIPAT, and for the activity "Following Tami's tracks". It can also be used interactively by means of tablets or other digital supports, allowing visitors to interact with this extraordinary palaeontological tracksite.</td></tr>
</table>

References

1. Gould, S.J. *Time's Arrow, Time's Cycle: Myth and Metaphor in the Discovery of Geological Time*; Harvard University Press: Cambridge, UK, 1987.
2. Rudwick, M.J.S. *Worlds Before Adam: The Reconstruction of Geohistory in the Age of Reform*; University of Chicago Press: Chicago, IL, USA, 2010.
3. Sande, A.; Burr, J.J.; Chiment, W.D.; Allmon, J.; Keith, R. A Problematic Fossil Brings Paleontology to the Classroom and the World. *J. Geo. Sci. Ed.* **2003**, *5*, 361–364.
4. Zafeiropoulos, G.; Drinia, H.; Antonarakou, A.; Zouros, N. From Geoheritage to Geoeducation, Geoethics and Geotourism: A Critical Evaluation of the Greek Region. *Geoscience* **2021**, *11*, 381. [CrossRef]
5. Bezzi, A. What is this thing called geoscience? Epistemological dimensions elicited with the repertory grid and their implications for scientific literacy. *Sci. Educ.* **1999**, *83*, 675–700.
6. Orion, N. A Holistic Approach for Science Education for All. *Eurasia J. Math. Sci. Technol. Ed.* **2007**, *3*, 111–118. [CrossRef]
7. Jolley, A.; Kennedy, B.M.; Brogt, E.; Hampton, S.J.; Fraser, L. Are we there yet? Sense of place and the student experience on roadside and situated geology field trips. *Geosphere* **2018**, *14*, 651–667.

8. Ozkaya de Juanas, S.; Barroso-Barcenilla, F.; Callapez, P.M.; Carvalho, P.C.S.; Lopes, F.C.; dos Santos, V.F. Descubriendo yacimientos del Cretácico de España y Portugal: Un proyecto educativo para el aprendizaje de las Ciencias de la Tierra fuera del aula. In *A Holistic View of Earth and Space in Its Human and Natural Aspects a Tribute to Professor Celeste Romualdo Gomes*; Centro de Investigação da Terra e do Espaço da Universidade de Coimbra: Coimbra, Portugal, 2020; Volume 2, pp. 613–634.
9. Ozkaya de Juanas, S.; Barroso-Barcenilla, F.; Callapez, P.M. Didactic and outreach possibilities of the Cretaceous palaeontological site Figueira da Foz (Portugal). *Comun. Geol.* **2021**, *108*, 125–130.
10. Blasi, I.C.; Pellicer, X.P.; Poch, J.; Galobart, À. Open-air paleontological sites: A dissemination opportunity and a risk to spoliation. A case study in the Orígens Geopark, Spain. *Geoconservation Res.* **2021**, *4*, 513–523.
11. Paik, I.S.; Huh, M.; Kim, H.J.; Kim, S.J.; Newsome, D. The Cretaceous fossil sites of South Korea identifying geosites, science and Geotourism. In *Geotourism: The Tourism of Geology and Landscape*; Newsome, D., Dowling, R.K., Eds.; Goodfellow Publishers: Oxford, UK, 2010.
12. Skelton, P.W.; Spicer, R.A.; Kelley, S.P.; Gilmour, I. *The Cretaceous World*; Cambridge University Press: Cambridge, UK, 2003.
13. Callapez, P.M.; Barroso-Barcenilla, F.; Soares, A.F.; Segura, M.; dos Santos, V.F. On the co-occurrence of Rubroceras and Vascoceras (Ammonoidea, Vascoceratidae) in the upper Cenomanian of the West Portuguese Carbonate Platform. *Cretac. Res.* **2018**, *88*, 325–336. [CrossRef]
14. Callapez, P.M.; Brandão, J.M.; dos Santos, V.F.; Gomes, C.R. Between history and contemporaneous geology: Revisiting a "classical" (geo) site from the Upper Cretaceous of Portugal. *Rev. Soc. Geol.* **2013**, *26*, 5–12.
15. Ozkaya de Juanas, S.; Alcalde-Fuentes, M.R.; Audije-Gil, J.; Barroso-Barcenilla, F.; Berrocal-Casero, M.; Callapez, P.M.; Pérez-Valera, J.A.; dos Santos, V.F.; Segura, M. Discovering the Cretaceous Life at the Didactic Area of the Palaeontological and Archaeological Interpretation Centre of Tamajón, CIPAT (Guadalajara, Spain). *Lucas Mallada* **2021**, *23*, 123–124.
16. Barroso-Barcenilla, F.; Goy, A.; Segura, M. Ammonite zonation of the Upper Cenomanian and Lower Turonian in the Iberian Trough, Spain. *Newsl. Stratigr.* **2009**, *43*, 139–164. [CrossRef]
17. Barroso-Barcenilla, F.; Callapez, P.M.; Ferreira Soares, A.; Segura, M. Cephalopod assemblages and depositional sequences from the upper Cenomanian and lower Turonian of the Iberian Peninsula (Spain and Portugal). *J. Iber. Geol.* **2011**, *37*, 9–28.
18. Barroso-Barcenilla, F.; Audije-Gil, J.; Berrocal-Casero, M.; Callapez, P.M.; Carenas, B.; Comas-Rengifo, M.J.; García Joral, F.; García-Hidalgo, J.F.; Gil-Gil, J.; Goy, A.; et al. El Cenomaniense-Turoniense de Tamajón (Guadalajara, España): Contexto geológico, contenido fósil e interpretación paleoambiental. *Bol. R. Soc. Esp. Hist. Nat. Sec. Geol.* **2017**, *111*, 67–84.
19. Berrocal-Casero, M. Los ecosistemas del Cretácico de Guadalajara: De la costa al mar. **2022** *Serv. Pub. Dip. Prov. Guadalajara. in press.*
20. Aguilar, M.J.; Ramírez del Pozo, J.; Riba, O. Algunas precisiones sobre la sedimentación y paleoecología del Cretácico inferior en la zona de Utrillas-Villarroya de los Pinares (Teruel). *Estud. Geol.* **1971**, *27*, 497–512.
21. Segura, M.; Barroso-Barcenilla, F.; Berrocal-Casero, M.; Castanera, D.; García-Hidalgo, J.F.; dos Santos, V.F. A new Cenomanian vertebrate tracksite at Tamajón (Guadalajara, Spain): Palaeoichnology and palaeoenvironmental implications. *Cretac. Res.* **2016**, *57*, 508–518. [CrossRef]
22. Vilas, L.; Mas, J.R.; García, A.; Arias, C.; Alonso, A.; Meléndez, N.; Rincón, R. Ibérica Suroccidental. In *El Cretácico de España*; García, Á., Ed.; Universidad Complutense de Madrid: Madrid, Spain, 1982; pp. 457–514.
23. Floquet, M.; Alonso, A.; Meléndez, A. El Cretácico Superior de Cameros-Castilla. In *El Cretácico de España*; García, Á., Ed.; Universidad Complutense de Madrid: Madrid, Spain, 1982; pp. 387–456.
24. Dos Santos, V.F.; Alcalde-Fuentes, M.R.; Audije-Gil, J.; Barroso-Barcenilla, F.; Berrocal-Casero, M.; Callapez, P.M.; Ozkaya de Juanas, S.; Pérez-Valera, J.A.; Segura, M. "Galloping Crocs" in the Upper Cretaceous of Tamajón (Guadalajara, Spain)? *Res. XXIV Bien. R. Soc. Esp. Hist. Nat. Sec. Geol.* **2021**, *1*, 208–210.
25. Barroso-Barcenilla, F.; Alcalde-Fuentes, M.R.; Audije-Gil, J.; Berrocal-Casero, M.; Callapez, P.M.; Ozkaya de Juanas, S.; Pérez-Valera, J.A.; dos Santos, V.F.; Segura, M. Discovering a Project for the Development of Geotourism in Rural Areas: The Palaeontological and Archaoelogical Intepretation Centre of Tamajón (CIPAT, Guadalajara. Spain). *Land* **2022**, *11*, 444. [CrossRef]
26. Sutton, M.D.; Rahman, I.A.; Garwood, R.J. *Techniques for Virtual Palaeontology*; Wiley-Blackwell: Oxford, UK, 2014.
27. Lyons, P.D.; Rioux, M.; Patterson, R.T. Application of a three-dimensional color laser scanner to paleontology: An interactive model of a juvenile Tylosaurus sp. basisphenoid-basioccipital. *Palaeontol. Electron.* **2000**, *3*, 4A.
28. UNESCO. Arouca Declaration. Arouca Geopark, Documentation. 2011. Available online: http://aroucageopark.pt/es/documentacion/ (accessed on 25 November 2022).
29. Stoffelen, A.; Vannaste, D. An integrative geotourism approach: Bridging conflicts in tourism landscape research. *Tour. Geogr.* **2015**, *17*, 544–560. [CrossRef]
30. Peppoloni, S.; Di Capua, G. Geoethics: Ethical, social, and cultural values in geosciences research, practice, and education. In *Geoscience for the Public Good and Global Development: Toward a Sustainable Future*; Wessel, G.R., Greenberg, J.K., Eds.; Geological Society of America: Boulder, CO, USA, 2016; pp. 17–23.
31. Vasconcelos, C.; Torres, J.; Vasconcelos, L.; Moutinho, S. Sustainable development, and its connection to teaching Geoethics. *Episodes* **2016**, *39*, 509–517. [CrossRef]
32. Vasconcelos, C.; Ribeiro, T.; Cardoso, A.; Orion, N.; Ben-Shalom, R. Educational Theoretical framework underpinning geoethical resources. In *Teaching Geoethics: Resources for High Education*; U. Porto Edições: Porto, Portugal, 2020.
33. Remondino, F.; Rizzi, A.; Girardi, S. 3D Ichnology-Recovering Digital 3D Models of Dinosaur Footprints. *Photogramm. Rec.* **2010**, *25*, 266–282. [CrossRef]

34. Razzolini, N.L.; Vila, B.; Castanera, D.; Falkingham, P.L.; Barco, J.L.; Canudo, J.I.; Manning, P.L.; Galobart, A. Intra-trackway morphological variations due to substrate consistency: The El Frontal Dinosaur Tracksite (Lower Cretaceous, Spain). *PLoS ONE* **2014**, *9*, e93708. [CrossRef] [PubMed]
35. Orion, N. Earth Science education: From theory to practice, implementation of new teaching strategies in different learning environments. In *Geociências nos Currículos dos Ensinos Básico e Secundário*; Universidade de Aveiro: Aveiro, Portugal, 2001; pp. 93–114.
36. Gardner, H. *Frames of Mind: The Theory of Multiple Intelligences*; Basic Books: New York, NY, USA, 1993.
37. Ozkaya de Juanas, S.; Barroso-Barcenilla, F. Paleontología y su didáctica en Primaria: Diseño y aplicación de actividades basadas en yacimientos cretácicos y sus fósiles. *Bol. R. Soc. Esp. Hist. Nat. Sec. Aula.* **2019**, *6*, 95–113. [CrossRef]
38. Brown, J.S.; Collins, A.; Duguid, P. Situated cognition and the culture of learning. *Educ. Res.* **1989**, *18*, 32–42. [CrossRef]
39. Hendricks, C. Teaching causal reasoning through cognitive apprenticeship: What are results from situated learning? *J. Educ. Res.* **2001**, *94*, 302–311. [CrossRef]
40. Dewey, J.D. *How We Think*; D.C. Heath & CO: Boston, MA, USA, 1910.
41. DeWitt, J.; Osborne, J. Supporting teachers on science-focused school trips: Towards an integrated framework of theory and practice. *Int. J. Sci. Educ.* **2007**, *29*, 685–710. [CrossRef]
42. Paris, S.G. Situated motivation and informal learning. *J. Mus. Educ.* **1997**, *22*, 22–27. [CrossRef]
43. Ausubel, D.P. *Educational Psychology: A Cognitive View*; Holt, Rinehart & Winston: New York, NY, USA, 1968.
44. Krathwohl, D.R. A Revision of Bloom´s Taxonomy: An Overview. *Theory Pract.* **2022**, *41*, 212–218. [CrossRef]
45. EC—European Commission. Directorate-General for Education, Youth, Sport and Culture. In *Key Competences for Lifelong Learning*; Publications Office of the European Union: Luxembourg, Luxembourg, 2019.
46. UN—United Nations. Sustainable Development Goals. 2022. Available online: https://www.un.org/sustainabledevelopment/sustainable-development-goals (accessed on 2 December 2022).
47. Ozkaya de Juanas, S.; Audije-Gil, J.; Barroso-Barcenilla, F.; Berrocal-Casero, M.; Callapez, P.M.; Segura, M. "Paleosentidos": Descubriendo el Cretácico Superior de Tamajón (Guadalajara, España). In *Libro de Resúmenes del XX Simposio de la Enseñanza de la Geología*; AEPECT: Menorca, Spain, 2018; pp. 135–142.
48. Wood, D.J.; Bruner, J.S.; Ross, G. The role of tutoring in problem solving. *J. Child Psychol. Psychiatry* **1976**, *17*, 89–100. [CrossRef]
49. Askew, S. *Feedback for Learning*; Routledge: London, UK, 2000.
50. Friedrich, F.; Beutel, R.G. Micro-computer tomography and a renaissance of insect morphology. *Proc. SPIE Int. Soc. Opt. Eng.* **2008**, *7078*, 545–550.
51. Kuzminsky, S.C.; Gardiner, M.S. Three-Dimensional Laser Scanning: Potential Uses for Museum Conservation and Scientific Research. *J. Archaeol. Sci.* **2012**, *39*, 2744–2751. [CrossRef]

Article

Prospective Study on Geosciences On-Line Education: UNESCO Global Geoparks in Spain and Portugal

Jesús Enrique Martínez-Martín [1,*], Pilar Ester Mariñoso [1], Emmaline M. Rosado-González [2] and Artur A. Sá [2]

1   Facultad de Educación, Universidad Camilo José Cela, 28692 Villanueva de la Cañada, Spain
2   UNESCO Chair on Geoparks, Sustainable Regional Development and Healthy Lifestyles,
   and Pole of the Geoscience Centre (CGeo), Department of Geology,
   University of Trás-os-Montes e Alto Douro, 5001-801 Vila Real, Portugal
*   Correspondence: jemartinez@ucjc.edu

Abstract: UNESCO Global Geoparks (UGGps) stand out as territories of excellence for the development of educational activities in the international arena. Their didactic potential, their multidisciplinarity and their importance for the development of non-formal and informal teaching activities have drawn the attention of institutions, organizations and governments of many countries. This, to such an extent, that the number of UGGps continues to increase year after year, having currently reached 177 territories spread over 46 countries. All of them work every day developing different activities and educational proposals aimed at the creation of the so-called "Quality Education", the fourth objective of the SDGs of the 2030 Agenda. The diversity of didactic plans, their adaptability and their accessibility mean that each UGGp is unique and different from the rest, maintaining the key values that make up this group of territories. This study describes the different educational proposals and activities that the Spanish and Portuguese UGGps show on their official websites, with the aim of analyzing their level of visibility before visiting the territories and highlighting their relevance in the education development framework.

Keywords: education; geosciences; geoparks; sustainability; online

Citation: Martínez-Martín, J.E.; Ester Mariñoso, P.; Rosado-González, E.M.; Sá, A.A. Prospective Study on Geosciences On-Line Education: UNESCO Global Geoparks in Spain and Portugal. *Geosciences* **2023**, *13*, 22. https://doi.org/10.3390/geosciences13020022

Academic Editors: Jesus Martinez-Frias and Assimina Antonarakou

Received: 1 December 2022
Revised: 30 December 2022
Accepted: 11 January 2023
Published: 19 January 2023

1. Introduction

Current educational models are in a state of constant evolution and adaptation. The events that have occurred in the last decade and the progress of information and communication technologies have forced experts to reassess the system and transform traditional elements into new hybrid tools, which maintain classic values and implement new educational methodologies. Each step that the educational system takes translates into a renovation that affects the teaching team, the centers, the students, the infrastructures and even the methods and the proper concept of "education" itself, making it behave almost like a living being that adapts to the moment depending on the need. In this context, the UNESCO Global Geoparks (UGGps) stand out, with an educational plan based on geotourism and directly related to sustainable development [1–3]. There is no doubt that, within these territories, there is a different and effective educational model that combines informal and non-formal methods to offer an accessible and adapted education for all which helps us understand our Planet [4,5]. The perfect union between science and society exists within the UGGps and it is necessary to live the experience to understand why it is so important for the social moment in which we find ourselves [6]. The variety of activities that are carried out based on sustainable education, such as the 'European Week of Geoparks' [7] or the 'Geoconvivencia' event [8], among many others, enjoy international relevance and have demonstrated, on many occasions, their educational value aimed at all kinds of public [1]. Even so, it is true that these activities have a lack of visibility within the scientific community, if we base our evaluation on existing publications in high-impact journals. Most of the educational projects are accessible if we review the number of abstracts that appear in international meetings, but this is outside the most important databases and, therefore, hidden behind the large number of publications that relate the UGGps to other subjects.

It is evident that the recent situation caused by the COVID-19 pandemic has led to a paradigm shift and an almost immediate update of the educational program in order to continue developing it continuously [9] and, for this reason, territories such as the UGGps are so important and even examples of good practices [10]. Their model and everything they encompass make them safe places to complement education in geosciences and learn about society, the environment and natural-heritage protection [11].

Nowadays, the terms "visibility", "online branding" or "personal brand" are very important elements for the development of projects on the web, since they offer a natural appeal that draws the attention of the public and generates sensations such as trust or confidence in the existence of a minimum of quality [12]. Parallel to these issues, web design, information, documents or audiovisuals contained therein can be a window through which people can obtain an idea of the content they can find within a UGGps in the fastest and most effective way possible. It is important to highlight this in the study, since it represents a fundamental role in how social networks and the Internet in general work and, in turn, how society reacts to content and interacts with it [13].

This study tries to collect and analyze the amount of educational information that is shown on the web pages of the Spanish and Portuguese UGGps themselves. The idea is to ask ourselves what the approach is that teachers, coordinators and curious people follow when it comes to finding out about the educational activities that are organized and carried out in these territories. Can I visit the UGGps with my students? Are educational programs able to complement formal educational programs taught at school? Can you help me with the guides on the routes? Do you have activities for students of all levels? All these questions are reflected in a series of variables that have been quantified to obtain visual results of the situation of education, in all its variants, on the UGGps web pages. By collecting information and segmenting it into variables, we can observe the current framework and the situation in which the UGGps find themselves in relation to education. Based on the data obtained, the possibilities and opportunities offered by the advancement of new technologies, the updating of educational systems and their implementation in these territories for sustainable development and their proper functioning are discussed.

The presence of education on the UGGps websites is a key factor promoting all the activities that these territories organize throughout the year and demonstrating their effectiveness as sustainable and safe educational alternatives, aimed at heritage protection, the learning of Earth Sciences and the search for a more aware and, ultimately, more sustainable society [14]. The good educational practices that take place in the UGGps and the integration of the SDGs within environmental education directly related to these territories have been reflected in numerous educational models, books and scientific publications [15–18]. Considering that, currently, online learning is at the center of our lives as one of the most important sources of information for our cognitive development, it is important to know the educational proposal that places such as the UGGps offer through this particular information media. For this reason, their study can shed some light on the adaptability of these territories to the digital world and their usefulness as an alternative method of education, not only in the field, but also online.

2. Objectives

With this research, we intend to achieve, as a main objective, the framing of the presence and visibility of the concept "education" and its methodologies in the Spanish and Portuguese UGGps web pages, from which three specific objectives arise:

- Promote the UGGps as essential and safe territories for the current educational model.
- Enhance the visibility and presence of the educational potential of these territories within the educative and scientific community.
- Understand the current panorama and discuss possible paths towards the educational future of the UGGps.

3. Context and Current Situation

The concept of "education" itself has undergone numerous modifications throughout history. Logically, an idea with such a wide range of perspectives and with such a complex network of ways of doing things, opinions and methodologies is impossible to conceive as a single element. Education is a living concept which is shaped based on society, culture and the moment in which we find ourselves, but it does not stop there; rather, each region and even each person understands and adapts their conception of education to such an extent that it increases the difficulty of understanding its evolution, making it complicated to define [19].

In addition to all this, the values and skills that we teach and intend to spread with the educational method change along with it and become variables to consider when applying different models or ways of teaching society. Currently, and thanks to the advancement of new technologies, online education, which used to be something that could be viable in the future but still had a lot to improve, has gained immense strength within educational plans and become a valid alternative for studies of all levels. The opportunities it offers, the ease, and the accessibility and adaptability options it generates have made it an indispensable tool for the educational system in just a few years [20]. The introduction of ICT and audiovisual elements in the classroom, such as the digital whiteboard, tablets, laptops or mobile phones, suggested that adaptation to the digital environment was imminent and, at present, it is difficult to find courses that do not offer this possibility in their educational offers [21,22].

All this educational adaptation has required an extra effort on the part of teachers, coordinators, students and families to be able to carry it out, but it has culminated in the creation of numerous "safe" initiatives and in proposals that had never previously arisen, but that have served to deal with this situation we find ourselves in. This is the case for outdoor activities, education in natural environments and the endless hours online that teachers from all over the world have had to teach, live and recorded. This process of conceptual change has only been possible thanks to the evolution of educational methods and concepts that, in parallel, have been including and using ICT as a valid learning method. Thanks to the fact that today's society remains constantly connected by computers, mobile phones and, obviously, the Internet, it has been possible to carry out the massive monitoring of the progress of students within the academic year, which has allowed the evaluation of concepts to be something possible at a time when difficulties were real and very complicated for teaching teams to face [23]. In fact, teachers, researchers and coordinators have also had to adapt themselves, leading to a massive entry of professionals into the technology sector who, surely, had previously only partially considered the use of these tools as educational enhancers in their classrooms [24,25]. A door has been opened to society that has seen that the Internet is a valid learning resource and an essential element for the dissemination of knowledge at all academic levels, which teachers themselves can use to liven up their classes and facilitate the understanding of content. These tools have been fundamental in the educational adaptation that has occurred in recent years due to COVID-19. Teachers, researchers and, in general, teams at all educational levels have had to act quickly to reduce the repercussion of the crisis as much as possible and, thus, be able to continue with their proposed didactic plans [26]. All this, framed by the idea of sustainability, amends the educational cycle and modernizes it to, once again, intertwining leisure with learning and improving the cognitive process.

In relation to the educational concept, sustainability and sustainable development should be highlighted as two of the most important key points of recent years. Society is increasingly aware of the search for a model that allows the co-existence of people and the Planet in harmony. The first definition of the concept "sustainable development" was in the Brundtland Report of the World Commission on Environment and Development entitled "Our common future", in 1987. This term referred to: "The satisfaction of the needs of the present generation without compromising the ability of future generations to meet their own needs". There are records of its evolution in 1992, during the "Earth Summit" or "Rio Summit" in Rio de Janeiro. "Benayas, Calvo and Gutiérrez, referenced by OREALC

(2009:9), represents a global regulation of the strategies that mediate between environment and development relations." Starting in 1997, the idea of sustainable development was established. Then, in the "United Nations Decade of Education for Sustainable Development (2005–2014): International Application Plan", UNESCO focuses on the term together with its three basic pillars: the environment, society and the economy. This concept, although like sustainability, differs from it because it is the set of means and processes to achieve sustainable objective, which are the same as a "long-term goal".

Education is a key element and one of the fundamental nuclei of sustainable development, as important as the social or economic component could be. According to UNESCO in 2015, "Education for Sustainable Development (ESD)" acts in a way that "equips students with the necessary capacity to make informed decisions and carry out responsible activities in favor of environmental integrity, economic viability and social justice, for current and future generations, with due respect for cultural diversity". Likewise, it is a fickle concept that influences all the sustainable development goals (SDG) of the 2030 Agenda, even having one of its own, goal number 4, which is defined as "Ensuring inclusive, equitable and quality education and promote lifelong learning opportunities for all" (Table 1).

Table 1. Disaggregated list of the targets and indicators of SDG4 of the UNESCO 2030 Agenda.

SDG4 Targets	Indicators
4.1 By 2030, ensure that all girls and boys complete free, equitable and quality primary and secondary education leading to relevant and effective learning outcomes.	Proportion of children and young people (a) in grades 2/3; (b) at the end of primary; and (c) at the end of lower secondary achieving at least a minimum proficiency level in (i) reading and (ii) mathematics, by sex and completion rate (primary education, lower secondary education, upper secondary education).
4.2 By 2030, ensure that all girls and boys have access to quality early childhood development, care and pre-primary education so that they are ready for primary education.	Proportion of children aged 24–59 months who are developmentally on track in health, learning and psychosocial well-being, by sex and participation rate in organized learning (one year before the official primary entry age).
4.3 By 2030, ensure equal access for all women and men to affordable and quality technical, vocational and tertiary education, including university.	Participation rate of youth and adults in formal and non-formal education and training in the previous 12 months, by sex.
4.4 By 2030, substantially increase the number of youth and adults who have relevant skills, including technical and vocational skills, for employment, decent jobs and entrepreneurship.	Proportion of youth and adults with information and communications technology (ICT) skills, by type of skill.
4.5 By 2030, eliminate gender disparities in education and ensure equal access to all levels of education and vocational training for the vulnerable, including persons with disabilities, indigenous peoples and children in vulnerable situations.	Parity indices (female/male, rural/urban, bottom/top wealth quintile and others such as disability status, indigenous peoples and conflict-affected, as data become available) for all education indicators on this list that can be disaggregated.
4.6 By 2030, ensure that all youth and a substantial proportion of adults, both men and women, achieve literacy and numeracy.	Proportion of population in a given age group achieving at least a fixed level of proficiency in functional (a) literacy and (b) numeracy skills, by sex.
4.7 By 2030, ensure that all learners acquire the knowledge and skills needed to promote sustainable development, including, among others, through education for sustainable development and sustainable lifestyles, human rights, gender equality, promotion of a culture of peace and non-violence, global citizenship and appreciation of cultural diversity and of culture's contribution to sustainable development.	Extent to which (i) global citizenship education and (ii) education for sustainable development are mainstreamed in (a) national education policies; (b) curricula; (c) teacher education and (d) student assessment.
4.a Build and upgrade education facilities that are child, disability and gender sensitive and provide safe, non-violent, inclusive and effective learning environments for all.	Proportion of schools offering basic services, by type of service.

Table 1. *Cont.*

SDG4 Targets	Indicators
4.b By 2020, substantially expand globally the number of scholarships available to developing countries, in particular least developed countries, small-island developing states and African countries, for enrolment in higher education, including vocational training and information and communications technology, technical, engineering and scientific programs, in developed countries and other developing countries.	Volume of official development assistance flows for scholarships by sector and type of study.
4.c By 2030, substantially increase the supply of qualified teachers, including through international cooperation for teacher training in developing countries, especially least developed countries and small-island developing states.	Proportion of teachers with the minimum required qualifications, by education level.

The UNESCO Global Geoparks (UGGps) have managed to mark a before and after in sustainable development from an educational point of view. "UNESCO Global Geoparks are unique and unified geographical areas in which sites and landscapes of international geological importance are managed with a holistic concept of protection, education and sustainable development" [6]. Under this definition, there is a direct link with society and the people that make up these territories called Geoparks. Using geological heritage as a resource and education as a tool, the UGGps carry out a territorial development strategy that involves the local population to maximize their potential in a sustainable way. Taking advantage of geoheritage and defending the natural environment, the UGGps help us to learn in a multidisciplinary way about the territory and everything it contains. That is why the UGGps are authentic open-air laboratories [27] where, thanks to earth sciences, we can learn about history, culture and everything that is hidden within the territory. In addition, the union between non-formal and informal education, methods such as in-situ education or the autonomy of visiting them, make them potential natural classrooms where students and teachers of all academic levels can learn, enjoy and enhance values as educational keys in an entertaining and simple way [28]. Moreover, thanks to the particularity of their geological heritage, the UGGps have been used to study other planets, not only geologically speaking, but also training astronauts and professionals for future space trips to Mars and other planetary bodies [29]. This is the case of the Lanzarote and Chinijo Islands UGGps, whose similarity to the Martian soil and landscape has served as an educational tool for the European Space Agency (ESA) to prepare future missions to the red planet [30].

"The walks through the almost Martian landscape of the Canary Island of Lanzarote have allowed the students of the Pangaea course to interpret the geological phenomena to understand the history of the formation of the island. Its goal is to help astronauts choose the best places to explore and collect rock samples" (ESA, 2022).

This characteristic not only recognizes the educational capacity of the UGGps as educational territories, but also highlights the importance of the nature that these territories contain and their relevance at a socio-cultural level [31].

The UGGps are not only open to visits from anywhere in the world, but also involve the educational centers of the territory, creating a sense of unity that motivates students and teaches the importance of caring for the environment and the planet in which we live [32]. This is a very relevant factor now, where the situation of the concept of sustainability and environmental protection have taken shape and are found in the vast majority of the international organizations and institutional plans. In addition, the advancement of information and communication technologies has allowed the widespread dissemination of the message, bombarding society from all possible media such as television, internet or radio [33,34].

It is evident that, in this situation, the UGGps model has attracted attention, being destinations visited by millions of people each year. At the same time, the number of

UGGps has been increasing and, with them, the number of studies and amount of research dedicated to understanding these territories in innumerable ways, whether from the pure sciences, the social sciences, geography or history [35]. For this study, a sample composed of the Spanish and Portuguese UGGps has been selected. Currently, Spain has 15 UGGps (Table 2), being the second country in the world with the largest number of Geoparks, after China. Portugal has five UGGps spread throughout the country (Table 3). Each UGGps is different and has its own characteristics, not only because of the different geological environments that we can find, but also because of the way each one applies the concepts and educational methods. However, the common elements unite the model, and the fundamental pillars that govern the values of these areas allow us to understand the UGGps as a whole and not as individual territories. For this reason, the analysis carried out did not consider the differences, but rather the common factors that the UGGps present. In this case, the on-Line visibility of educational programs, methods, activities and events on the web pages of each UGGps were studied. As mentioned above, the boom in information and communication technologies has been exponential and has marked the last decade, changing the way we had of understanding society until now.

Table 2. List of current Spanish UGGps ordered in chronological order based on the date of their Global Geopark Network designated year.

Spanish UGGps	Declaration Year
Maestrazgo Cultural Park	2000
Cabo de Gata—Nijar	2001
Sierras Subbéticas	2006
Sobrarbe—Pirineos	2006
Basque Coast	2010
Sierra Norte de Sevilla	2011
Villuercas-Ibores-Jara	2011
Cataluña Central	2012
Molina & Alto Tajo	2014
El Hierro	2014
Lanzarote and Chinijo Islands	2015
Las Loras	2017
Origens	2018
Courel Mountains	2019
Granada	2020

Table 3. List of the current Portuguese UGGps arranged in chronological order based on the date of their Global Geopark Network designated year.

Portuguese UGGps	Declaration Year
Naturtejo	2006
Arouca	2009
Azores	2013
Terras de Cavaleiros	2014
Serra de Estrela	2019

Visibility is a key factor within the field of web technologies and the current social moment [36]. We live in an era where people receive significant amounts of information in the palm of their hand thanks to smart phones and the advancement of the internet. This information goes through our subconscious automatically and, in a few seconds, can help us decide, for example, if we choose a specific product, find out about a specific topic or, in this case, choose a destination to visit [37].

Tourism is the basis of the UGGps and what allows their constant operation, so having a good presentation with a website in good condition and of good quality is something necessary for these territories. Going further into the matter, one of the fundamental pillars of the UGGps is education in all its forms and variables, and that is why the online exhibition of the activities, plans, methods or courses that are developed within they themselves is so important [38]. Not being visible on the internet currently translates into not existing for society, which has become accustomed to making quick searches on the net and opting for the most attractive links.

How does a website become attractive and attract the confidence of the Internet user? Factors such as the design, the layout of the elements, the source or the images and audiovisuals that we show on them are sometimes just as important as the content included in them [39,40]. The relationship between the variety of content, the constant updating of a website or the resources found on it through videos, image frames, PDFs or educational documents can give the exact key that the visitor to the web was looking for. The idea, drawn from marketing and advertising theories, of creating an identity or personal brand is not so far from the concept of on-Line visibility, and is influential by providing originality and character to a specific concept [41]. As Pérez stated in 2008 [40]: "Specifically, the personal brand could be understood as a combination of attributes transmitted through a name or a symbol, which influences the thinking of a certain public and creates value for its owner". This means that having a "personal brand" translates into the generation of trust and expectations in the target audience and, therefore, it is essential to unite the online model of the UGGps to show a common message that represents the values of each UGGps. All this goes through a process in which the fusion of pillars and base values, the objectives set by the territories, the target audience, the study of the current situation and the design of a positioning and representation strategy allow the cognitive centralization of the brand and the final representation as a whole, and not as a denomination with infinite subdivisions. Cantone, in 2011 [42], pointed out a series of advantages of working on branding successfully that were summarized by Climent-Rodriguez and Navarro-Abal in 2017 [43] as:

a. Being able to be known by your targets, that is, by your potential audience, the people who want to be known.
b. Able to be differentiated from the rest of the professionals who may compete with you.
c. Have the possibility of positioning yourself as an expert in your specialty.
d. Favor the perception by others as a leader and facilitate the possession of effective networking.
e. Help find partners and collaborators for new projects.
f. Create many new opportunities related directly or not to your profession.
g. Find new lines of professional activity, or recent partners and clients.

However, focusing on the educational factor as such, the context of visibility varies slightly and deviates from the base meaning. There is a variation of the concept outlining a simpler version that does not necessarily take into account SEO, positioning or personal branding [44]. It is about the ease of locating educational content within the website in a simple way for the visitor. Moreover, information and communication technologies allow us to camouflage the educational content, so they can be named as leisure, and simplify the way of obtaining knowledge, teaching without the public noticing they are learning [45]. This is the case, for example, of the educational audiovisuals and promotional videos that we can find about the UGGps. In most of them, we can see fascinating and immersive images that show us the genuine geological heritage of the territory, its cultures, customs and landscapes. This feeling of fascination creates a motivation to continue learning that, naturally, opens the viewer to the territory. Thanks to those few minutes, people are able to relate tourism inside the UGGps with three key factors: nature, culture and society [34,46]. The presence of this type of resource is absolutely necessary, since it introduces the environment to the target audience and motivates future browsing of the website. Simple menus, easy-to-find sections or information boxes are useful tools for locating the educational activities that are generated in the UGGps and can be easily implemented on web pages. These territories are authentic educational engines with respect to the field of natural sciences and general knowledge about everything that the territory contains; however, what can we find and how can we approach our passage through the

UGGps so that it is productive? That is the fundamental question that many teachers, interested and curious people, or professionals in the sector ask themselves when they directly access the website of a UGGps [47].

4. Methodology

The process followed for the formalization of the graphs and the subsequent analysis was based on quantifying the presence of the variables mentioned below from a numerical rule 1–0, indicating with the value 1 the presence on the website and with the value 0 the absence. One by one, all the web pages of the Spanish and Portuguese UGGps were reviewed, giving a total of 20 study samples. All the data was arranged progressively in an Excel sheet that shows the set of variables and the analyzed UGGps. The quantification of qualitative variables is not only a useful tool for preparing studies, but also makes it easy to visualize factors and variables such as concepts, situations or connections which are generally impossible to observe. For this reason, and since "Education" is such a complex term to represent, we opted for its diversification and adaptation to the study environment, in this case, the UGGps. Throughout the analysis, the data were treated as anonymous and only the total results of the study were mentioned, grouped according to the country to which these territories correspond. Finally, for a better understanding of the educational situation on the websites of the UGGps analyzed, different graphs were created with the results obtained. For their design, the presence (value 1) or non-presence (value 0) of the categories mentioned below was represented. Giving a short example: Is there information about a defined educational project on the official UGGps website? If the response was positive, it was defined with the value "1" and reflected in the graphics as "verifiable". If not, the category was defined with the value "0" and reflected as "not verifiable" in the graphical representations.

5. Categories Description

This section lists and defines the variables selected for the study of educational presence within the web pages of the Spanish and Portuguese UGGps. With the idea of portraying the educational reality of these websites, online places such as educational centers, museums or the UGGps themselves were used as a reference, which are able to show all their educational potential on the network. All the variables are of great importance when it comes to drawing a model of how these territories work and not visiting them blindly [8]. Finally, 13 variables subdivided into 15 were defined, taking into account that number 10: "Primary, Secondary and University education programs" was analyzed separately and represented according to the presence or absence of said plans on the websites (Table 4). The positioning of the information within the website, the internal menus or the links were not analyzed as such, even though they are an important aspect, since they facilitate the search for information for Internet users. In this case, it was decided to dispense with said analysis since it would lead to an environment derived from the visibility of the information that would blur the focus of the study and would speak more about web design than about the educational framework itself.

5.1. Defined Educational Project

Showing a well-defined educational project is essential when capturing the attention of teachers, counselors or curious people who access the UGGps website. The first impression is a key factor and, in the case of education, setting well-structured objectives, a conceptual map or an outline of the educational process during the visit will make a difference when choosing to carry out an activity or a visit alone with students or with family and friends. Being territories based on education, the first step to show is having a well-prepared educational project [48].

Table 4. Summary of the quantified and chosen variables to carry out the research about the visibility of education on the web pages of the Spanish and Portuguese UGGps. Variable number 10 "Primary, secondary and university education programs" was unified for its definition, since it represents the presence and visibility of the different plans or strategies within the same educational environment at different levels.

Quantified Variables
1.   Defined educational project
2.   Educational programs
3.   Schools go to the UGGps
4.   The UGGps goes to schools
5.   UGGps pet/mascot
6.   Educational methodologies
7.   Complementary educational initiatives
8.   Downloadable educational documents
9.   Educational audiovisuals
10. Primary, Secondary and University education programs
11. Background—pastcourses
12. Programs for Geosciences
13. Educational programs on other topics

5.2. Educational Programs

From the teaching point of view, this is one of the most important categories, since it allows the preparation of the visit not only for the moment of its realization but also before and after it in the classroom. The activity program, the schedules and the key dates are essential in every academic section of the UGGps. Within the web pages, we can find them in posters, brochures or even in the text itself, as a subsection [49].

5.3. Schools Go to the UGGps

This category is based on the digital sample of the possibility of organizing school or educational groups and visiting the UGGp [50]. Within it, we can include facilities that the territory can provide such as organization, guides, and routes for the different academic groups or means of transport. Obviously, it is important to show this category visually, either in a web subsection, group images or explanatory videos on how to organize visits.

5.4. The UGGps Goes to Schools

This sector is of vital importance for the educational field, since it allows you to "visit" any UGGp without leaving the classroom. It is a key tool for centers to prepare thematic classes and future excursions or for others that do not have the possibility of traveling to the territory at a specific time [12]. It is important to highlight this possibility on the web pages to open the range and allow academic centers to approach the UGGps without having to visit them.

5.5. UGGps Pet/Mascot

Having a pet/mascot can be an interesting educational resource for young children. This could be reflected in the websites with activities, mini-games and didactic images that provide information on the adaptation of the concepts of the UGGps for all audiences. Many UGGps already make use of this tool to promote geosciences and sustainable values among the smallest of the house [51,52].

5.6. Educational Methodologies

It is evident that the methodologies that we can find in these territories are a key factor within the mentality of teachers and those curious about deciding to visit these territories, either alone or with students. If the working methods are not attractive or not considered valid for the purpose of the visit, it will be very difficult to attract new visitors and promote educational plans. Fortunately, the combination of informal and non-formal education that exists in the UGGps makes them very complete educational experiences [53].

5.7. Complementary Educational Initiatives

The presence of activities complementary to those normally carried out within the UGGps can be a claim not only to attract new visitors, but also encourage people who have visited the UGGps and have been amazed by its heritage and its people to come back and make the trip with a different approach. In addition, the presence of activities such as meetings, presentations, special events or characteristic days celebrations are synonymous with quality educational activity and, therefore, it is interesting when it is reflected on the web pages.

5.8. Downloadable Educational Documents

PDF documents, educational videos, sheets, images, maps, brochures, etc. All downloadable educational material is positive when it comes to complementing the activities carried out in the UGGps and extrapolating everything learned to the classroom and to personal leisure. Resources such as these facilitate the understanding of concepts and enhance the educational experience of visitors [54].

5.9. Educational Audiovisuals

The advancement of information and communication technologies has made possible, with relatively few resources, the availability of incredible audiovisual materials to show the peculiarities of the UGGps to the whole world. The presence of an educative promotional video about what we can find in the Geopark on the website is something very necessary nowadays [55].

5.10. Primary, Secondary and University Education Programs

The presence of these educational offers on the web pages of the UGGps can guide teachers of all educational levels about the adaptability of the routes or the workshops that the Geoparks offer to their educational visits. The existence of these plans can be seen in past or future activities, talks or downloadable brochures, where the content can be evaluated and it decided whether it is suitable for the level of the students who wish to visit the territory [1,56].

5.11. Background—Past Courses

Showing publicly how past courses or events in old editions have been developed can be a sign of quality and a reason to consider visiting a territory, attend an event or develop a course with students. If there is a reference to how the activities have progressed in previous years, it will be easier for those interested to obtain an idea from the website of how the Geopark works, specifically regarding to the environment and involvement that it is capable of creating during the realization of an educational proposal.

5.12. Programs for Geosciences

UGGps are territories strongly focused on Geotourism and Geosciences. What kind of programs do you carry out within these subjects? Do you have geological guides of the Geopark? Are there specific events depending on the date of the visit? These and other questions are the ones that connoisseurs and curious people ask themselves when visiting the UGGps web pages to find out more about their content. It is important, even if it is clear from the beginning, to highlight the activities carried out in the territories and publicize them to promote social participation and the generation of new events and educational opportunities.

5.13. Educational Programs on Other Topics

Although the UGGps focus their educational activity on geology and earth sciences, we can also find a perfect connection between society and the environment. That is why the history of the territory, traditions or gastronomy play a fundamental role within the UGGps themselves and provide them their own identity. It is a reality that geology is essential in the UGGps, but without the people who live in them every day, said Geopark could not exist as itself. For this reason, the elaboration of routes, conferences, meetings or talks aimed at identifying the traditions and ways of life of the society that makes up the territory make it interesting to understand the social situation and the reason for the panorama that exists within the UGGps.

6. Results

The results presented below are divided into two sections according to the UGGps analyzed and the countries in question. The first section, showing the data from the Spanish UGGps, comprises the first 13 variables (Figure 1), while the second shows the last three (Figure 2). In the same way, Figures 3 and 4 correspond to the first and second sections of the Portuguese UGGps. All the data was organized and defined in graphs to offer a visual perspective of the situation of education in the analyzed web pages.

Finally, a comparative graph of all the variables was drawn up where the differences in the visibility of education in the online offer of the Spanish and Portuguese UGGps can be clearly observed (Figure 5). The obtained results give us a broad view of the situation of education in the web pages of the Spanish and Portuguese UGGps. There is no doubt that the sample differs considerably, since we are comparing the results obtained from 15 Spanish UGGps directly with those obtained from the five Portuguese UGGps.

In the representation of the results, the visibility of content was not taken into account in any of its meanings (either in the nature of SEO or in the ease of finding information within websites). Although they were taken into account for the discussion and conclusions, quantifying and representing these variables could guide the study to other fields that are outside the educational framework. It is important to emphasize that, although they are essential elements, they do not influence the presence or not of said information and, therefore, they were excluded from the variables analyzed. All the results are shown anonymously to represent a general framework of how we find the situation of education in the different UGGps websites. The intention of this study is not to make a critical representation of the online educational environment, but, with the utmost respect possible, to promote these territories as great educational alternatives and improve the current model with data, proposals and possible solutions.

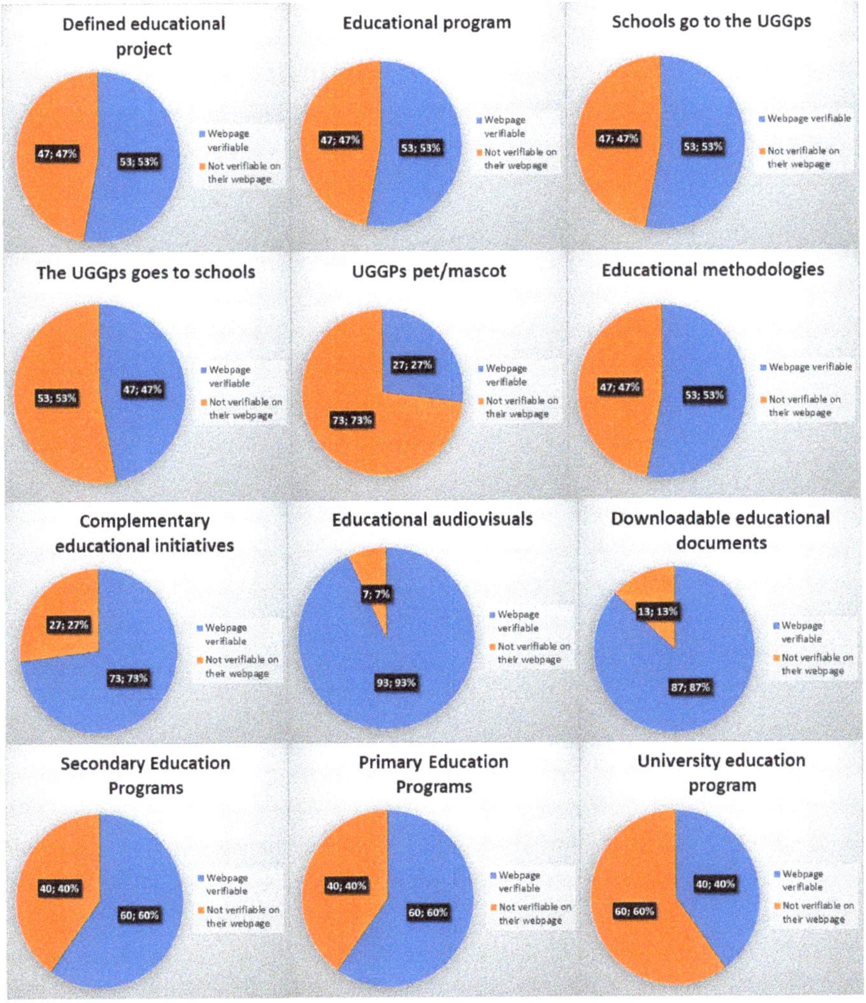

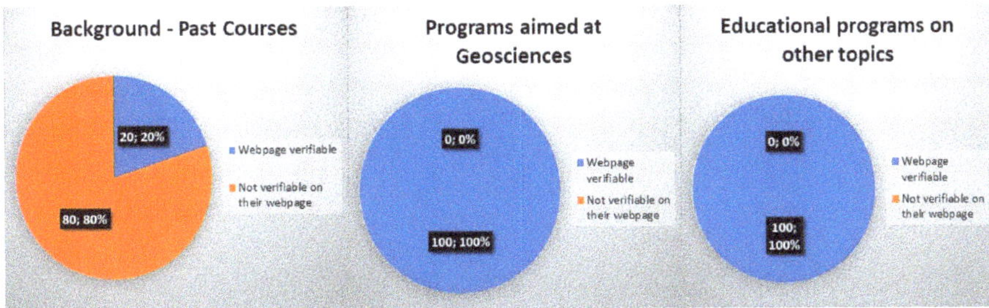

Figure 1. Graphic representation of the first 13 variables analyzed with respect to the visibility of education on the web pages of the Spanish UGGps.

Figure 2. Graphic representation of the last three variables analyzed with respect to the visibility of education on the web pages of the Spanish UGGps.

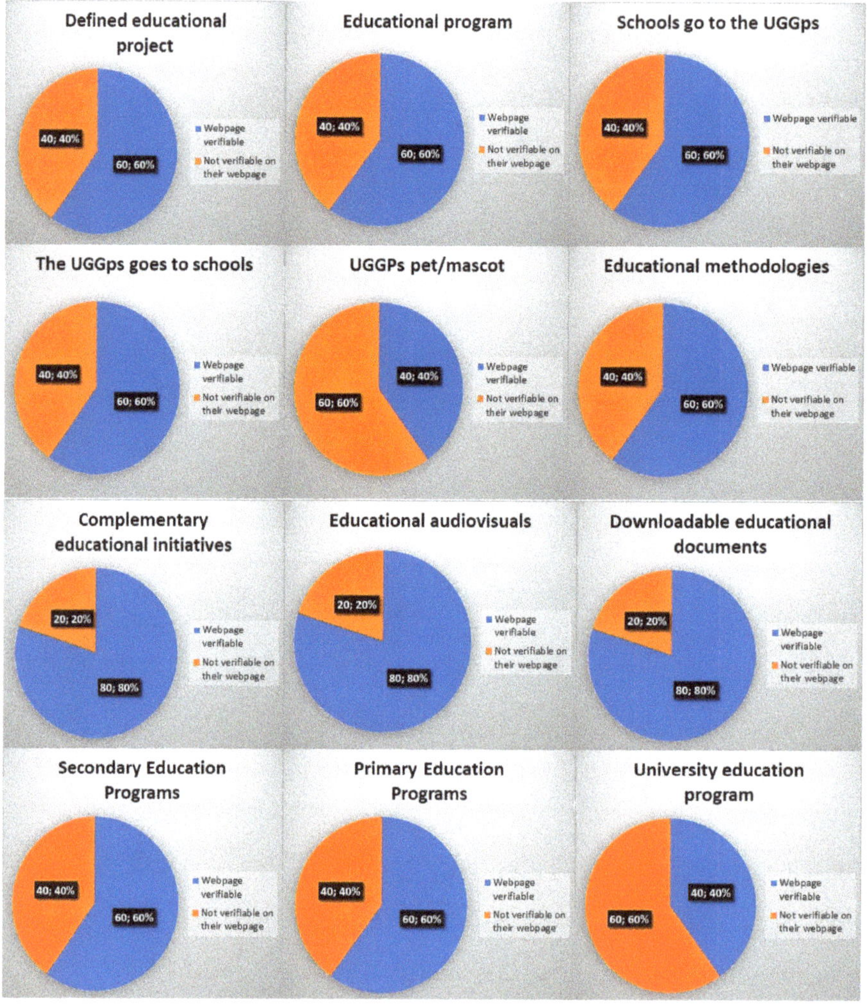

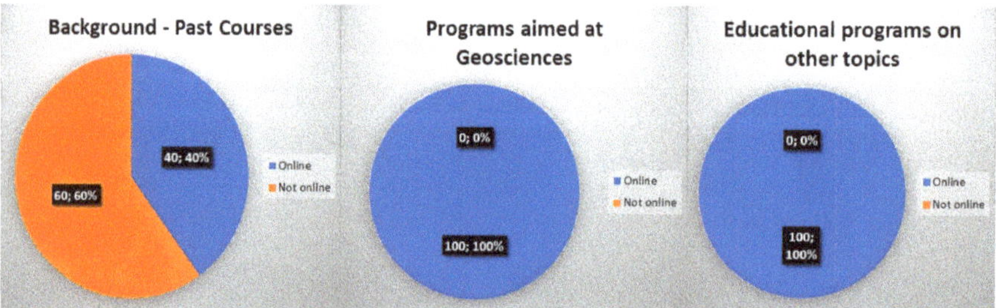

Figure 3. Graphic representation of the first 13 variables analyzed with respect to the visibility of education on the web pages of the Portuguese UGGps.

Figure 4. Graphic representation of the last three variables analyzed with respect to the visibility of education on the web pages of the Portuguese UGGps.

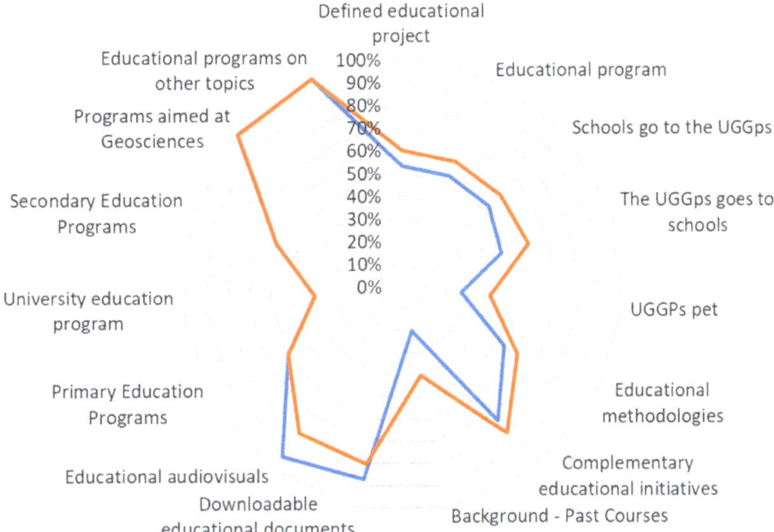

Figure 5. Graphic comparison of the visibility of education in the online offer of the Spanish and Portuguese UGGps.

7. Discussion

Online visibility is an important concept nowadays, where the Internet and the advancement of information and communication technologies have marked a new way of working, promoting and teaching [57]. Developing models and methodologies have evolved drastically until reaching a point where the audiovisual stands out and attracts everyone's attention. Society has become accustomed to having all possible information in the palm of its hand [58] thanks to smart phones or tablets, and tries to find out as much as possible about destinations, routes and territories before preparing a trip [59]. Destinations should be promoted in such a way that the number of visitors increases and, in the alternative context of the concept itself, it should be encouraged that web design facilitates the search for educational information. It is necessary that content is easy to locate within the website, adapted and accessible to everyone. This not only falls within the fundamental educational pillars of the UGGps, but also in the branch of objective number 4 of the SDGs of the 2030 Agenda. Updating web pages and positioning information strategically can be a fundamental factor for visitors when selecting a territory as a future destination. That is why the need to show all the possibilities has become a long-distance race of continuous updating, in which all the existing sectors participate every day, many times, even without realizing it. In the results obtained from the web pages analyzed, a reality can be highlighted: It is difficult to know what activities or projects are carried out daily in these territories simply by visiting their website, which is the largest showcase that can be used for promotion on the Internet. It would be fine to have menus or examples of how the UGGps works in the activities inserted in the website. Being territories focused on education, and this being an essential pillar for its operation, it is interesting to observe these results and realize that, in the eyes of the Internet and what is shown from their own web pages, many of the activities carried out in the UGGps remain totally invisible and overshadowed by other factors. The result culminates in the image that the UGGps are places with a remarkable geological heritage where you can enjoy nature, but nothing more. This is something completely incorrect and totally differs from the very definition of the UGGps concept. The UGGps are essential places for the future of society. Acting as laboratories and classrooms in the open air [27], they teach us in a multidisciplinary way every loophole that the territory hides. Science, history, and tradition, etc. all this is condensed and shown to the public from the UGGps. In addition, they are territories with tremendous potential when it comes to teaching not only in a sustainable way, but also the very concept of sustainability and how to apply it. In short, the UGGps are essential outdoor experiences for the future and, observing the results, we realize that many of these characteristics, which could be shown from the point of view of education, remain hidden, are difficult to find or, directly, are not shown on their official pages. According to the statistics obtained, there are many web pages of the Spanish and Portuguese UGGps in which we cannot find an educational program, background, or even ways to visit the Geopark or tools to move the Geopark to the classrooms. It is true that there are variables that stand out from others, such as the existence of promotional audiovisuals. This is something necessary, as mentioned above, and very attractive in view of the heritage panorama and the visibility of geology and Earth Sciences; however, on many of the websites, it is not possible to find an educational variant, or it is unclear with regards to the importance given to the environment and the landscape. It is also true that, while some web pages are very complete and detailed, with a large number of the variables visible and easily navigable, there are many others that are outdated and need an update of content to be closer to the current quality web standards. The key lies in building trust through the quality of the website, and this is an essential factor in the age we live in. It is important to underline the importance of visibility, since no matter how well the UGGps acts or how innovative the projects carried out are, if they are not visible, they do not exist for the majority of society. In addition, during the investigation, it was detected that the UGGps websites were very different from each other. It is logical that each UGGps has its style, since, although the fundamental pillars are shared by all UGGps, their differences are the key factor for allowing each one to be different faces of the same model. However, and following the theme of web visibility and the importance of generating trust in the visiting public, it is important to consider branding as a tool to attract public attention. The creation of a common domain, a brand-new model or the use of a similar structure for web pages could be a start to carry out this unification and, thus, present a brand style of the UGGps. If we put ourselves in the shoes of a teacher who needs to know how the educational activities of a UGGps work, how they have been changing in recent years or how they adapt to the academic curriculum that he intends to teach, we find certain difficulties within the website of some UGGps to locate the vast majority of the educational content that is generated within these territories. It is understandable that, if there really is this interest in visiting the Geopark, the visitor obtains extensive information, not only by searching the website, but also by calling the interpretation centers or the places they wish to visit. However, neglecting something as important as the web page of the territories is something that not only eliminates visiting opportunities, but also gives a bad impression when it comes to knowing and expanding knowledge about the territory. Internet is a spectacular promotional tool and, right now, essential in the times in which we live. Places as interesting and useful for society from an educational point of view as the UGGps should be promoted and deserve to be presented as a visible educational offer within networks to enhance their operation and maximize the tourism that occurs within them. Nowadays, we are experiencing a technological explosion that we can take advantage of to further emphasize the importance of quality education and create a socio-planetary synergy through which we can live together sustainably [60]. For this, it is necessary to enter the world of new technologies, such as virtual reality, and create a "Virtual Geopark" [61]. This not only allows tours without getting up from the sofa but can be used as an educational method which allows UGGps to connect with part of society that does not have the possibility of traveling for visits and creates an educational experience adapted and accessible to all [62–64]. This is just a small example of what new technologies can contribute. In fact, it can be already found small traces of the presence of UGGps in this area, such as the free app "Geotours," which allows you to explore the "European Atlantic Geotourism Route" in a simple and attractive way using mechanics of augmented reality [65] or some tests of gamifying the UGGps method [66]. In the app, we can download routes, digitally visit several of the most emblematic Geosites of the UGGps and win collectibles and medals that make its use more enjoyable and easier. Finally, UGGps stand out as educational territories in the field of Earth Sciences and sustainability. Their innovative methods and their mix of non-formal and informal educational programs make them true educational engines which constitute an alternative, attractive and original scenario within the current academic *curricula*. The transition to online education produced by the advancement of new technologies and the events that have occurred in recent years in relation to COVID-19 pandemic [10,67] have made the UGGps incredible educational alternatives which provide a safe, enjoyable and simple educational environment for every level. That is why the work must continue to make all kinds of activities visible in order to continue growing and spreading the message of quality, accessible and adapted education for all. The UGGps teach us to unravel the hidden message behind the landscape, to discover incredible cultures, traditions and places and that is why their daily work needs to be promoted and their potential demonstrated. The idea of creating a sustainable model in which society, nature and the economy can coexist and complement each other is closer every day and, in order fully achieve it, the UGGps educational work is essential.

8. Conclusions

The UGGps stand out as educational territories focused on the development and protection of natural heritage. Little by little, they have been building a reputation thanks to the activities, courses, plans and methods they use to communicate everything that is contained in the territory. Thanks to science, history, traditions and the spectacular nature of the environment that we enjoy in these places, the UGGps exist by and for the people, this being their fundamental objective. Web pages are equivalent to small windows that society can visit to echo what is happening in certain places and quickly determine whether it is of interest to them. Therefore, it is essential that these websites show reality in a clear, simple and adapted way for everyone. Being one of the most interesting digital showcases today, the UGGps must channel all their educational content and capture it in their web pages to show why they are ideal places to visit for teachers, those interested, those curious or anyone who has an interest in learning about a specific territory. That is why the possibility of unifying the web pages of the UGGps should be studied, not only in the educational field, but also in the visual field. Network visibility and the correct presence of content, as well as branding and personal image, are essential and differentiating factors on the network to stand out and communicate our message effectively. Virtual reality, digital applications, downloadable documents and educational audiovisuals are just some of the resources that are beginning to be used in the educational field and that, step by step, UGGps are beginning to implement within their toolkit to update, adapt and compile information that reaches everyone. For now, numerous paths and new opportunities are opening in the education sector that imply educational quality and accessibility at all levels, and the UGGps are continuously adapting to stay on trend. In short, the UGGps are and will continue to be relevant places in terms of education, and their adaptation to the digital medium is a process that has already begun and must continue, to highlight their importance in achieving a future that is more respectful of the planet and the people who live in it and, ultimately, more sustainable for everyone.

Author Contributions: Writing—original draft, J.E.M.-M.; Supervision, P.E.M., E.M.R.-G. and A.A.S. All authors have read and agreed to the published version of the manuscript.

Funding: This research received no external funding.

Data Availability Statement: Not applicable.

Acknowledgments: This study has been financially supported by the UCJC-Santander scholarship of excellence for the promotion of research.

Conflicts of Interest: The authors declare no conflict of interest.

References

1. Fernández Álvarez, R. Los geoparques como recurso para la enseñanza-aprendizaje del espacio geográfico en Educación Primaria: El paisaje de las áreas de montaña. *Didáctica Geográfica* **2019**, *20*, 27–53. [CrossRef]
2. Rocha Da Silva, E.M. *The Contribution of the European UNESCO Global Geoparks for the 2030 Agenda for Sustainable Development—A Study Based on Several Data Sources*; Universidade NOVA de Lisboa: Lisbon, Portugal, 2020.
3. Rosado-González, E.M.; Sá, A.A.; Palacio-Prieto, J.L. UNESCO Global Geoparks in Latin America and the Caribbean, and Their Contribution to Agenda 2030 Sustainable Development Goals. *Geoheritage* **2020**, *12*, 36. [CrossRef]
4. Henriques, M.H.; Tomaz, C.; Sá, A.A. The Arouca Geopark (Portugal) as an educational resource: A case study. *Episodes* **2012**, *35*, 481–488. [CrossRef]
5. Rodrigues, J. Pedagogical Geosciences tools to explain Naturtejo Geopark in both non-formal and formal environments. In *Libro de Ponencias II Conferencia del Proyecto Geoschools: Geología y Sociedade: Alfabetización Geocientífica*; Publicaciones del Seminário de Paleontologia de Zaragoza; Universidad de Zaragoza: Zaragoza, Spain, 2012. [CrossRef]
6. UNESCO. What is a UNESCO global geopark? 2022. Available online: http://www.globalgeopark.org/aboutGGN/6398.htm (accessed on 19 February 2022).
7. Jones, C. History of Geoparks. *Geol. Soc. Lond. Spéc. Publ.* **2008**, *300*, 273–277. [CrossRef]
8. Blanco, C.M.-C.; Castro, A.B.S. El muestreo en la investigación cualitativa. *NURE Investig. Rev. Científica Enfermería* **2007**, *27*, 1–4.
9. Bonilla-Guachamín, J.A. Las dos caras de la educación en el COVID-19. *CienciAmérica* **2020**, *9*, 89–98. [CrossRef]
10. Martini, B.G.; Zouros, N.; Zhang, J.; Jin, X.; Komoo, I.; Border, M.; Watanabe, M.; Frey, M.L.; Rangnes, K.; Van, T.T.; et al. UNESCO Global Geoparks in the "World after": A multiplegoals roadmap proposal for future discussion. *Episodes* **2022**, *45*, 29–35. [CrossRef]
11. Henriques, M.H.; Canales, M.L.; García-Frank, A.; Gomez-Heras, M. Accessible Geoparks in Iberia: A Challenge to Promote Geotourism and Education for Sustainable Development. *Geoheritage* **2019**, *11*, 471–484. [CrossRef]
12. Reuben, R. The Use of Social Media in Higher Education for Marketing and Communications: A Guide for Professionals in Higher Education. 2008. Available online: https://rachelreuben.com/2008/08/19/social-media-uses-higher-education-marketing-communication/ (accessed on 22 April 2022).
13. Maignan, I.; Lukas, B.A. The Nature and Social Uses of the Internet: A Qualitative Investigation. *J. Consum. Aff.* **1997**, *31*, 346–371. [CrossRef]
14. Aires, L.; Dias, P.; Azevedo, J.; Rebollo, M.; García-Perez, R. Education, digital inclusion and sustainable online communities. In *E-Learning and Education for Sustainability*; Azeiteiro, U.M., Leal Filho, W., Caeiro, S.S., Eds.; Peter Lang D. International Academic Publishers: Bern, Switzerland, 2014; pp. 263–273. [CrossRef]
15. Hellqvist, M. Teaching Sustainability in Geoscience Field Education at Falun Mine World Heritage Site in Sweden. *Geoheritage* **2019**, *11*, 1785–1798. [CrossRef]
16. Locke, S.; Libarkin, J.; Chang, C.-Y. Geoscience Education and Global Development. *J. Geosci. Educ.* **2012**, *60*, 199–200. [CrossRef]
17. Stewart, I. Sustainable geoscience. *Nat. Geosci.* **2016**, *9*, 262. [CrossRef]
18. Vasconcelos, C. (Ed.) *Geoscience Education*; Springer International Publishing: Berlin/Heidelberg, Germany, 2016. [CrossRef]
19. Gardner, H. How education changes. In *Globalization: Culture and Education in the New Millennium*; University of California Press: Oakland, CA, USA, 2004; pp. 235–258.
20. Sun, A.; Chen, X. Online Education and Its Effective Practice: A Research Review. *J. Inf. Technol. Educ. Res.* **2016**, *15*, 157–190. [CrossRef]
21. Abarzúa, A.; Cerda, C. Integración curricular de TIC en educación parvularia. *Rev. Pedagog.* **2011**, *32*, 13–43.
22. García Sánchez, M.d.R.; Reyes Añorve, J.; Godínez Alarcón, G. Las Tic en la educación superior, innovaciones y retos [The ICT in higher education, innovations and challenges]. *RICSH Rev. Iberoam. Cienc. Soc. Hum.* **2017**, *6*, 299–316. [CrossRef]
23. García-Peñalvo, F.J.; Corell, A.; Abella-García, V.; Grande, M. La evaluación online en la educación superior en tiempos de la COVID-19. *Educ. Knowl. Soc.* **2020**, *21*, 26. [CrossRef]
24. Adell Segura, J.; Sales Ciges, A. El profesor online: Elementos para la definición de un nuevo rol docente. In Proceedings of the IV Congreso de Nuevas Tecnologías de la Información y de la Comunicación para la Educación. Nuevas Tecnologías en la Formación Flexible y a Distancia, Seville, Spain, 14–17 September 1999; Universidad de Sevilla; Secretariado de Recursos Audiovisuales y Nuevas Tecnologías: Seville, Spain, 1999; pp. 1–15.
25. Ferrada-Bustamante, V.; González-Oro, N.; Ibarra-Caroca, M.; Vergara-Correa, D.; Castillo-Retamal, F.; Ried-Donaire, A. Formación docente en TIC y su evidencia en tiempos de COVID-19. *Rev. Saberes Educ.* **2021**, *6*, 144. [CrossRef]

26. Sandoval, C.H. La Educación en Tiempo del Covid-19 Herramientas TIC: El Nuevo Rol Docente en el Fortalecimiento del Proceso Enseñanza Aprendizaje de las Prácticas Educativa Innovadoras. *Rev. Tecnológica* **2020**, *9*, 24–31. [CrossRef]

27. Silva, E.; Sá, A.A. Educational challenges in the Portuguese UNESCO Global Geoparks: Contributing for the implementation of the SDG 4. *Int. J. Geoheritage Park.* **2018**, *6*, 95–106. [CrossRef]

28. González Tejada, C.; Girault, Y. Los Geoparques Mundiales de la UNESCO en España: Entre divulgación científica y desarrollo turístico. *Cuad. Geográficos* **2021**, *60*, 255–274. [CrossRef]

29. Martínez, J.E.; Mariñoso, P.E. Timanfaya lava flows geosite: A historical and educational approach. *Int. J. Earth Sci.* **2020**, *109*, 2697–2698. [CrossRef]

30. Martínez-Frías, J.; Mateo, E. Lanzarote: Mars on Earth. In *Lanzarote and Chinijo Islands Geopark: From Earth to Space*; Mateo, E., Martínez-Frías, J., Vegas, J., Eds.; Springer International Publishing: Berlin/Heidelberg, Germany, 2019; pp. 143–148. [CrossRef]

31. Martínez Frías, J.; Mateo Mederos, M.; Lunar Hernández, R. Los geoparques como áreas de investigación, geoeducación y geoética en geociencias planetarias: El geoparque de Lanzarote y Archipiélago Chinijo. *Geotemas* **2016**, *16*, 343.

32. Catana, M.M.; Brilha, J.B. The Role of UNESCO Global Geoparks in Promoting Geosciences Education for Sustainability. *Geoheritage* **2020**, *12*, 1. [CrossRef]

33. Farsani, N.T.; Coelho, C.; Costa, C.; Neto de Carvalho, C. (Eds.) *Geoparks and Geotourism—New Approaches to Sustainability for the XXI Century*; BrownWalker Press: Boca Raton, FL, USA, 2012.

34. Fassoulas, C.; Nikolakakis, E.; Staridas, S. Digital Tools to Serve Geotourism and Sustainable Development at Psiloritis UNESCO Global Geopark in COVID Times and Beyond. *Geosciences* **2022**, *12*, 78. [CrossRef]

35. Farsani, N.T.; Coelho, C.O.A.; Costa, C.; Amrikazemi, A. Geo-knowledge Management and Geoconservation via Geoparks and Geotourism. *Geoheritage* **2014**, *6*, 185–192. [CrossRef]

36. Rowley, J. Online branding. *Online Inf. Rev.* **2004**, *28*, 131–138. [CrossRef]

37. Bansal, M.; Sharma, D. Improving webpage visibility in search engines by enhancing keyword density using improved on-page optimization technique. *Int. J. Comput. Sci. Inf. Technol.* **2015**, *6*, 5347–5352.

38. Fernández-Cavia, J.; Fabra, B.U.P.; Castro, D. Comunicación y branding en los sitios web nacionales de turismo. *Cuad. Info* **2015**, *37*, 167–185. [CrossRef]

39. Berry, L.H. Cognitive effects of web page design. In *Instructional and Cognitive Impacts of Web-Based Education*; IGI Global: Hershey, PA, USA, 2000; pp. 41–55.

40. Pérez-Ortega, A. *Marca Personal: Cómo Convertirse en la Opción Preferente*; ESIC: New Delhi, India, 2008.

41. Powell, T.A.; Jones, D.L.; Cutts, D.C. *Web Site Engineering: Beyond Web Page Design*; Prentice-Hall, Inc.: Hoboken, NJ, USA, 1998. [CrossRef]

42. Cantone, D. Como Elaborar tu Estrategia de Personal Branding Paso a Paso. 2011. Available online: Http://Davidcantone.Com/About (accessed on 16 March 2022).

43. Climent-Rodríguez, J.-A.; Navarro-Abal, Y. Branding y reputación: Pilares básicos de la visibilidad online del profesor de educación superior. *Rev. Iberoam. Educ. Super.* **2017**, *8*, 66–76. [CrossRef]

44. Vanegas, C.A.; Pinzón Núñez, S.A. Accesibilidad en el diseño de páginas web. *Vínculos* **2012**, *9*, 177–190. [CrossRef]

45. Cuberos, R.C.; Sánchez, M.C.; Ortega, F.Z.; Garcés, T.E.; Martínez, A.M. Videojuegos activos como recurso TIC en el aula de Educación Física: Estudio a partir de parámetros de ocio digital-Active Videogames as ICT tool in Physical Education classroom: Research from digital leisure parameters. *Digit. Educ. Rev.* **2016**, *29*, 112–123.

46. Brilha, J.B.R.; Legoinha, P.A.R.; Butler, J. The Internet and the Geology teaching in Portugal. *Comput. Geosci.* **1999**, *25*, 205–206.

47. Pereira, D.I.; Ramos, I.; Medina, W. Are Geoparks webpages attractive to potential tourists? Some results of an evaluation procedure. In Proceedings of the International Congress Arouca 2011 Geotourism in Action, Arouca, Portugal, 9–13 November 2011; 2011; pp. 159–161. Available online: http://hdl.handle.net/1822/15202 (accessed on 10 April 2022).

48. Adams, D. Defining educational quality. *Improv. Educ. Qual. Proj. Publ.* **1993**, *1*, 1–24.

49. Glumac, O.; D'Arcy, G.; de Sousa Morais, M.R.C. Co-creation and Co-design of Educational Programmes with Young People: A Comparative Study Between Dublin and Porto. In *Perspectives on Design II*; Springer: Berlin/Heidelberg, Germany, 2022; pp. 61–76. [CrossRef]

50. Lin, J.-C.; Su, S.-J. Environmental Education for Geoparks—Practices and Challenges. In *Geoparks of Taiwan*; Springer: Berlin/Heidelberg, Germany, 2019; pp. 105–118. [CrossRef]

51. Del Mazo, I.; Bonilla Leo, A.; Carnicer Corchero, M.d.P.; Caro López, M.C.; García Iglesias, S.; Gil Pacheco, M.I.; Gordillo Valenzuela, J.; Jara Álvarez, P.; Muñoz Calderón, C.; Parejo Sánchez, A.; et al. Geopaca Folk: Rustic Performance on the Road. 2020. Available online: https://hdl.handle.net/11162/206523 (accessed on 16 March 2022).

52. Velaz Sánchez, M.; Gándara Carretero, I.; Bonilla Leo, A. Geopaca: De Mascota a Influencer. 2019. Available online: https://hdl.handle.net/11162/201744 (accessed on 13 May 2022).

53. Rodrigues, J.; de Carvalho, C.N.; Ramos, M.; Ramos, R.; Vinagre, A.; Vinagre, H. Geoproducts—Innovative development strategies in UNESCO Geoparks: Concept, implementation methodology, and case studies from Naturtejo Global Geopark, Portugal. *Int. J. Geoheritage Parks* **2021**, *9*, 108–128. [CrossRef]

54. Kim, J.Y.; Fienup, D.M. Increasing Access to Online Learning for Students with Disabilities During the COVID-19 Pandemic. *J. Spéc. Educ.* **2022**, *55*, 213–221. [CrossRef]

55. Pattier, D. Teachers and Youtube: The use of video as an educational resource. *Ric. Pedagog. Didatt. J. Theor. Res. Educ.* **2021**, *16*, 59–77. [CrossRef]
56. Wang, J.; Zouros, N. Educational Activities in Fangshan UNESCO Global Geopark and Lesvos Island UNESCO Global Geopark. *Geoheritage* **2021**, *13*, 51. [CrossRef]
57. Pujadas, F.J.C. Impacto de las nuevas tecnologías en la masificación de la educación. *Rev. Sci.* **2019**, *4*, 173–187. [CrossRef]
58. Correa, R.I. La sociedad mesmerizada: Medios, nuevas tecnologías y conciencia crítica en educación. In *La Sociedad Mesmerizada*; Universidad de Huelva: Huelva, Spain, 2019.
59. Varela-Ordorica, S.A.; Valenzuela-González, J.R. Uso de las tecnologías de la información y la comunicación como competencia transversal en la formación inicial de docentes. *Rev. Electrónica Educ.* **2020**, *24*, 1–20. [CrossRef]
60. Marte-Espinal, R. Uso de las tecnologías en la educación. In *Revista: Atlante. Cuadernos de Educación y Desarrollo*; Universidad de Málaga: Malaga, Spain, 2017; pp. 1–9.
61. Cortés, F.P.; Rebolledo, A.M.; Zamorano, P.G.; Ossandón, J.M.; Montaña, A.; Obreque, M.K.; Castro, G.R.; Oñate, J.J.; Ulloa, M.E.; Kuschel, P.H. El geo-parque virtual y los módulos de aprendizaje: Una propuesta de aprendizaje sobre los riesgos naturales. In *Nuevas Ideas en Informática Educativa Memorias del XVII Congreso Internacional de Informática Educativa*; TISE: Santiago, Chile, 2012; pp. 434–440.
62. Hincapié Montoya, M.; Valencia-Arias, A.; Fernandez, J.; Cuéllar Rojas, Ó. Aplicaciones de Realidad Aumentada y Realidad Virtual en Geociencias: Un análisis bibliométrico. In *Un Análisis de los Campos de la Ingeniería: Usos y Aplicaciones*; Americana: Barranquilla, Colombia, 2020; pp. 95–109.
63. Lege, R.; Bonner, E. Virtual reality in education: The promise, progress, and challenge. *JALT CALL J.* **2020**, *16*, 167–180. [CrossRef]
64. Flores Tena, M.; Ortega Navas, M.C.; Sánchez Fuster, M.C. Las nuevas tecnologías como estrategias innovadoras de enseñanza-aprendizaje en la era digital. *Rev. Electrónica Interuniv. Profr* **2021**, *24*, 29–42. [CrossRef]
65. Elmqaddem, N. Augmented Reality and Virtual Reality in Education. Myth or Reality? *Int. J. Emerg. Technol. Learn.* **2019**, *14*, 234. [CrossRef]
66. Pereira, D.; da Gama, K.S.; Silveira, C.B.d.M.; Dias e Cordeiro, I. Creación de un prototipo y test de una aplicación para la gamificación de la visita al geoparque de Araripe (Ceará-Brasil). *Estud. Perspect. Tur.* **2019**, *28*, 1021–1031.
67. Monasterio, D.; Briceño, M. Educación mediada por las Tecnologías: Un desafío ante la coyuntura del COVID-19. *Obs. Conoc.* **2020**, *5*, 100–108.

Review

The Status of Geoethical Thinking in the Educational System of Greece: An Overview

Spyros Mosios *, Efthymios Georgousis and Hara Drinia *

Department of Geology and Geoenvironment, National and Kapodistrian University of Athens, 15784 Athens, Greece
* Correspondence: spymosi@geol.uoa.gr (S.M.); cntrinia@geol.uoa.gr (H.D.); Tel.: +30-210-7274394 (H.D.)

Abstract: In recent years, the concern about geoethics in Greece has begun to grow. This review aims to present the current geoethical thinking in Greece's educational system through a thorough research of all educational levels' curricula and the actions developed on this topic in the Greek geoscientific community. In the Greek educational reality, geoeducation is not widespread, and geoethics is used in limited school curricula. The research highlighted a significant lack of initiatives to promote geoethical thinking and the values related to geological heritage and the need to protect it.

Keywords: geoenvironmental education; geoethics; awareness; Greece

1. Introduction

In recent years, the geological research community has shifted towards interdisciplinary and socially relevant topics, one of which is the study and inventory of geological heritage. The study of geological heritage represents an emerging field of geosciences that involves, among other things, the investigation of the scientific, educational, aesthetic, and cultural values of the Earth's natural and cultural features, including geological and geomorphological features, fossils, minerals, landscapes, and other natural resources. The management of geological heritage is guided by ethical considerations, such as the respect for the rights of indigenous people, the protection of biodiversity, and the promotion of sustainable development [1].

Geoethics, on the other hand, is a relatively new and interdisciplinary field that addresses the ethical implications of the use and management of the Earth's resources and environment. It encompasses ethical considerations related to geosciences and the environment, such as the protection of geological heritage, the management of natural resources, and the mitigation of natural hazards [2]. Furthermore, it includes the ethical implications of the use of geotechnology, such as the use of geothermal energy and the handling of geological waste. It aims to ensure that the needs and well-being of present and future generations are considered when making decisions regarding the use and management of natural resources, including geoheritage [3].

Geoethics is often considered to be a subset of environmental ethics, as it deals specifically with the ethical issues related to the Earth sciences and their applications. However, it is also considered to be a distinct field of study, with its own unique focus and set of concerns. Environmental ethics is a branch of philosophy that considers the moral and ethical relationship between human beings and the natural environment. It examines questions such as how human activities impact the environment and what moral obligations we have to protect and preserve the natural world [4]. Geoethics, on the other hand, is an interdisciplinary field that focuses on the ethical and societal implications of the Earth sciences and their applications. It encompasses the study of ethical and social issues related to the use and management of natural resources, the impacts of human activities on the Earth's system, the conservation of biodiversity, and the mitigation of natural hazards.

Citation: Mosios, S.; Georgousis, E.; Drinia, H. The Status of Geoethical Thinking in the Educational System of Greece: An Overview. *Geosciences* **2023**, *13*, 37. https://doi.org/10.3390/geosciences13020037

Academic Editors: Jesús F. Jordá Pardo and Karoly Nemeth

Received: 5 December 2022
Revised: 24 January 2023
Accepted: 26 January 2023
Published: 29 January 2023

Geoethics also deals with the ethical responsibilities of geoscientists and other Earth science professionals [5].

Two documents provide guidelines for ethical decision-making in the field of geosciences but are developed by different organizations and adopted at different conferences: The International Declaration on Geoethics is a document that was adopted by the International Union of Geological Sciences (IUGS) at the 35th International Geological Congress in Florence, Italy, in 2004. The International Declaration on Geoethics is a set of guidelines that provide ethical principles and values for geoscientists and other stakeholders involved in the management and use of the Earth's resources [6].

The Cape Town Statement on Geoethics, also known as the Cape Town Declaration, is a document that was adopted by the International Association for Promoting Geoethics (IAPG) at the 2nd International Conference on Geoethics in Cape Town, South Africa, in 2013. The statement lays out the principles and values of geoethics and provides guidance for the ethical practice of geosciences [7].

Geoethical thinking refers to the process of considering and applying ethical principles and values to the use, management, and conservation of the Earth's resources and environment [8]. The goal of geoethical thinking is to ensure that decisions regarding the use and management of the Earth's resources are guided by ethical considerations, considering the needs and well-being of present and future generations, as well as the impacts on non-human species and ecosystems. It also involves the consideration of cultural, social, and economic aspects. It is interdisciplinary, involving input from various fields, such as geology, environmental science, philosophy, and sociology, among others [9].

The protection and preservation of geoheritage is a key aspect of geoethics. This includes not only the preservation of geological and geomorphological features but also the protection of fossils and minerals, the conservation of landscapes and the management of other natural resources [10]. Education and awareness-raising are also important components of geoethics in relation to geoheritage. This includes not only educating the public about the scientific and cultural value of geoheritage but also raising awareness of the importance of its protection and preservation [11].

1.1. The Necessity of Geoethics and Geoethical Education

Geoethics is an important field that addresses the ethical considerations surrounding the use and management of the Earth's resources and environment. It encompasses research and reflection on the values that should guide appropriate behavior and practice at the intersections of human activities and the geosphere [6]. It provides a framework for making ethical decisions related to the use and management of the Earth's resources and environment, promotes sustainability in the use and management of natural resources, protects and preserves geological heritage, promotes transparency and accountability in the use and management of natural resources and the environment, and raises awareness of the responsibility of humans as a powerful geological force [12].

The field of geoethics has also turned to education in recent years, with a focus on reflecting on the way humans relate to the geosphere and how geologists work during their academic and professional activities. Geoethical education, which incorporates ethical considerations related to the Earth's resources and environment into education, is an important aspect of promoting geoethical thinking in society. This can include integrating ethical considerations into the curriculum of relevant educational programs, such as geology, environmental science, and earth science, and promoting geoethical education through outreach programs, educational resources, and community engagement initiatives [13]. Informal or formal "geoeducation" is a key area of interest as part of education for sustainable development and the promotion of geoheritage values, in order to achieve the implementation of geoconservation objectives and to ensure effective management of geoheritage [14].

1.2. Objectives of the Review

Greece is in the northeastern part of the Mediterranean basin (Figure 1), which is a tectonically active region characterized by complex geodynamic processes. The geodynamic setting of Greece is shaped by the interaction of several major tectonic plates, including the African, Eurasian, and Aegean plates, which have resulted in the formation of a diverse range of geological features, including islands, mountains, and volcanic activity [15–18]. This setting makes Greece a unique and interesting place to study geology and geodynamic processes and is home to a rich and diverse geoheritage, which includes a wide range of geological and geomorphological features, fossils, minerals, landscapes, and other natural resources [1].

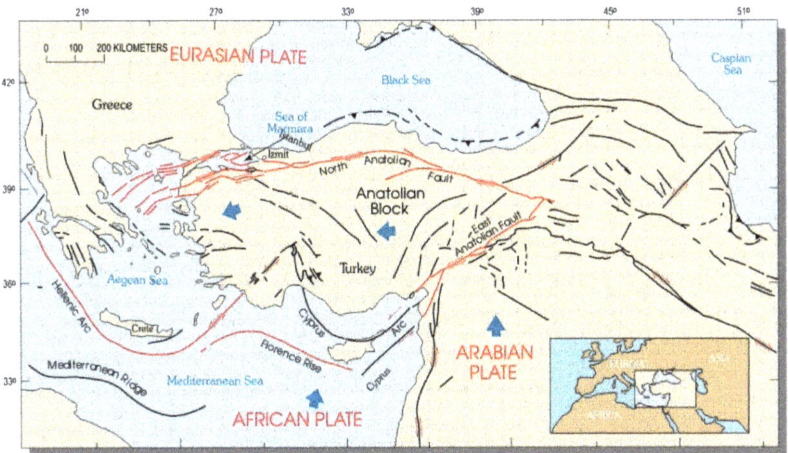

Figure 1. The tectonic map of Eastern Mediterranean Basin, including the North Anatolian fault, East Anatolian fault, and Hellenic and Florence trenches (by U.S. Geological Survey, National Earthquake Information Center—U.S. Geological Survey, National Earthquake Information Center, Public Domain, https://commons.wikimedia.org/w/index.php?curid=33751528 (accessed on 5 December 2022)).

In Greece, the advancement of geoethical thinking has been gaining increasing attention in recent years. However, geoethical thinking is still not fully integrated into the educational system and curricula, but there are some efforts to incorporate it in the universities' curricula and research programs [19]. The focus of this review will be to examine the extent to which geoethical considerations have been integrated into higher education curricula and research activities. The main research question that we will seek to answer is to what degree has geoethical thinking been incorporated into educational programs and what studies have been conducted in this field.

2. Methodology

We have employed a multifaceted approach to evaluate the state of geoethical thinking in the educational system of Greece. This approach includes:

- A review of the existing literature on the topic of geoethics in the educational system of Greece, including academic journals, reports, and other publications, to gain an understanding of what has been studied and written on the topic and to identify areas where further research is needed.
- An analysis of the curriculum of relevant educational programs in Greece to determine the extent to which geoethical considerations are incorporated into the curriculum, providing insight into the current state of the field, and identifying areas where further attention is needed.

- Examination of case studies of specific educational programs or initiatives that have incorporated geoethical considerations in Greece, which provides detailed information on how geoethical thinking has been applied in practice and highlights best practices and lessons learned.

To measure the level of geoethical research in Greece, we have used a variety of metrics, including:

- Tracking the number of research papers and publications on the topic in academic journals, conference proceedings, and books.
- Monitoring the amount of funding allocated to geoethical research in Greece.
- Counting the number of research centers and groups in Greece that focus on geoethical issues.
- Tracking the number of students who complete their Ph.D. or master's degrees in the field of geoethics.
- The number of workshops and conferences organized in Greece on the topic of geoethics.

It is important to note that the use of a combination of these methods is essential to obtain a comprehensive understanding of the status of geoethical thinking in the educational system of Greece.

3. Results

3.1. The Concept of Geoethics in the Greek Reality

When searching for the word "geoethics" in Greek on the internet, the results are scarce and refer mostly to general concepts, typically without any references. These results focus mainly on the first two decades of the 21st century, especially after 2012. Greece is a national section of the IAPG network, which, as is known, has a mission "to coordinate efforts in promoting Geoethics and enlarging the IAPG network in the country, by encouraging the participation of geoscientists in their activities on the basis of equal opportunities and favoring the exchange of information among its members through newsletters, publications or other suitable tools such as a website and social networks" [20]. Therefore, the Cape Town Statement on Geoethics is available in Greek. Furthermore, The National and Kapodistrian University of Athens-Applied Philosophy Research Laboratory (NKUA-APRL), which was established with a purpose to facilitate research in the area of applied philosophy and to conduct empirical and/or theoretical research in all areas of philosophy, signed with the International Association for Promoting Geoethics (IAPG), Memorandum of Understanding (MoU) (2019), with the aim of cooperation in developing empirical and/or theoretical research in the areas of geoethics and bioethics and to coordinate activities aimed at promoting the discussion on the ethical, social, and cultural implications of geosciences and biosciences [21,22]. However, the concept of geoethics has not yet gained widespread acceptance within the Greek educational system or among the scientific community in Greece [10].

3.2. Geoethics in the Greek Educational System

It is important to note that the inclusion of geoethics in curricula is a relatively new field, and there is limited information available about it, especially for specific countries such as Greece. However, in order to investigate the inclusion of geoethics in Greek curricula, we contacted the Earth and Environmental Science departments and the Geography departments of universities or colleges to inquire about the inclusion of geoethics in their curricula. We searched for academic journals that focus on geoethics and education, such as the *Journal of Environmental Education*, the *Journal of Geography in Higher Education*, and the *International Journal of Geoethics*, to find articles about the inclusion of geoethics in curricula. We searched for online resources, including websites of professional societies, such as the Greek Geological Society, that provide information and resources on geoethics in education. We searched for government reports on education and the environment, which may provide information on the inclusion of geoethics in curricula. Finally, we reached out to experts in the field of geoethics and education, such as professors or researchers, who may be able to provide information on the inclusion of geoethics in curricula.

The conducted survey revealed that, unfortunately, there are no specific programs or courses on geoethics offered in Greek higher education institutions. However, Greek universities and research centers may have professors and researchers who specialize in environmental ethics and sustainable development, who may also be interested in geoethics. It is worth noting that, while it is not a common subject, there are some universities in Greece that have a Department of Environmental Studies, Environmental Engineering, and other related fields, where environmental ethics and sustainable development are being taught, and these departments could also be a good starting point for students who are interested in pursuing studies in geoethics.

It is also important to note that, as an interdisciplinary field, it is often taught through collaboration between different departments, such as geology, environmental science, ethics, and philosophy. Therefore, while it may not be a common subject in Greek higher education, there are opportunities for students interested in geoethics to explore the field through various disciplines and through international events. Moreover, it should be noted that, until now, in the Greek higher education community, there does not seem to be a particular reflection, equivalent to the international one, as to whether geoethics should be established in the field of environmental ethics or as something clearly different based on the foundations of the professional ethics of applied geosciences, although it seems that the most promising interface of geoethics with adjacent fields is likely the relation to the field of environmental ethics [23].

As far as the primary or secondary education in Greece is concerned, as in higher education, there are no specific curriculum or programs on geoethics offered. However, environmental education and sustainable development are often included in the primary and secondary education curriculum in Greece, and these subjects may include some aspects of geoethics. It is important to note that, while it is not a common subject, environmental education and sustainable development constitute a good starting point for students to be exposed to the ethics and societal implications of the Earth sciences and their applications.

However, it should be claimed that the education of students in geosciences is incomplete in the curricula of the geology–geography subject of the lower secondary education (Gymnasium) and the concept of the geoenvironment, as a witness of geological phenomena, is absent and rarely consists of a subject matter of the educational programs of environmental education. As a result, students are unable to perceive values of the geological heritage. In contrast to Greek geosciences curricula programs, in other countries, the didactics of geosciences studies aims to students' knowledge of the principles of geoethics [24].

The European Geosciences Union (EGU) has developed a geoethics initiative that aims to promote the integration of geoethics in geoscience education and research across Europe, which includes didactics of geosciences. Additionally, some universities in European countries, such as the UK, have started to offer courses or modules on geoethics, and some professional societies, such as the American Geological Institute, have also developed resources and guidelines for incorporating geoethics into education and training programs.

From the results of recent research concerning the new curriculum on environment and education for sustainable development, it is arguable whether the concepts will lead to the development of values in the field of geoethics since the basic concepts are mainly biodiversity, protection of nature and land, the environment in general, ethical behavior, and, secondarily, geological and geomorphological knowledge, geodiversity, and geoheritage [19].

3.3. Overview of the Current Geoethical Research in Greece

Geoethical research aims to understand and address the ethical dimensions of environmental and Earth science issues. Current geoethical research encompasses a wide range of topics, including sustainable development, climate change, environmental justice (meaning the research on the distribution of environmental burdens and benefits among different communities and the ethical implications of these disparities), biodiversity conservation, resource management, and research on the ethical dimensions of the relationship between science and society, such as the responsibilities of scientists and policymakers and the communication of scientific information to the public.

Overall, geoethical research is an interdisciplinary field that uses philosophical, sociological, legal, and other perspectives to understand and address ethical issues related to the Earth and the environment.

In terms of geoethical research in Greece, it is a relatively new field with limited information available. Our survey, however, revealed that the amount of geoethical research being conducted in Greece today is still small, but growing.

The scientific and research activity of the last few years at the level of postgraduate and doctoral theses reflects the current research being conducted in the field of geoethics in Greece and the interest in promoting geoethical thinking primarily in the Greek scientific community (Table 1). A common theme of these research contributions is the recognition of the importance of geoethics concerning the relationship of the earth sciences to society and the Earth.

Table 1. Master's and Doctoral Theses on Geoethics (unpublished).

Master and Doctoral Theses	Title	Affiliation	Scientific Domain (According to Affiliation)
Tsikripis, N.	Geoethics and Environmental Management	Department of Geology and Geoenvironment, UoA	environment management; natural disasters
Georgousis, E.	Man and Geoenvironment: An Interdisciplinary Approach to Environmental Ethics and Geoethics	Department of Geology and Geoenvironment, UoA	geoethics; environmental ethics; traditional indigenous environmental knowledge
Mosios, S.	Geoenvironmental education and its geoethical dimension at national, European and international level	Department of Geology and Geoenvironment, UoA	geoethics; geoeducation
Zafeiropoulos, G.	The importance of geo-environmental education in understanding geological heritage and geo-ethical awareness. The case of the Dodecanese barren islands	Department of Geology and Geoenvironment, UoA	geoethics; geoeducation; geocultural heritage
Koupatsiaris, A.	Geoethics as a factor of environment-friendly behaviour through a geo-environmental education programme for primary and secondary schools	Department of Geology and Geoenvironment, UoA	geoethics; geoeducation

The master's research paper, "Geoethics and Environment Management", highlights the contribution of geoethics in the management of natural hazards and disasters in the context of sustainable development and in communication among the scientific community, citizens, and the media. Aside from postgraduate theses, four doctoral theses are currently being written at the National and Kapodistrian University of Athens (UoA) Faculty of Geology and Geoenvironment, covering a wide range of geoethical research.

The contribution of the doctoral dissertation research titled "Man and geoenvironment: an interdisciplinary approach to environmental ethics and geoethics" lies in the connection of geoethics with the field of environmental ethics, the possibility of using traditional natural resources management practices, traditional indigenous and local populations environmental knowledge and their perceptions on environmental risks, as well as the integration of the traditional knowledge and value system in the field of geoethics. The research titled "Geoenvironmental education and its geoethical dimension at national, European and international level" outlines the current state of geoeducation and geoethics in Greece, Europe, and worldwide as well. It focuses on exploring the geoethical dimensions in the educational system, which as a miniature society, prepares future responsible citizens for the environmental challenges, and presents the reasons and the need for the integration of geoenvironmental education not only in the curricula of the educational system but also in the actions and activities of non-formal education for the benefit of society. The Ph.D. thesis, "The importance of geoenvironmental education in the understanding of geological heritage and geoethical awareness. The case of the Dodecanese barren line islands", highlights the importance of geoenvironmental education in the understanding of geological heritage and geoethical awareness. Finally, the thesis, "Geoethics as a factor of environment-friendly behaviour through a geo-environmental education programme for primary and secondary schools", aims to show that geoethics in the context of geoeducation contributes to raising awareness and fostering values and responsibility and that geocultural heritage and geoethics can strengthen the links between people and their place and between their places of origin and their memories.

Various articles in peer-reviewed international journals address a range of topics surrounding the concept of ethics in geosciences (Table 2) and provide a much-needed basis for discussion to promote geoethical thinking in education. All the efforts are isolated but promising and should be enriched with the ultimate goal of integrating geoethics as a central part of all geoscience courses.

Table 2. Papers and conference presentations on Geoethics.

Paper Reference	Title	Affiliation	Scientific Domain (According to Affiliation)
Georgousis, E.; Savelides, S.; Mosios, S.; Holokolos, M.-V.; Drinia, H.	The Need for Geoethical Awareness: The Importance of Geoenvironmental Education in Geoheritage Understanding in the Case of Meteora Geomorphes, Greece.	Department of Geology and Geoenvironment, UoA	geoethics; values; geocultural heritage; geoeducation; quantitative analysis
Georgousis, E.; Savelidi, M.; Savelides, S.; Holokolos, M.-V.; Drinia, H.	Teaching Geoheritage Values: Implementation and Thematic Analysis Evaluation of a Synchronous Online Educational Approach.	Department of Geology and Geoenvironment, UoA	geocultural heritage; geoethical values, perception, awareness; qualitative thematic analysis
Zafeiropoulos, G.; Drinia, H.; Antonarakou, A.; Zouros, N.	From Geoheritage to Geoeducation, Geoethics and Geotourism: A Critical Evaluation of the Greek Region.	Department of Geology and Geoenvironment, UoA; Department of Geography, University of Aegean	geoheritage; geoconservation; geoeducation; geotourism; sustainable development
Georgousis, E.; Savelidi, M.; Savelides, S.; Mosios, S.; Holokolos, M.-V.; Drinia, H.	How Greek Students Perceive Concepts related to Geoenvironment: A Semiotics Content Analysis.	Department of Geology and Geoenvironment, UoA	geodiversity; geoheritage; geoethics; geotourism; geoeducation; semiotics content analysis
Savelides, S.; Georgousis, E.; Fasouraki, R.; Papadopoulou, G.; Drinia, H.	"Storm Tossed Sea Rocks in Pelion": An environmental synchronous online education program.	Department of Geology and Geoenvironment, UoA	environmental education; synchronous online education
Georgousis, E.; Savelidi, M.; Savelides, S.; Mosios, S.; Holokolos, M.-V.; Drinia, H.	The inclusion of Geoethical Values in the Design of Educational Policy for the Next Decade: The Case of the Greek Educational System	Department of Geology and Geoenvironment, UoA	geoethics; values; educational policy; Greek educational system
Georgousis, E.; Mosios, S.; Savelides, S.; Holokolos, M.-V.; Drinia, H.	The New Curriculum on Environment and Education for Sustainable Development and its Potential for Shaping Attitudes and Behaviors towards the Geoenvironment and Cultivating Geoethical values	Department of Geology and Geoenvironment, UoA	geoethics; geodiversity; geoheritage; geoeducation; environment; sustainable development
Mosios, S.; Georgousis, E.; Savelides, S.; Holokolos, M.-V.; Drinia, H.	Geodiversity and Geoheritage: The geoethical dimensions of a Geoeducation Program in the context of an experimental Lower Secondary Education school club	Department of Geology and Geoenvironment, UoA	geodiversity; geoheritage; geotourism; geoethics; geoeducation

3.4. Overview of the Initiatives to Promote Geoethics in Greece

In recent years, there have been several international conferences and workshops on geoethics held around the world, with a focus on topics such as the ethical considerations related to natural resource management, the protection of geological heritage, and the integration of geoethics into education. As such, we mention, indicatively International Geoethics Day, an event that is celebrated worldwide every year, GOAL workshops, and IAPG's participation in the annual General Assembly of EGU. These events have brought together experts from various fields, including geology, sociology, philosophy, law, and education, to discuss and share their perspectives on the field of geoethics. In Greece, the 16th International Congress of the Geological Society of Greece in 2022 marked the first instance of such an event where a sub-session regarding geoethics (Sub-session 1: Geoheritage, geoconservation, geoeducation, geoethics of the special session: Geological heritage for education and sustainable development) was included. Unfortunately, the number of presentations on geoethics was quite limited.

In the frame of the events organized by IAPG–Greece, a lecture titled Modern Geoethical Issues was presented by Dr. Gerassimos Papadopoulos (IAPG–Greece coordinator) on 6 March 2019, at the Department of Geology and Geoenvironment of the University of Athens. Additionally, on 8 November 2022, IAPG–Greece and SafeGreece organized an online event on the topic "Public communication of seismic hazard issues and geoethical dilemmas". It is likely that there will be similar conferences and workshops on geoethics in Greece in the future.

As far as research funding is concerned, at the moment there are no funding opportunities and eventually no research centers and groups that focus on geoethical issues, except for the research conducted through environmental education programs.

4. Discussion

The necessity of geoethics is to ensure ethical decision-making, sustainability, protection of geological heritage, transparency, accountability, and social responsibility in the use and management of natural resources and the environment.

Geoethical education is important for fostering a culture of ownership and environmental responsibility in society and for ensuring that future generations of geoscientists and decision-makers are equipped with the knowledge and skills needed to make informed and ethical decisions related to the Earth's resources and environment.

Effective teaching tools aimed at developing geoethics awareness are needed, enabling young people to become conscious and active citizens [24]. That is why the planning and implementation of geoenvironmental education programs for lower secondary education school (Gymnasium) students with the aim of empowering them in geocultural values is essential for the Greek school reality [25–27]. In contrast to the prevailing situation in schools and curricula [26,27], there are positive examples of planning and implementation of educational activities organized for elementary and high school students, mainly in the context of "Educational Programs of School Activities" and "Skills workshops" [28]. Such activities contribute to the awareness and empowerment of students to the values of geoethics [25,29,30] and consequently to the development of appropriate behavior regarding the interaction of human activities with the Earth system [10,25,28,29,31].

According to the literature review and our opinion, the following activities, that were implemented recently, could be used as a guideline. The program "Geoethics in Indoor Learning Environments: The Storm Tossed Sea Rocks in Pelion" is an environmental synchronous online education program in which the scenario attempts to empower students with knowledge, attitudes, and values that promote geodiversity and geocultural heritage [25]. Students should develop an ethical code and sense of responsibility for the environment protection and preservation [32] of the coasts and especially the storm-tossed sea rocks. Through assignments, presentations, and public debates, they develop and spread geoethics attitudes and values of sustainability [25].

Additionally, a geoeducation program was undertaken by 2 teachers and 45 students at an experimental lower secondary education school who expressed the desire to participate as members of the respective Creativity and Innovation Group. It was designed with the aim of broadening students' ethical concerns concerning the recognition of geodiversity's intrinsic value, which essentially means that people do not have the right to reduce geodiversity and that students are expected to realize their personal values' framework, which may signal their transition to a higher stage of ethical thinking [26,33]. In the context of "Skills workshops", teachers may choose the workshop titled "Geological heritage" and give students of secondary education the opportunity to learn about geocultural heritage, become familiar with virtual tour routes, and enhance observation and critical skills.

Greece boasts a unique position on the European geological map with numerous significant geotopes, including paleontological remains, rare geomorphological formations, and thousands of caves. This is why Greece has been a member of the UNESCO European Network of Geoparks since 2015, with seven designated geoparks (Lesvos Petrified Forest, Psiloreitis Geopark in Crete, Chelmos–Vouraikos Geopark, Vikos–Aoos Geopark, Sitia Geopark in Crete, Grevena–Kozani Geopark, and Kephalonia–Ithaka Geopark) among the 94 geoparks in 28 European countries [1]. This rich geological heritage highlights the importance of considering Geoethics in outdoor learning environments in Greece.

The positive examples of designing and implementing environmental education programs and educational activities in geologically protected areas organized for primary and lower secondary education students can contribute to the need to educate students in geosciences [34–36]. The geoethics elements can be incorporated into almost any geoscience fieldtrip [37]. Educational fieldtrips can enhance students' interest and awareness in various geoethics topics. They can be prepared accordingly to the geoethics dilemmas existing in the schools' area and aligned with a specific curriculum. It also links the instinct to learn with the need to teach and to learn the conceptual contents without the need to memorize them without understanding. By integrating the development of knowledge and competencies and by increasing the motivation to learn, field trips can be a powerful educational strategy to teach geoethics in higher education. More relevantly, the domain of fieldtrips has the potential to achieve GOAL project aims and an awareness geoethics learning [37]. A large body of research carried out over the last three decades shows the positive learning outcomes that field studies have had in education and within the positive and environmental cognitive fields [38–41]. To help and motivate teachers to plan and implement such activities successfully, it would be useful to organize training courses and prepare educational materials.

Geoethical education is not implemented in geoscience curricula; however, it is important to disperse knowledge and provide guidelines for its application in higher education [42] because such education will help aspiring future geoscientists and young people to become more sensitive to their natural environment, geohazards, energy, and conservation of natural resources, etc. [43]. Furthermore, students will recognize their personal, communal, and professional responsibilities to society and the planet, and for all these reasons, it is obviously unethical not to provide the necessary geoethical knowledge in schools and universities [44]. It should be noted that many university geoscience departments lack in incorporating considerations of ethics or geoethics into their strategies for student development, curriculum, or research efforts [45].

In conclusion, as abovementioned, geoethics should be the primary focus of all geoscience courses, provided that it is combined with the development of critical thinking skills and effective educational practices.

5. Conclusions

The aim of this review has been based on the system in force to demonstrate the current situation regarding the promotion and utilization of geoethics and its values in the Greek educational reality. The review outcomes showed that there is a significant lack in the implementation of geoethics in school and university curricula. That calls for geoethical awareness and the easiest way to achieve this goal is through the inclusion of geoeducation at all levels of education and through the promotion of geoethical values among Greek society since there are many challenges that still need to be addressed in Greece. These include a lack of enforcement of environmental laws and regulations, insufficient funding for environmental protection, and a lack of public awareness and participation in environmental decision-making. In conclusion, the advancement of geoethical thinking in Greece is crucial for addressing these challenges and ensuring sustainable development for present and future generations.

Author Contributions: Conceptualization, S.M., E.G. and H.D.; methodology, S.M. and E.G.; investigation, S.M. and E.G.; resources, S.M., E.G. and H.D.; writing—original draft preparation, S.M. and E.G.; writing—review and editing, S.M., E.G. and H.D.; supervision, H.D. All authors have read and agreed to the published version of the manuscript.

Funding: This research received no external funding.

Data Availability Statement: The data presented in this study are available on request from the corresponding author.

Acknowledgments: The authors gratefully thank the journal editor and the three reviewers for their thorough consideration of this paper.

Conflicts of Interest: The authors declare no conflict of interest.

References

1. Zafeiropoulos, G.; Drinia, H.; Antonarakou, A.; Zouros, N. From Geoheritage to Geoeducation, Geoethics and Geotourism: A Critical Evaluation of the Greek Region. *Geosciences* **2021**, *11*, 381. [CrossRef]
2. DeMiguel, D.; Brilha, J.; Meléndez, G.; Azanza, B. Geoethics and geoheritage. In *Teaching Geoethics: Resources for Higher Education*; Vasconcelos, C., Schneider, S., Peppoloni, S., Eds.; U. Porto Edições: Porto, Portugal, 2020; pp. 57–72, ISBN 978-989-746-254-2.
3. Druguet, E.; Passchier, C.W.; Pennacchioni, G.; Carreras, J. Geoethical education: A critical issue for geoconservation. *Episodes* **2013**, *36*, 11–18. [CrossRef]
4. Brennan, A.; Lo, N. Environmental Ethics. In *the Stanford Encyclopedia of Philosophy*; Edward, N.Z., Ed.; Stanford University: Stanford, CA, USA, 2022. Available online: https://plato.stanford.edu/archives/sum2022/entries/ethics-environmental/ (accessed on 5 December 2022).
5. Bobrowsky, P.; Cronin, V.S.; Di Capua, G.; Kieffer, S.W.; Peppoloni, S. The emerging field of geoethics. In *Scientific Integrity and Ethics in the Geosciences*; Gundersen, L.C., Ed.; AGU Special Publication; Wiley: Hoboken, NJ, USA, 2017; pp. 175–212. [CrossRef]
6. International Declaration on Geoethics: AGID Working Group for Geoethics, Pribram, Czech Republic, October 2011. Available online: http://tierra.rediris.es/Geoethics_Planetary_Protection/AGID_Geoethics_International_Declaration.htm (accessed on 10 November 2022).
7. Di Capua, G.; Peppoloni, S.; Bobrowsky, P.T. The Cape Town Statement on Geoethics. *Ann. Geophys.* **2017**, *60*, 6. [CrossRef]
8. Bohle, M. Geoethical thinking and wicked socio-environmental systems. *Geophys. Res. Abstr.* **2018**, *20*, 37. [CrossRef]
9. Bohle, M. (Ed.) *Exploring Geoethics: Ethical Implications, Societal Contexts, and Professional Obligations of the Geosciences*; Palgrave Pivot: Cham, Switzerland, 2019.
10. Peppoloni, S.; Di Capua, G. The Meaning of Geoethics. In *Geoethics: Ethical Challenges and Case Studies in Earth Sciences*; Peppoloni, S., Wyss, M., Eds.; Elsevier: Amsterdam, The Netherlands, 2015; pp. 3–14, ISBN 9780127999357.
11. Mogk, D.W.; Geissman, J.W.; Bruckner, M.Z. Teaching Geoethics Across the Geoscience Curriculum: Why, when, what, how, and where? In *Scientific Integrity and Ethics: With Applications to the Geosciences*; Gundersen, L.C., Ed.; American Geophysical Union: Washington, DC, USA, 2018.
12. Nwankwoala, H.O. Geoethics As An Emerging Discipline: Perspectives, Ethical Challenges And Prospects. *Earth Sci. Malays.* **2019**, *3*, 1–8. [CrossRef]
13. Imbernon, R.A.L.; Castro, P.d.T.A.; Mansur, K.L. Geoethics in the Scenario of the Geological Society in Brazil. *Geosciences* **2021**, *11*, 462. [CrossRef]
14. Andrăşanu, A. Basic Concepts in Geoconservation. In *Mesozoic and Cenozoic Vertebrates and Paleoenvironments–Tributes to the Career of Dan Grigorescu*; Csiki, Z., Ed.; Ars Docendi: Bucharest, Romania, 2006; pp. 37–41, ISBN (10) 973-558-275-9. Available online: https://www.academia.edu/10715520/Basic_concepts_in_Geoconservation (accessed on 10 November 2022).
15. Makris, J. Geophysical investigation of the Hellenides, Hamburger Geophys. *Einzelschriften* **1977**, *34*, 124.
16. McKenzie, D. Some remarks on the development of sedimentary basins. *Earth Planet. Sci. Lett.* **1978**, *40*, 25–32. [CrossRef]
17. Papazachos, B.C.; Comninakis, P.E. Geotectonic significance of the deep seismic zones in the Aegean Area. *Thera Aegean World* **1978**, *1*, 121–129.
18. Pichon, X.L.; Angelier, J. The Hellenic Arc and Trench System: A key to the neotectonic evolution of the Eastern Mediterranean. *Tectonophysics* **1979**, *60*, 1–42. [CrossRef]

19. Georgousis, E.; Mosios, S.; Savelides, S.; Holokolos, M.-V.; Drinia, H. The New Curriculum on Environment and Education for Sustainable Development and its Potential for Shaping Attitudes and Behaviors towards the Geoenvironment and Cultivating Geoethical values. In Proceedings of the 16th International Congress of the Geological Society of Greece, Patras, Greece, 17–19 October 2022. Available online: https://gsg2022.gr/wp-content/uploads/2022/10/GSG2022_Book-_Of_Abstracts_V3 _compressed.pdf (accessed on 20 November 2022).
20. IAPG. National Sections. Available online: https://www.geoethics.org/sections (accessed on 12 November 2022).
21. IAPG. IAPGeoethics Newsletter 2019, 4. Available online: http://docs.wixstatic.com/ugd/5195a5_cdb634dd417049b1ac9929319 ad51c2c.pdf (accessed on 12 November 2022).
22. IAPG. Affiliations and Agreements. Available online: https://www.geoethics.org/affiliations-agreements (accessed on 12 November 2022).
23. Bohle, M.; Di Capua, G. Setting the Scene. In *Exploring Geoethics–Ethical Implications, Societal Contexts, and Professional Obligations of the Geosciences*; Bohle, M., Ed.; Palgrave Pivot: Cham, Switzerland, 2019; pp. 1–24. [CrossRef]
24. Vasconcelos, C.; Torres, J.; Vasconcelos, L.; Moutinho, S. Sustainable development and its connection to teaching geoethics. *Episodes* **2016**, *39*, 509–517. [CrossRef]
25. Savelides, S.; Georgousis, E.; Fasouraki, R.; Papadopoulou, G.; Drinia, H. "Storm Tossed Sea Rocks in Pelion" an environmental synchronous online education program. In Proceedings of the 13th Conference on Informatics in Education (13th CIE2021), Athens, Greece, 9–10 October 2021. Available online: http://events.di.ionio.gr/cie/images/documents21/CIE2021_OnLineProceedings/ CIE2021_Binder1.pdf (accessed on 12 November 2022).
26. Georgousis, E.; Savelidi, M.; Savelides, S.; Mosios, S.; Holokolos, M.-V.; Drinia, H. How Greek Students Perceive Concepts Related to Geoenvironment: A Semiotics Content Analysis. *Geosciences* **2022**, *12*, 172. [CrossRef]
27. Georgousis, E.; Savelidi, M.; Savelides, S.; Mosios, S.; Holokolos, M.-V.; Drinia, H. *The Inclusion of Geoethical Values in the Design of Educational Policy for the Next Decade: The Case of the Greek Educational System*; EGU General Assembly: Vienna, Austria, 2022. [CrossRef]
28. Georgousis, E.; Savelides, S.; Mosios, S.; Holokolos, M.-V.; Drinia, H. The Need for Geoethical Awareness: The Importance of Geoenvironmental Education in Geoheritage Understanding in the Case of Meteora Geomorphes, Greece. *Sustainability* **2021**, *13*, 6626. [CrossRef]
29. Georgousis, E.; Savelidi, M.; Savelides, S.; Holokolos, M.-V.; Drinia, H. Teaching Geoheritage Values: Implementation and Thematic Analysis Evaluation of a Synchronous Online Educational Approach. *Heritage* **2021**, *4*, 3523–3542. [CrossRef]
30. Savelidi, M.; Savelides, S.; Georgousis, E.; Papadopoulou, G.; Fasouraki, R.; Drinia, H. Microcontroller Systems in Education for Sustainable Development Service. A Qualitative Thematic Meta-Analysis. *Eur. J. Eng. Technol. Res.* **2022**, *CIE 2758*, 53–60. [CrossRef]
31. Peppoloni, S.; Di Capua, G. Geoethics as global ethics to face grand challenges for humanity. *Geol. Soc. Spec. Publ.* **2021**, *508*, 13–29. [CrossRef]
32. Fermeli, G.; Markopoulou-Diakantoni, A. Selecting Pedagogical Geotopes in Urban Environment. *Bull. Geol. Soc. Greece* **2004**, *36*, 649–658. Available online: https://ejournals.epublishing.ekt.gr/index.php/geosociety/article/view/16770 (accessed on 8 November 2022).
33. Mosios, S.; Georgousis, E.; Savelides, S.; Holokolos, M.-V.; Drinia, H. Geodiversity and Geoheritage: The geoethical dimensions of a Geoeducation Program in the context of an experimental Lower Secondary Education school club. In Proceedings of the 16th International Congress of the Geological Society of Greece, Patras, Greece, 17–19 October 2022. Available online: https://gsg2022. gr/wp-content/uploads/2022/10/GSG2022_Book-_Of_Abstracts_V3_compressed.pdf (accessed on 20 November 2022).
34. Zouros, N.; Valiakos, I. Geoparks management and assessment. *Bull. Geol. Soc. Greece* **2010**, *43*, 965–977. [CrossRef]
35. Fassoulas, C.; Zouros, N. Evaluating the influence of Greek Geoparks to the local communities. *Bull. Geol. Soc. Greece* **2010**, *43*, 896–906. [CrossRef]
36. Drinia, H.; Tsipra, T.; Panagiaris, G.; Patsoules, M.; Papantoniou, C.; Magganas, A. Geological Heritage of Syros Island, Cyclades Complex, Greece: An Assessment and Geotourism Perspectives. *Geosciences* **2021**, *11*, 138. [CrossRef]
37. Orion, N.; Ben-Shalom, R.; Ribeiro, T.; Vasconcelos, C. Geoethics in field-trips: A global geoethical perspective. In *Teaching Geoethics: Resources for Higher Education*; Vasconcelos, C., Schneider-Voß, S., Peppoloni, S., Eds.; Universidade do Porto: Porto, Portugal, 2020; pp. 111–133, ISBN 978-989-746-254-2. [CrossRef]
38. McKenzie, A.; White, R.T. Fieldwork in Geography and Long-Term Memory Structures. *Am. Educ. Res. J.* **1982**, *19*, 623–632. [CrossRef]
39. Orion, N. A Model for the Development and Implementation of Field Trips as an Integral Part of the Science Curriculum. *School Sci. Math.* **1993**, *93*, 325–331. [CrossRef]
40. Knapp, D. Memorable Experiences of a Science Field Trip. *School Sci. Math.* **2000**, *100*, 65–72. [CrossRef]
41. Knapp, D.; Barrie, E. Content Evaluation of an Environmental Science Field Trip. *J. Sci. Educ. Technol.* **2001**, *10*, 351–357. [CrossRef]
42. Drąsutė, V.; Corradi, S.; Peppoloni, S.; Di Capua, G. Geoethics and New Medias: Sharing Knowledge and Values. In *Virtual Conference Proceedings The Future of Education International Conference*; Filodiritto Editore: Bologna, Italy, 2020. Available online: https://conference.pixel-online.net/FOE/files/foe/ed0010/FP/6590-SCI4598-FP-FOE10.pdf (accessed on 20 November 2022).

43. Gogoi, K.; Deshmukh, B.; Mishra, M. Addressing Perception of Geoethics Through Geoscience Curriculum at IGNOU. *Educ. Quest Int. J. Educ. Appl. Soc. Sci.* **2017**, *8*, 503–508. Available online: https://ndpublisher.in/admin/issues/EQv8n3h.pdf (accessed on 15 November 2022). [CrossRef]
44. Promduangsri, P.; Crookall, D. Geoethics education: From theory to practice—A case study. In Proceedings of theEGU General Assembly 2020, Online, 4–8 May 2020. [CrossRef]
45. Cronin, V.S. *Geoethics as a Common Thread That Can Bind a Geoscience Department Together*; Special Publications; Geological Society: London, UK, 2020; Volume 508, pp. 55–65. [CrossRef]

 *geosciences*

Review

Worldwide Trends in Methods and Resources Promoting Geoconservation, Geotourism, and Geoheritage

Michael E. Quesada-Valverde and Adolfo Quesada-Román *

Laboratorio de Geografía Física, Escuela de Geografía, Universidad de Costa Rica, San Pedro 2060, Costa Rica
* Correspondence: adolfo.quesadaroman@ucr.ac.cr

Abstract: This study aims to provide a systematic analysis of the literature of methods and resources supporting geoconservation and geotourism worldwide, while identifying current and future trends in the field. This paper offers a comprehensive bibliometric analysis which comprises the period of 2011–2021 after an in-depth systematic literature review of 169 papers, using Web of Science. The volume of research on these topics is growing rapidly, especially in Italy, Poland, Brazil, Russia, and China; these constitute the most productive countries. The main identified geomorphological environments are sedimentary, volcanic, aeolian, coastal, fluvial, and karstic. We discovered that the main methods for evaluating geoconservation and geotourism are geomorphological mapping, the study of economic values for geotourism, field work as a research tool, geoheritage management, documentation, exploration, and inventories of geoheritage at a regional level. The main determined resources are UNESCO Geoparks, educational activities, digital tools, geomanagement, economic values, geoitineraries, and geoeducation programs. To our knowledge, this is the first study dealing with methods and resources publicizing geoconservation and geotourism, worldwide. Knowing about the most successful methods and resources for promoting geoconservation and geotourism can definitely be useful for future endeavors in countries where geoheritage studies are starting to be developed.

Keywords: geoheritage; geodiversity; geoconservation; geotourism; Web of Science; bibliometric analysis; co-citation analysis

Citation: Quesada-Valverde, M.E.;
Quesada-Román, A. Worldwide
Trends in Methods and Resources
Promoting Geoconservation,
Geotourism, and Geoheritage.
Geosciences **2023**, *13*, 39. https://
doi.org/10.3390/geosciences13020039

Academic Editors: Hara Drinia,
Karoly Nemeth and Jesus
Martinez-Frias

Received: 19 December 2022
Revised: 27 January 2023
Accepted: 28 January 2023
Published: 30 January 2023

1. Introduction

Over the last 30 years, growing scientific interest has emerged in geoconservation, geotourism, and geoparks [1,2]. In fact, it was at the International Symposium on the Conservation of Geological Heritage in 1991 that the term geoheritage was used for the first time, and numerous conceptualizations have been proposed and established from that moment on [3]. The concept relates to the preservation of the characteristics of the planet that are important for geosciences such as landforms, geological outcrops, and their main traits [4]. Hence, geoheritage constitutes the geological heritage of a site and a new paradigm for physical geography [5].

Geoheritage has been relegated to the background in international events such as the Earth Summit in Rio de Janeiro and Agenda 21, the Millennium Declaration. The same occurred in 2015 when the United Nations adopted the Sustainable Development Goals, yet geosciences were not included, leaving aside the importance of geodiversity for sustainable development. The World Tourism Organization (UNWTO) has followed the same path, with some resolutions that leave geotourism aside [1,6].

The panorama is different in international organizations such as the International Union for the Conservation of Nature (IUCN) and the United Nations Educational, Scientific, and Cultural Organization (UNESCO), which have proposed the topics of geoheritage/geodiversity in the program of their international forums. In 2014, a group of geoheritage specialists was established within the World Commission on Protected Areas.

In addition, UNESCO created the International Geoparks Program and other individual initiatives emerged such as the International Association for the Conservation of Geological Heritage (ProGEO) to promote the academic scientific development of geoconservation. Moreover, the International Association of Geomorphologists (AIG) presented a specific group of researchers working on geoheritage. Additionally, the International Union of Geological Sciences established the International Commission on Geoheritage [1].

The scientific interest in geoheritage, geoconservation, and geotourism studies has grown, due to the interest in its scientific, academic, historical, social, cultural, and aesthetic values [4]. Many of these values are exposed in the geodiversity worldwide. Geodiversity encompasses elements of abiotic nature such as geology, geomorphology, soils, and hydrology [7,8]. This concept has been widely applied in geoconservation and geoheritage contexts, in addition to the attributed values by society to aspects of the abiotic, natural environment, due to their historical importance. Moreover, geodiversity increases the quality of the relations between diverse processes and interrelations of the Earth system.

Geodiversity is considered the core of national geoheritage strategies, assessments, and geological conservation, which supports biodiversity and geosystem services such as geotourism [9–11]. As a global, regional, and local concept, geodiversity has contributed to the formation of new knowledge and new avenues of research and results [12]. Geosites are the representative elements of geodiversity and have been made known through geotourism and in initiatives such as the UNESCO Global Geoparks for a decade [13]. There is a logical succession from geology, science through geological heritage, and the identification of sites of geoheritage importance, to the determination of geosites or geopark establishment, then to geoconservation, which leads to geomanagement, geoeducation, and geotourism [14,15].

The interest in geoconservation has been increasing since the 1990s, and the IUCN has promoted initiatives to integrate geodiversity and geoheritage [16,17]. This is defined as the policies, approaches, and efforts aimed at geoconservation. It is a tradition that differs in time and space, since countries such as the United Kingdom and Australia have protected their geoheritage for 70 years, while in the rest of the countries it is a recent and partial process [18]. The main objectives of geoconservation are protection and sustainable use of exceptional elements of geodiversity [19,20].

The growing interest in geoconservation has been demonstrated by the numerous site inventories that have been carried out in different countries [21,22]. The scientific community has focused its efforts on the formulation of various qualitative and quantitative methodologies to evaluate geosites for conservation [23–25]. Geosite inventories and assessment provide a basis for the protection and use of geoheritage, and are considered basic steps in geoconservation strategies and a management support tool [26–28].

Geotourism started as a scientific discipline that emerged in the line of geological engineering in geotourism, and that migrated towards the study of geoheritage. It is in a phase of exponential growth of research, with scientific productivity and the diversification of information covering research trends such as geosites, geoheritage, and geoparks [20]. Geotourism is a specific form of nature tourism focused on the discovery of geology and geomorphology revealed through scientific research, promoting the protection of geodiversity and the awareness of visitors [29,30]. It is one of the most recent concepts within tourism studies today [31]. These studies focus on identifying, describing, and evaluating geoheritage and its geotourism potential on different scales. A small number of researchers are interested in tourists, local communities, and sustainable development [32,33]. It has therefore been positioned as a strategic route for the promotion of sustainable tourism [34,35]. It is also an effective method of bringing geosciences to a wider audience [36] and a way to generate social, economic, and environmental benefits [20,37,38], especially in rural areas of developing countries with strong human pressures relating to natural resources [39,40].

The growth of geotourism is evident with the expansion of the UNESCO World Network of Geoparks initiative [31]. Geotourism activities are promoted, with geoparks being the basis for the development of geotourism and new geoparks proposals around the world [39–41]. Geoparks expose and contain the geological, geomorphological, hydrographic and edaphic values, and the geological diversity, historical structures and traditional culture which are resources for many tourist activities [42]. Geo-education in the geological and environmental sciences, and sustainable regional development are key factors for integration into geotourism studies [30,43,44].

There are many bibliometric analyses and systematic literature reviews dealing with geoheritage [13,30], geodiversity [5], geoconservation [45,46], geoparks [47–49], and geotourism [13,31,32,38,50]. We hypothesize that, despite the many reviews of geoheritage that have been carried out, a systematic literature review of the methods and resources promoting geoconservation and geotourism has not been performed. Considering the global scientific growth of geoheritage, geoconservation, and geotourism, we present a systematic literature review of the main world trends in the methods and resources that promote geoconservation and geotourism. In addition, we present some data on how the investigative literature has been produced around geoheritage, geoconservation, and geotourism. As far as we know, this is the first study dealing with methods and resources publicizing geoconservation and geotourism.

2. Materials and Methods

Systematic literature reviews have demonstrated successful identifying trends, prospective study fields, research gaps, and unstudied geographic areas in several knowledge fields. The reviews have also discussed detailed analyses related to geoheritage [51–53], geodiversity [5,8,12], geoconservation [46], and geotourism [13,38].

We used the Web of Science Advanced Research Query Builder, specifically the All Web of Science Citation Index Expanded. We used the following query expression:

$$(ALL = (\text{"geotourism" AND "geoconservation" AND "geoheritage"})) \text{ AND } (LA == (\text{"ENGLISH"})) \tag{1}$$

The query was configured to show the last 10 years of studies; that is, from 2011 to January 2022 (including all of 2021). Therefore, the bibliographic sources and a tab-delimited text file (.txt) were obtained. We only used study cases, not including review papers. With the help of an Excel spreadsheet, a bibliographic database was designed with authors, title, journal of publication, keywords, abstract, country of study, corresponding author, corresponding country, publication year, Web of Science category, and research area. The methods, resources, and environments were extracted by reading all the abstracts of every paper. We split all the determined environments and summed them independently for each of their environments. From these data, frequency analyses were performed to determine the bibliometrics. Subsequently, using the VOSviewer version 1.6.18 program, the tab-delimited text file (.txt) was loaded to generate bibliometric maps of country of correspondence, country of study, environment, methods, year of publication, resources, journal, and authors. The results served to corroborate both the manual bibliometric analysis in Excel and the automated one from VOSviewer. Our goal is to map the current state of the art with the different tools used in recent years to promote geoconservation and geotourism. Our primary intention is to uncover hidden patterns and provide support to stakeholders in order to bring new directions and visualize the interconnectedness of specific subject areas. Rather than providing a critical evaluation of each research paper, we aim to simply showcase the current progress in terms of methods and resources promoting geoconservation and geotourism.

3. Results and Discussion

3.1. Methods to Promote Geoconservation and Geotourism

Global trends around methods that promote geoconservation and geotourism demonstrated three broad generalities (Figure 1). First of all, there is still a research trend and tradition related to the inventory and evaluation of geoheritage in general. Moreover, those evaluations for the promotion of geotourism have been made through SWOT analysis, the application of surveys to visitors and the description of geoheritage. In addition, the issue of geoconservation through proposals and management plans of geoheritage have been supported by the cartographic discipline and the inventory and evaluation of geosites for the promotion of geotourism.

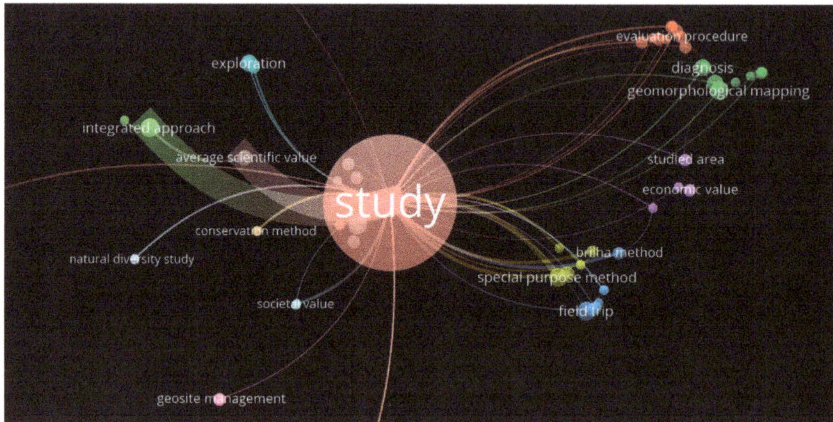

Figure 1. Global trends in methods for promoting geoconservation and geotourism.

The studies are divided into procedures for the evaluation of geotourism, geomorphological mapping, the study of economic values for geotourism, field work as a research tool, geoheritage management, documentation, exploration, and inventories of geoheritage at the regional level. In each of the commented-upon methods, the literature exposes proposals and new quantitative methods for geological interpretation. The geomorphological mapping support diagnoses that in turn allow the valuation of geotourism and later the geoconservation and monitoring plans in relation to the values of the geological heritage.

The methods and world resources that are the trend for promoting geoconservation and geotourism have not been worked on, and therefore the present academic exercise is novel, and it is difficult to develop a comparative analysis of the results with previous experiences at the level of bibliometric reviews of the literature. From the results, it is possible to identify the perspectives of some methods in authors such as Duarte et al. [38], who, using the Scopus database, indicate that the geoheritage inventory and evaluation processes are still scarce, but an excellent method is used in the first steps of promoting geoconservation and geotourism. The authors also highlighted studies that analyze and evaluate the experience of visitors in projects associated with geoconservation and geotourism. In this light, they become relevant to our results, as the SWOT analysis and the application of visitor surveys appear to be important methods for the evaluation of geoheritage and its relationship with visitors. Ólafsdóttir and Tverijonaite [32], using Scopus, Web of Science, and Science Direct indicated that, in research into geotourism management, new models and approaches to evaluating geosites and their forms of promotion regain importance. Farsani [49] highlighted the figure of the geopark as a resource and the need to inventory and evaluate geoheritage as a method, prior to these projects. Reynard [18] showed how the IAG (International Association of Geomorphologists) has created guidelines for mapping and making geoproducts, another of the results discussed in this research (Figure 1).

Numerous publications by the scientific community on geoheritage have indicated that UNESCO Geoparks are the main way to promote geotourism and geoconservation. Duarte et al. [38] identified numerous studies explaining the topic of geoparks, finding an important relationship with issues of socioeconomic development and local economy, which is discussed by Ólafsdóttir [31] in the case of the rural socioeconomic development associated with geotourism. Galvão [50] also found this association in the literature on the topic of geoparks, and Farsani [49] mentioned how geoparks are related to the issue of geoeducation and geoconservation of local communities. Moreover, social media has a huge impact on the decision-making of future visitors of different geotouristic destinations [54,55]. The integration of local/regional knowledge of geosites, the integration of local stakeholders, public and private sector decision-makers, and the proper use of technologies such as GIS and social media could enhance the promotion of geoheritage sites.

3.2. Resources in the Promotion of Geoconservation and Geotourism

The systematic literature review indicates that the global promotion of geosites is one of the most used in studies related to geoconservation and geotourism worldwide (Figure 2). Trends show that the main route for geotourism and geoconservation uses the concept of UNESCO Geoparks. These trends also relate to the educational activities in schools and universities that propose to integrate geosciences into the academic field to make people more aware of the importance of geoheritage and its protection. Geoparks can provide socio-economic benefits to local communities in line with sustainable development for both developed and developing countries [51].

Figure 2. Most used resources that expose the main trends in the promotion of geotourism and geoconservation.

Another trend we found has to do with the digital tools which become the key support for geomorphological and geological mapping to technologies in mobile devices that allow the promotion of educational and geotourism activities. Another major link in these conflicting trends contrasts with the geomanagement related to the proper administration of geosites that attract tourists around the world and that, without proper regulation, puts much of the world's geoheritage at risk. This also permeates the economic values and the development of geotourism in a sustainable way. Digital tools have improved geolocation, geointerpretation and geomodelling in the promotion of geoheritage and geoconservation worldwide [56].

The figure of modern geotourism supported by geoitineraries that make the most relevant geosites and their values available, allow geotourism to be promoted through field guides that expose geodiversity and, in turn, involve geoconservation at a local level, contributing to the protection of geosites. Geotourism projects are closely related to geoeducation programs through technologies such as information panels and web pages, which affect the popularization of geoheritage. At the academic level, the role of geotourism is very important in field research to obtain the most detailed information and to be able to carry out an adequate promotion of geotourism and geoconservation. Somma [57] has identified the fact that interactive, didactic earth-science activities can enhance geoconservation for a broader audience.

Some of the most important resources gaining prominence in the processes of promoting geotourism and geoconservation are technological and digital tools. These are related to processes such as geomorphological and geological mapping and technologies associated with mobile devices that allow projects to be promoted, as well as the different educational and geotourism activities. From this perspective, Williams [45] in interviews with global professional experts in geoconservation and geotourism in 2018, indicated that 25% of them used geographic information technologies to support decision-making and communication.

3.3. Country of Correspondence and Study

The research derived from an analysis of the countries with more than two appearances per frequency (Figure 3) of the main authors on topics related to geotourism, geoconservation, and geoheritage, are from countries such as Italy, Poland, Brazil, Russia, and China. We found an association of co-citation among the various countries, leaving out some countries such as Turkey, Greece, Romania, among others.

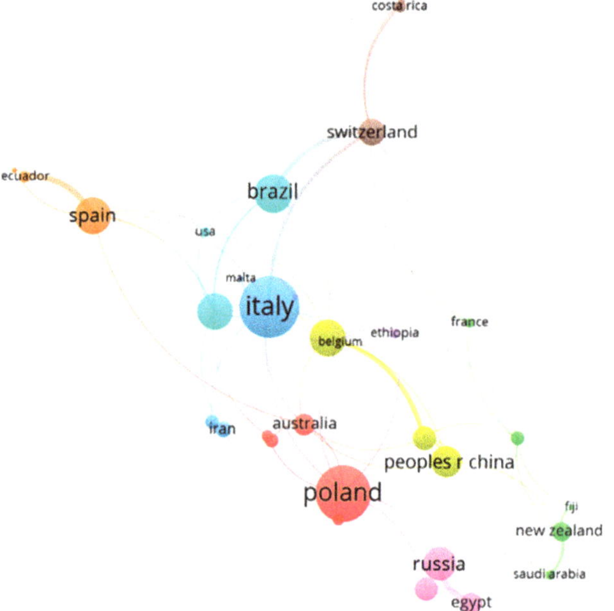

Figure 3. Network visualization in the VOSviewer program. The size of the circle and label are proportional to the frequency in number of studies per country of correspondence. The intensity of the lines shows the strength of the association between co-citations and the colors show the groups of co-citations among authors of each country.

As for the country of the study, the trend varies very little. We found that the countries with the most studies in their territory relating to geoheritage, geoconservation and geotourism are in descending order: Italy, Brazil, China, Poland, and Egypt (Figure 4). Russia is displaced by Egypt as the fifth country, mainly due to the great geological and archaeological interest that surrounds it. The complete data are located in the Supplementary Material. When it comes to the country of correspondence and the country of the study [38] of geotourism, we found high production in Brazil, Australia, and Italy. In terms of geotourism, most of the studies correspond to Europe, supporting the idea that Italy, followed by Poland, Serbia, the United Kingdom, and Slovakia are the countries of correspondence par excellence. The author also indicated that the case of the Asian continent is headed by Iran, followed by China. In Africa, Egypt leads and is followed by Morocco and Cameroon. In the case of Australasia, it is led by Australia followed by New Zealand. In America, Brazil and the United States stand out. Ruban [51] showed that geotourism has been researched mainly in the Middle East, Europe, South America, and East Asia, finding leading researcher communities in Italy, Brazil, China, and Poland. Herrera et al. [20], using the

Scopus database, also indicated that geotourism and geosites also coincide in countries such as Italy, Spain, China, Portugal, and Brazil. The future of geoconservation and geotourism will include the appearance of several countries with alluring places and geosites to be valued and promoted through evaluation, geoparks, and international recognition. Many developing countries and regions will find in geotourism an innovative, sustainable, and profitable way to generate income for their population.

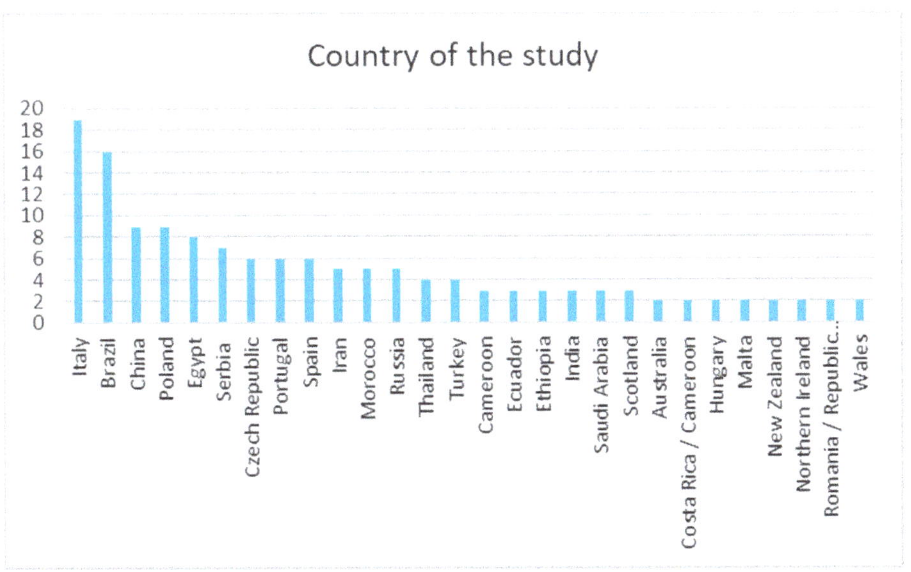

Figure 4. Graph showing the number of studies related to geoheritage, geoconservation and geotourism worldwide, with respect to the 169 articles analyzed.

3.4. Authors

From 661 authors, the authors who published the most on topics related to geoheritage, geoconservation and geotourism were Ruban Dmitry A, Kubalikova Lucie, Reynard Emmanuel, Coratza Paola, and Hose Thomas A (Figure 5), whereas the most cited are Hose Thomas A, Henriques María Helena, dos Reis Rui Pena, Kubalikova Lucie, and Reynard Emmanuel. In the case of the authors with more publications on topics related to geotourism, geoheritage and geoconservation, the results showed some variation, which may be due to the different search parameters and topics covered in each particularity. For example Duarte et al. [38], for authors associated with the topic of geotourism and territorial development, found in descending order those such as Hose, TA, Newsome, D, Dowling, R, Farsani, NT, Poiraud, A and Reynard, E Herrera et al. [20]. In the analysis of the literature regarding geoheritage and geosites they identified that among those associated with geotourism, Ruban D stands out, while in geoconservation it is Brilha J. Herrera et al. [20], indicating that the authors with most publications on geotourism, in descending order, are Ruban DA, Hose TA, Marković SB, Migoń P, and Farsani NT The complete data is located in the Supplementary Material. The number of authors working on geoheritage is growing rapidly, and these names may change in the near future, due to the specialization of researchers in different approaches of geoconservation and geotourism.

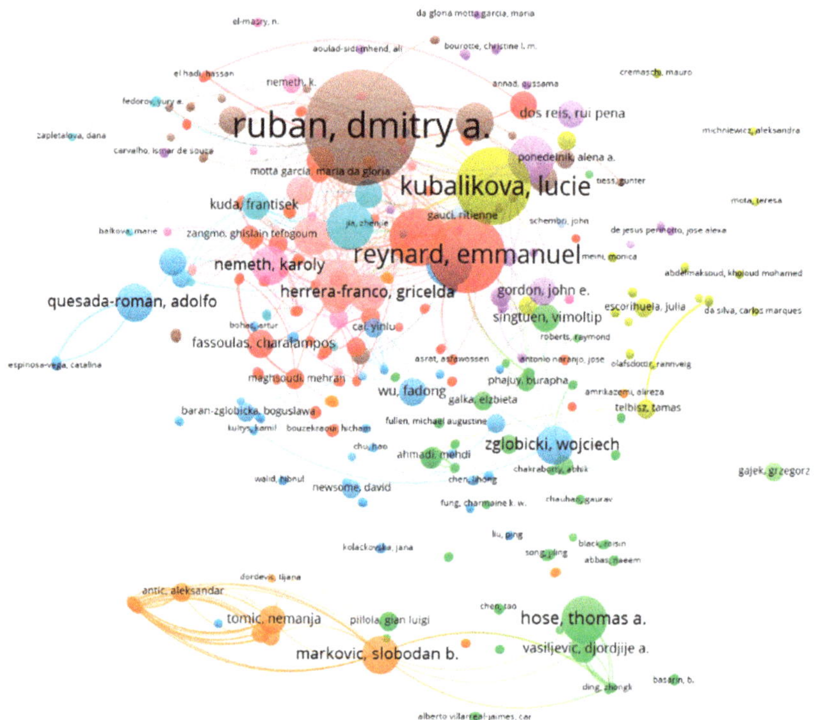

Figure 5. Most-productive authors dealing with methods and resources promoting geoconservation and geotourism.

3.5. Scientific Journals

The bibliometric analysis highlighted the fact that the first five most-used journals in publications related to geoheritage, geoconservation and geotourism issues are *Geoheritage*, *Sustainability, Quaternary International, Resources, and Proceedings of the Geologists Association*. The complete data are located in the Supplementary Material. The results obtained from the main scientific journals are comparable with those obtained by Herrera et al. [20] where *Geoheritage* journal tops the list, with the most publications and citations. The *Quaternary International* journal is second in terms of number of publications. Duarte et al. [38] found an association of geoheritage studies with the journal *Geoheritage*, followed by the journal *Vulcanology*, then the journals *Geotourism* and *Global Geotourism Perspectives*. Ruban et al. [53] in their article on the unique and climbing geology also found the *Geoheritage* journal to be the main journal for geoheritage issues. Ólafsdóttir and Tverijonaite [32] determined that the main journals related to the topics of geoheritage and geotourism are the *Geoheritage* journal, followed by the *Geojournal of Tourism and Geosites*. There is an increase in the number of journals publishing articles related to geoheritage worldwide. This opens up a plethora of options for publishing to authors worldwide and from different disciplines, enhancing the inclusion of different technologies, approaches, and methods to value, spread, promote, and protect geosites worldwide.

3.6. Geomorphological Environment

The study allowed us to identify the current and formative environments in which research on geoheritage, geoconservation and geotourism has been developed, particularly volcanic environments, valleys, and coasts (Figure 6). We used sedimentary environments because there were several environments in the same study. Sometimes, these sedimentary environments are linked to fluvial, glacial, gravitational, karstic, metamorphic, coastal, and even volcanic environments. Therefore, we find studies in sedimentary environments that largely involve deserts, valleys, and coasts around the world where fluvial, aeolian, and karstic geomorphology are well represented. The most frequently mentioned geographical features are sinkholes, karstic formations, deposits such as loess fields, mines, islands, and Jurassic coasts. Formations and environments of volcanic origin from both effusive and intrusive events are also well represented. Examples of these are plutonic morphologies or volcanoes.

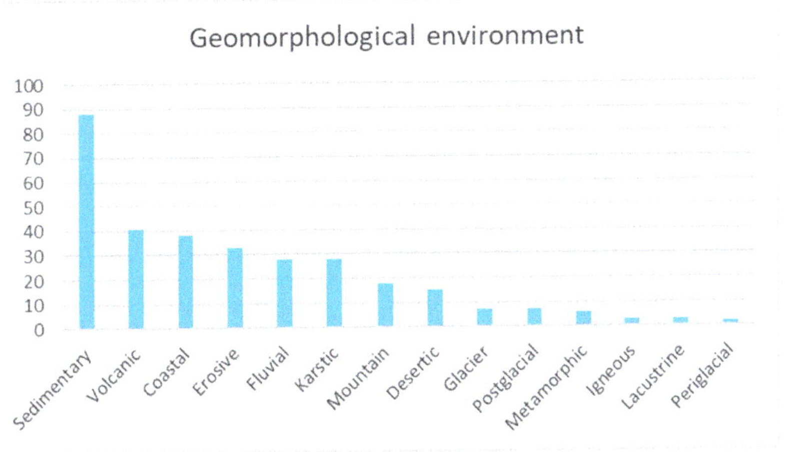

Figure 6. Main study environments of geoheritage, geoconservation and geotourism around the world.

The studies referring to the geomorphological environments analyzed by various scientific publications do not differ significantly from those found in our research. In the case of Ólafsdóttir and Tverijonaite [32], volcanic-formation environments, mountainous areas, followed by urban geoheritage, quarries, and mining areas are the main contexts where research on geotourism and geoconservation has focused. The authors added that in mountain areas, geomorphosite evaluations predominate. These areas require particular mapping techniques in order to generate good baseline information for geoheritage purposes [58]. In the case of caves, the main avenue of research is to examine tourist motivations, and geotourism management is analyzed in the volcanic and loess contexts. Quesada et al. [59] determined that in Central America volcanic environments, coastal areas, karstic environments, glacial, and fluvial environments are the most common in the region. The opportunity to enhance different geomorphic environments in order to study geosites, is immense. There are large regions of the world where geoheritage studies are still scarce, which possess incredible landforms, processes, and landscapes.

4. Conclusions

We have presented a systematic literature review of methods and resources supporting geoconservation and geotourism worldwide. We used 169 scientific articles based on the Web of Science between 2011 and 2021. We found that Italy, Poland, Brazil, Russia, and China are the most productive countries for these topics. We determined that the main methods used to promote geoheritage are procedures for the evaluation of geotourism, geomorphological mapping, the study of economic values for geotourism, field work as a research tool, geoheritage management, documentation, exploration, and inventories of geoheritage at the regional level. The main defined resources for socializing and supporting geosite visualization and their use are UNESCO Geoparks, educational activities, digital tools, geomanagement, economic values, geoitineraries, and geoeducation programs. Future studies should include less-studied regions, through collaborative studies with those countries and their researchers. A huge number of alluring geosites and potential geoparks are invisible to national and international audiences. The inclusion of innovative methods and resources to promote geoconservation and geotourism will definitely generate an economic production chain which will generate more and more diversified incomes on a local and regional scale in both developed and developing countries, with a sustainable approach such as is geoheritage.

Supplementary Materials: The following are available online at https://www.mdpi.com/article/10.3390/geosciences13020039/s1.

Author Contributions: Conceptualization, A.Q.-R.; methodology, A.Q.-R.; software, M.E.Q.-V.; validation, M.E.Q.-V.; formal analysis, M.E.Q.-V.; investigation, M.E.Q.-V.; resources, A.Q.-R.; data curation, M.E.Q.-V.; writing—original draft preparation, M.E.Q.-V. and A.Q.-R.; writing—review and editing, M.E.Q.-V. and A.Q.-R.; visualization, M.E.Q.-V. and A.Q.-R.; project administration, A.Q.-R.; funding acquisition, A.Q.-R. All authors have read and agreed to the published version of the manuscript.

Funding: Research Project "Geoturismo en América Central", C1211, Vicerrectoría de Investigación, Universidad de Costa Rica.

Data Availability Statement: Data will be available in the Supplementary Material and on request to the authors.

Acknowledgments: A special thanks to Manuel Peralta who assisted in some editing tasks of the paper, and Soll Kracher for her very useful corrections in English syntax, which greatly improved the final version of the manuscript.

Conflicts of Interest: The authors declare no conflict of interest.

References

1. Reynard, E.; Brilha, J. Geoheritage: A multidisciplinary and applied research topic. In *Geoheritage*; Elsevier: Amsterdam, The Netherlands, 2018; pp. 3–9.
2. Coratza, P.; Vandelli, V.; Fiorentini, L.; Paliaga, G.; Faccini, F. Bridging terrestrial and marine geoheritage: Assessing geosites in Portofino Natural Park (Italia). *Water* **2019**, *11*, 2112. [CrossRef]
3. Németh, B.; Németh, K.; Procter, J.N.; Farrelly, T. Geoheritage conservation: Systematic mapping study for conceptual synthesis. *Geoheritage* **2021**, *13*, 45. [CrossRef]
4. Shekhar, S.; Kumar, P.; Chauhan, G.; Thakkar, M.G. Conservation and sustainable development of geoheritage, geopark, and geotourism: A case study of Cenozoic successions of Western Kutch, India. *Geoheritage* **2019**, *11*, 1475–1488. [CrossRef]
5. Claudino-Sales, V. Geodiversity and geoheritage in the perspective of geography. *Bull. Geography. Phys. Geogr. Ser.* **2021**, *21*, 45–52. [CrossRef]
6. Gray, M.; Crofts, R. The potential role of the geosciences in contributing to the UN's Sustainable Development Goals. In *Parks Stewardship Forum*; University of California: San Diego, CA, USA, 2022; Volume 38.
7. Ibanez, J.J.; Brevik, E.C. Divergence in natural diversity studies: The need to standardize methods and goals. *Catena* **2019**, *182*, 104110. [CrossRef]
8. Dos Santos, F.M.; de La Corte Bacci, D.; Saad, A.R.; da Silva Ferreira, A.T. Geodiversity index weighted by multivariate statistical analysis. *Appl. Geomat.* **2020**, *12*, 361–370. [CrossRef]
9. Gray, M.; Gordon, J.E. Geodiversity and the '8Gs': A response to Brocx & Semeniuk. *Aust. J. Earth Sci.* **2020**, *67*, 437–444.
10. Kubalíková, L.; Drápela, E.; Kirchner, K.; Bajer, A.; Balková, M.; Kuda, F. Urban geotourism development and geoconservation: Is it possible to find a balance? *Environ. Sci. Policy* **2021**, *121*, 1–10. [CrossRef]
11. Elkaichi, A.; Errami, E.; Patel, N. Quantitative assessment of the geodiversity of M'Goun UNESCO Geopark, Central High Atlas (Morocco). *Arab. J. Geosci.* **2021**, *14*, 1–19. [CrossRef]
12. Gray, M. Geodiversity: A significant, multi-faceted and evolving, geoscientific paradigm rather than a redundant term. *Proc. Geol. Assoc.* **2021**, *132*, 605–619. [CrossRef]
13. Herrera-Franco, G.; Montalván-Burbano, N.; Carrión-Mero, P.; Apolo-Masache, B.; Jaya-Montalvo, M. Research trends in geotourism: A bibliometric analysis using the scopus database. *Geosciences* **2020**, *10*, 379. [CrossRef]
14. Brocx, M.; Semeniuk, V. The '8Gs'—A blueprint for Geoheritage, Geoconservation, Geo-education and Geotourism. *Aust. J. Earth Sci.* **2019**, *66*, 803–821. [CrossRef]

15. Quesada-Román, A.; Pérez-Umaña, D. Tropical paleoglacial geoheritage inventory for geotourism management of Chirripó National Park, Costa Rica. *Geoheritage* **2020**, *12*, 58. [CrossRef]

16. Pérez-Umaña, D.; Quesada-Román, A.; De Jesús-Rojas, J.C.; Zamorano-Orozco, J.J.; Dóniz-Páez, J.; Becerra-Ramírez, R. Comparative analysis of geomorphosites in volcanoes of Costa Rica, Mexico, and Spain. *Geoheritage* **2018**, *11*, 545–559. [CrossRef]

17. Mucivuna, V.C.; da Garcia, M.D.G.M.; Reynard, E. Criteria for assessing geological sites in National Parks: A study in the Itatiaia National Park, Brazil. *Geoheritage* **2022**, *14*, 1–19. [CrossRef]

18. Reynard, E.; Coratza, P. Scientific research on geomorphosites. A review of the activities of the IAG working group on geomorphosites over the last twelve years. *Geogr. Fis. Dinam. Quat.* **2013**, *36*, 159–168.

19. Moura, P.; da Glória Motta Garcia, M.; Brilha, J. Guidelines for Management of Geoheritage: An Approach in the Sertão Central, Brazilian Northeastern Semiarid. *Geoheritage* **2021**, *13*, 42. [CrossRef]

20. Herrera-Franco, G.; Carrión-Mero, P.; Montalván-Burbano, N.; Caicedo-Potosí, J.; Berrezueta, E. Geoheritage and Geosites: A Bibliometric Analysis and Literature Review. *Geosciences* **2022**, *12*, 169. [CrossRef]

21. Santos, D.S.; Reynard, E.; Mansur, K.L.; Seoane, J. The specificities of geomorphosites and their influence on assessment procedures: A methodological comparison. *Geoheritage* **2019**, *11*, 2045–2064. [CrossRef]

22. Pérez-Umaña, D.; Quesada-Román, A.; Tefogoum, G.Z. Geomorphological heritage inventory of Irazú volcano, Costa Rica. *Int. J. Geoheritage Park.* **2020**, *8*, 31–47. [CrossRef]

23. Mucivuna, V.C.; Garcia, M.D.G.M.; Reynard, E. Comparing quantitative methods on the evaluation of scientific value in geosites: Analysis from the Itatiaia National Park, Brazil. *Geomorphology* **2022**, *396*, 107988. [CrossRef]

24. Bollati, I.M.; Zerboni, A. The Po Plain Loess Basin (Northern Italy): Scientific Values, Threats, and Promotion Opportunities. *Geoheritage* **2021**, *13*, 74. [CrossRef]

25. Mucivuna, V.C.; Reynard, E.; Garcia, M.D.G.M. Geomorphosites assessment methods: Comparative analysis and typology. *Geoheritage* **2019**, *11*, 1799–1815. [CrossRef]

26. Carrión-Mero, P.; Borja-Bernal, C.; Herrera-Franco, G.; Morante-Carballo, F.; Jaya-Montalvo, M.; Maldonado-Zamora, A.; Berrezueta, E. Geosites and geotourism in the local development of communities of the Andes mountains. A case study. *Sustainability* **2021**, *13*, 4624. [CrossRef]

27. Santos, D.S.; Mansur, K.L.; Seoane, J.; Mucivuna, V.C.; Reynard, E. Methodological proposal for the inventory and assessment of geomorphosites: An integrated approach focused on territorial management and geoconservation. *Environ. Manag.* **2020**, *66*, 476–497. [CrossRef]

28. Tefogoum, G.Z.; Quesada-Román, A.; Pérez-Umaña, D. Geomorphosites inventory in the Eboga Volcano (Cameroon): Contribution for geotourism promotion. *Géomorphologie* **2020**, *26*, 19–33. [CrossRef]

29. Bussard, J.; Reynard, E. Heritage Value and Stakeholders' Perception of Four Geomorphological Landscapes in Southern Iceland. *Geoheritage* **2022**, *14*, 89. [CrossRef]

30. Santangelo, N.; Valente, E. Geoheritage and geotourism resources. *Resources* **2020**, *9*, 80. [CrossRef]

31. Ólafsdóttir, R. Geotourism. *Geosciences* **2019**, *9*, 48. [CrossRef]

32. Ólafsdóttir, R.; Tverijonaite, E. Geotourism: A systematic literature review. *Geosciences* **2018**, *8*, 234. [CrossRef]

33. Quesada-Román, A.; Pérez-Umaña, D. State of the art of geodiversity, geoconservation, and geotourism in Costa Rica. *Geosciences* **2020**, *10*, 211. [CrossRef]

34. Reyes, C.A.R.; Amorocho-Parra, R.; Villarreal-Jaimes, C.A.; Meza-Ortíz, J.A.; Castellanos-Alarcón, O.M.; Madero-Pinzon, H.D.; Carvajal-Díaz, J.D. Geotourism in regions with influence from the oil industry: A study case of the Middle Magdalena Valley Basin (Colombia). *Geoheritage* **2021**, *13*, 1–26. [CrossRef]

35. Kharbish, S.; Henaish, A.; Zamzam, S. Geodiversity and geotourism in Greater Cairo area, Egypt: Implications for geoheritage revival and sustainable development. *Arab. J. Geosci.* **2020**, *13*, 451. [CrossRef]

36. Zoboli, D.; Pillola, G.L. The Funtana Morimenta Ichnosite (Sardinia, Italy): A Potential Geotourist Attraction. *Geoheritage* **2021**, *13*, 30. [CrossRef]

37. Štrba, Ľ.; Kolačkovská, J.; Kudelas, D.; Kršák, B.; Sidor, C. Geoheritage and geotourism contribution to tourism development in protected areas of Slovakia—Theoretical considerations. *Sustainability* **2020**, *12*, 2979. [CrossRef]

38. Duarte, A.; Braga, V.; Marques, C.; Sá, A.A. Geotourism and territorial development: A systematic literature review and research agenda. *Geoheritage* **2020**, *12*, 65. [CrossRef]

39. Kubalíková, L. Assessing geotourism resources on a local level: A case study from Southern Moravia (Czech Republic). *Resources* **2019**, *8*, 150. [CrossRef]

40. Quesada-Román, A.; Tefogoum, G.Z.; Pérez-Umaña, D. Geomorphosites comparative analysis in Costa Rica and Cameroon volcanoes. *Geoheritage* **2020**, *12*, 90. [CrossRef]

41. Skibiński, J.; Kultys, K.; Baran-Zgłobicka, B.; Zgłobicki, W. Geoparks in SE Poland as areas of tourism development: Current state and future prospects. *Resources* **2021**, *10*, 113. [CrossRef]

42. Özgeriş, M.; Karahan, F. Use of geopark resource values for a sustainable tourism: A case study from Turkey (Cittaslow Uzundere). *Environ. Dev. Sustain.* **2021**, *23*, 4270–4284. [CrossRef]

43. Stoffelen, A.; Groote, P.; Meijles, E.; Weitkamp, G. Geoparks and territorial identity: A study of the spatial affinity of inhabitants with UNESCO Geopark De Hondsrug, The Netherlands. *Appl. Geogr.* **2019**, *106*, 1–10. [CrossRef]

44. Németh, B.; Németh, K.; Procter, J.N. Visitation rate analysis of geoheritage features from earth science education perspective using automated landform classification and crowdsourcing: A geoeducation capacity map of the auckland volcanic field, New Zealand. *Geosciences* **2021**, *11*, 480. [CrossRef]
45. Williams, M.A.; McHenry, M.T.; Boothroyd, A. Geoconservation and geotourism: Challenges and unifying themes. *Geoheritage* **2020**, *12*, 1–14. [CrossRef]
46. Quijas, S.; Romero-Duque, L.P.; Trilleras, J.M.; Conti, G.; Kolb, M.; Brignone, E.; Dellafiore, C. Linking biodiversity, ecosystem services, and beneficiaries of tropical dry forests of Latin America: Review and new perspectives. *Ecosyst. Serv.* **2019**, *36*, 100909. [CrossRef]
47. Stoffelen, A. Where is the community in geoparks? A systematic literature review and call for attention to the societal embedding of geoparks. *Area* **2020**, *52*, 97–104. [CrossRef]
48. Farsani, N.T.; Coelho, C.O.; Costa, C.M.; Amrikazemi, A. Geo-knowledge management and geoconservation via geoparks and geotourism. *Geoheritage* **2014**, *6*, 185–192. [CrossRef]
49. Galvão, A.; Mascarenhas, C.; Marques, C.; Braga, V. Geotourism as Promoter of Sustainability Development: A Systematic Review and Research Agenda. *Econ. Manag. Geotourism* **2022**, *2022*, 1–18.
50. Ruban, D.A.; Mikhailenko, A.V.; Yashalova, N.N.; Scherbina, A.V. Global geoparks: Opportunity for developing or "toy" for developed? *Int. J. Geoheritage Park.* **2022**, *11*, 54–63. [CrossRef]
51. Zakharovskyi, V.; Németh, K. Systematic Literature Review of the Natural Environment of the Coromandel Peninsula, New Zealand, from a Conservation Perspective. *Conservation* **2021**, *1*, 21. [CrossRef]
52. Németh, B.; Németh, K.; Procter, J.; Farrelly, F. Application of advanced data mining tools for better defining the geoheritage values of volcanic fields. In *European Geoscience Union General Assembly*; EGU: Vienna, Austria, 2018; pp. EGU2018-18652.
53. Ruban, D.A.; Ermolaev, V.A. Unique geology and climbing: A literature review. *Geosciences* **2020**, *10*, 259. [CrossRef]
54. Frey, M.L. Geotourism—Examining tools for sustainable development. *Geosciences* **2021**, *11*, 30. [CrossRef]
55. Salamzadeh, A.; Tajpour, M.; Hosseini, E.; Salamzadeh, Y. Geotourism and Destination Brand Selection: Does Social Media Matter? In *Economics and Management of Geotourism*; Springer: Berlin/Heidelberg, Germany, 2022; pp. 105–124.
56. Fassoulas, C.; Nikolakakis, E.; Staridas, S. Digital Tools to Serve Geotourism and Sustainable Development at Psiloritis UNESCO Global Geopark in COVID Times and Beyond. *Geosciences* **2022**, *12*, 78. [CrossRef]
57. Somma, R. The Inventory and Quantitative Assessment of Geodiversity as Strategic Tools for Promoting Sustainable Geoconservation and Geo-Education in the Peloritani Mountains (Italy). *Educ. Sci.* **2022**, *12*, 580. [CrossRef]
58. Campos, N.; Quesada-Román, A.; Granados-Bolaños, S. Mapping Mountain Landforms and Its Dynamics: Study Cases in Tropical Environments. *Appl. Sci.* **2022**, *12*, 10843. [CrossRef]
59. Quesada-Román, A.; Torres-Bernhard, L.; Ruiz-Álvarez, M.A.; Rodríguez-Maradiaga, M.; Velázquez-Espinoza, G.; Espinosa-Vega, C.; Toral, J.; Rodríguez-Bolaños, H. Geodiversity, Geoconservation, and Geotourism in Central America. *Land* **2022**, *11*, 48. [CrossRef]

geosciences

Article

Nisyros Aspiring UNESCO Global Geopark: Crucial Steps for Promoting the Volcanic Landscape's Unique Geodiversity

Paraskevi Nomikou [1,*], Dimitrios Panousis [1], Elisavet Nikoli [1], Varvara Antoniou [1], Dimitrios Emmanouloudis [2,3], Georgios Pehlivanides [4], Marios Agiomavritis [5], Panagiotis Nastos [1], Emma Cieslak-Jones [1] and Aris Batis [5]

[1] Department of Geology and Geoenvironment, School of Science, National and Kapodistrian University of Athens, Panepistimioupoli, 15784 Athens, Greece

[2] Department of Forest and Natural Environment Sciences, School of Geotechnical Sciences, International Hellenic University, 66100 Drama, Greece

[3] UNESCO Chair on Conservation and Ecotourism of Riparian and Deltaic Ecosystems, 66100 Drama, Greece

[4] Hands-On Studio, Research and Art Direction, Branding, UX/UI Design, Project Management, 54655 Thessaloniki, Greece

[5] Econtent Systems P.C, Software, Website and Mobile Application Development, 15784 Athens, Greece

* Correspondence: evinom@geol.uoa.gr

Abstract: Nisyros Geopark, an island geopark in the Southeastern Aegean Sea, Greece, is here presented as an official candidate for the UNESCO Global Geoparks designation, featuring outstanding geological, natural and cultural characteristics tightly connected to its volcanic origin. It covers a total area of 481 km^2 and includes Nisyros, an active volcano and the main island, the surrounding islets of Pachia, Strongyli, Pergousa, Kandeliousa and the marine region among them. It features 24 geosites and a network of well-established walking trails. Furthermore, there are two internationally designated Natura 2000 areas covering its entire surface and also exceptional archaeological and cultural sites, including fortresses, remnants of ancient habituations and numerous churches and monasteries. It is the only area in the broader region of the Eastern Mediterranean that hosts all these features within such a restricted area. The initial efforts of the management body of Nisyros Geopark and its scientific team to promote its unique geodiversity included the complete design, construction and launch of the official website, the mobile application "Nisyros Volcano App", a modern informative leaflet regarding the region of the hydrothermal craters (Lakki), a Geopark guidebook and a series of panels and signs for the geosites.

Keywords: Nisyros; geopark; island; volcano; UNESCO; geodiversity; digital; traditional; website; application

Citation: Nomikou, P.; Panousis, D.; Nikoli, E.; Antoniou, V.; Emmanouloudis, D.; Pehlivanides, G.; Agiomavritis, M.; Nastos, P.; Cieslak-Jones, E.; Batis, A. Nisyros Aspiring UNESCO Global Geopark: Crucial Steps for Promoting the Volcanic Landscape's Unique Geodiversity. *Geosciences* **2023**, *13*, 70. https://doi.org/10.3390/geosciences13030070

Academic Editors: Karoly Nemeth and Jesus Martinez-Frias

Received: 27 January 2023
Revised: 21 February 2023
Accepted: 27 February 2023
Published: 1 March 2023

1. Introduction

Throughout the history of Earth, complex geological processes gradually shaped every aspect of the environment of a living, dynamic world that became home to countless life forms and, eventually, humans. As the evolution of geosciences worldwide uncovers the scientific importance of the different stages of the planet's evolution, it has become apparent that geology and the landscape provided the seeds for humanity to flourish and develop. Thus, the need for protection, conservation and promotion of the many geological and geomorphological monuments that are scattered all across the globe has slowly initiated new forms of geology-related education, tourism and other environmentally responsible activities seeking to uncover and preserve the roots of modern-day societies. Geotourism, a form of alternative, sustainable and responsible tourism that focuses on discovering a region's geological and geomorphological heritage, is recently becoming more and more sought after, not only by earth science enthusiasts, but also by the broader public [1,2]. It offers opportunities for exploring the history of the Earth through geosites, sites of unique geological and geomorphological value, and geotrails that promote an area's geodiversity, thus attracting the interest of a significant number of tourists in an area [3]. At the same time, it is helping a new type of environmentally aware and geo-culturally respectful tourist to emerge, through proper educational activities based on simple and understandable scientific information dissemination methods, both traditional and digital [4–7]. Thus, geotourism, along with all the educational value it has to offer, has become a powerful tool for the sustainable development of local and regional communities [1]. The best way to enhance and further diversify the geological value of an area, with geotourism and education, while incorporating local communities and all the other aspects of local societies in a sustainable development manner, is through the initiative of UNESCO Global Geoparks (UGGps) [8–10].

According to [11], 'UNESCO Global Geoparks are single, unified geographical areas where sites and landscapes of international geological significance are managed with a holistic concept of protection, education and sustainable development'. Having the area's geological heritage at their core, they form uniquely diverse societies where topics regarding the sustainable use of the Earth's resources, the mitigation of the effects of climate change and the reduction of the impact of natural disasters are enhanced and connected to all other aspects of the area's cultural and natural heritage [12,13]. Hence, they strengthen shared cultural bonds among their communities, create resilience regarding possible geological threats that may affect those communities, and enhance the quality and quantity of economic activities, education and geotourism, all while the geoheritage of the area is protected. UGGps were first established in 2015 [14], although the initiative for the promotion and preservation of geological features through a holistic approach involving biological, cultural, tangible and intangible heritage, as well as the local communities, had already been around since the mid-90s [2,5,15]. They are established as a bottom-up process that involves all relevant local and regional stakeholders and authorities, thus demanding a strong and lasting commitment of the local communities, as well as local multiple partnerships with long-term support by all parts of the local society, both public and political. Thus, UGGps, along with the other two UNESCO site designations (Biosphere Reserves and World Heritage Sites), provide a complete visibility and conservation schema for the world's most outstanding and endangered cultural, biological and geological diversity [11]. Today, there are 177 officially recognized UGGps in 46 countries spread across all continents (as of December 2022), each one been connected to regional and continental networks, which are the European Geoparks Network—EGN, the Asia-Pacific Geoparks Network—APGN, the Latin America and the Caribbean Geoparks Network—LACGN and the newest African UNESCO Global Geoparks Network (AUGGN) [16].

For an area to become a UGGp, a series of crucial steps needs to be taken in order to prepare the application dossier and then function as a de facto UGGp for at least one year, before eventually undergoing its evaluation process. During this period, the area is referred to as an Aspiring UNESCO Global Geopark (aUGGp). It must adhere to the four essential values that a UGGp incorporates, which include geological heritage of outstanding international value, as it is implied by thorough research and scientific publications, a strong management body with legal existence under a national legislation, enhanced visibility through sustainable geotourism and geoeducation, as well as strong networking capabilities [2,11]. Gradually, an aspiring Geopark must incorporate within its local communities, local stakeholders, political entities and facilities a management plan focusing on the ten most important topics within a UGGp: the sustainable development of all its inhabitants, as well as the wise exploitation of the area's natural resources; the promotion of awareness of possible geological hazards and climate change, their impact and their mitigation; the encouragement of new and innovative scientific research and public dissemination of its results on topics regarding the geological heritage of the Geopark, geoconservation, as well as its cultural history, local and indigenous knowledge and the empowerment of women through the development of their cooperatives [2,5,9,11,12].

Of particular interest is the aspect of visibility of a UGGp, which is mainly achieved not only by its network, but also through its geotourism initiatives that stem from a strong educational plan primarily focusing on sustainability. The development and promotion of educational activities for all kinds of audiences is a substantial factor that can determine a Geopark's positive impact both on local societies and tourists, thus offering a better understanding of the importance of the geological, biological, cultural, tangible and intangible heritage of the area. This understanding will then provide the roots for a further appreciation and respect for the Geopark's natural and human environment, which will eventually guarantee the conservation of geoheritage [8]. It will also play a decisive role in the UGGp's contribution to the fulfillment of the 17 United Nations Sustainable Development Goals (SDGs), that comprise the Agenda 2030 for Sustainable Development adopted by the United Nations [17,18]. Thus, sustainable educational activities within Geoparks should always be supported by local schools, academic institutions and universities, and include training activities not only for students, locals and tourists, but also for teachers and special tour guides, in order to establish responsible promoters of the area's geo-cultural and natural environment. The involvement of the geoscientific community, that comprises the core of the activities provided by a UGGp, is crucial for the achievement of these goals [19].

The promotion and communication of scientific information through innovative educational initiatives has started shifting from traditional means, such as books, leaflets and classic informative panels, towards more digital approaches that further improve the educational and geotouristic value of a UGGp. During the past few years, many Geoparks have adopted initiatives regarding the implementation of digital technologies (GIS mapping, websites and mobile applications, UAVs) in many aspects of their provided services, particularly for geoeducational purposes and to enhance the management of their geoheritage, thus providing innovative and diversified geotourism opportunities [13,20]. Those include examples of digital 3D virtual geo-routes [21], virtual field trips and virtual reality environments [22,23]. In this way, accessibility to almost every aspect of a Geopark is gradually made far more possible in a world where modern everyday life is primarily carved by technology, thus opening a window to some of nature's most beautiful and unique places worldwide that the UGGps represent, and communicating their social and environmental goals [24]. However, as more and more people, especially younger audiences, refer to technological developments like computers, smartphones and tablets both for recreational and educational activities, the need to spread the concept and goals of UGGps more easily to the wider public has gradually led to the development of geotouristic mobile applications [13,25]. Such applications would have a greater educational impact on younger audiences and inspire them to discover what a UGGp has to offer.

The aim of this paper is to present the initial efforts of the management body of Nisyros aUGGp, Greece, toward the goal of achieving the UNESCO Global Geopark status, through a series of both traditional and digital developments focusing on the introduction of the Geopark's extraordinary features to the global geotourism and geoeducational scene. These included the design, construction and installation of informative panels and signs on the designated geosites, as well as signs indicating possible hazardous areas at the most important geological attractions of the main island of Nisyros. A new, informative leaflet regarding the hydrothermal area of Lakki was also produced, with a unique and innovative foldable, poster-like design, as well as a comprehensive, plain language Nisyros Geopark guidebook, describing all aspects of the Nisyros Geopark. Finally, two digital products were also highlighted, the official Nisyros Geopark website and the first free informative mobile application for the touristic exploration and geo-interpretation of the area of Lakki within the volcano's caldera.

2. Nisyros Aspiring UNESCO Global Geopark

The Nisyros aspiring UNESCO Global Geopark (Nisyros Geopark) is an island complex located between Kos and Tilos islands (Dodecanese prefecture) in the southeastern part of the Aegean Sea in Greece. It includes Nisyros island, the youngest—and still active—volcano of the South Aegean Volcanic Arc, together with the surrounding volcanic islets of

Strongyli, Pachia, Pergousa, the non-volcanic islet of Kandeliousa as well as the marine area between them (Figure 1). It occupies a total land and sea area of 481 km².

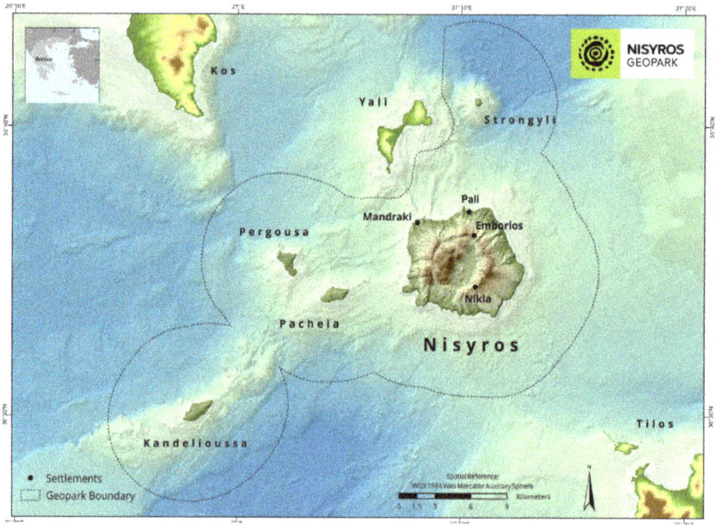

Figure 1. Geotouristic map of Nisyros aspiring UNESCO Global Geopark.

Nisyros Geopark is an active volcanic field located in the South-Eastern Aegean Sea, and is part of the South Aegean Volcanic Arc (also known as the Hellenic Volcanic Arc), a chain of active Quaternary volcanoes that stretches from mainland Greece (Sousaki, Methana) to the southeastern corner of the Aegean (Milos, Santorini Volcanic Field and Kos-Yali—Nisyros Volcanic Field) [26,27]. Being part of the Dodecanese group in the eastern Aegean Sea, and located between the islands of Kos and Tilos, it lays within a prehistoric volcanic field that generated the largest volcanic eruption in the Eastern Mediterranean Sea (Kos Plateau Tuff), 161,00 years ago [28–32]. With only 160,000 years of volcanic activity, following the tremendous eruptive cycles of Kos Plateau Tuff, Nisyros is considered Greece's youngest active volcanic edifice [33].

From a geodynamic perspective, Nisyros is located within one of the most tectonically complex regions of the eastern Mediterranean, featuring important geodynamic events that eventually shaped the area and gave birth to the volcano [34–36]. The dominant geodynamic events that characterize the region are the ongoing northward subduction of the African plate beneath the Eurasian plate, along with an extensional geodynamic regime that has been established within the entire Aegean Sea region since the Late Miocene, leading to the thinning of the continental crust accompanied by the development of a series of fault systems that accommodate deformation (Figures 2 and 3). As a result of the above processes, the entire region of the eastern Aegean Sea has been subjected to the development of graben—horst systems, bounded by faults. In the southeastern Aegean, the submarine area between Kos and Tilos islands constitutes a regional graben, through which hot molten rock produced due to the subduction rises to upper lithospheric levels and finally to the surface, thus creating volcanism [37]. Over the past 160,000 years, volcanic products have filled the regional graben, giving birth to the Nisyros volcanic complex.

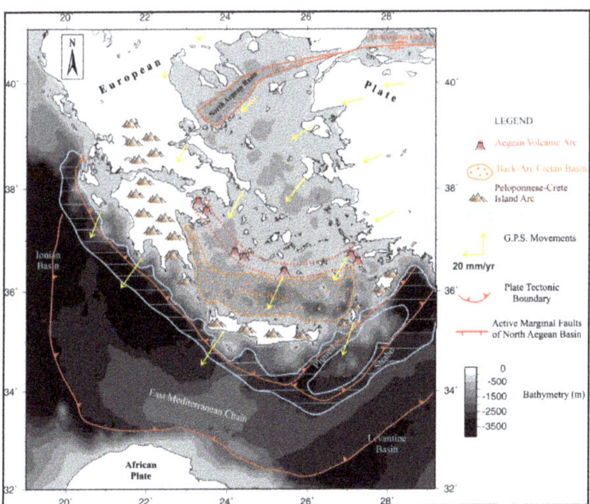

Figure 2. Geotectonic setting of the Hellenic Volcanic Arc (modified from [27]).

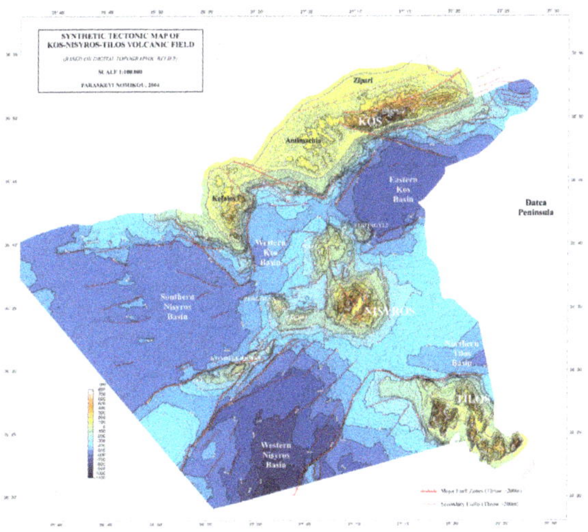

Figure 3. Synthetic tectonic map of the Kos-Nisyros-Tilos volcanic field [31].

Systematic research has been taking place in the region for many years, yielding important results regarding the volcano both from onshore and offshore observations, and in many fields of geosciences, including volcanology, geochemistry, geophysics, oceanography and seismology [38–48].

Thanks to that research, it is now known that Nisyros and the surrounding islets of the Geopark are entirely composed of volcanic rocks, apart from Kondelioussa, which consists of Mesozoic limestones and forms a small-scale regional horst between Kos and Tilos [49]. The volcanic evolution of Nisyros is divided into five eruptive cycles, each one of which produced volcanic products such as lava flows of variable composition, pyroclastic deposits such as tuff and pumice successions, domes, the caldera, hydrothermal craters and faults. They consist of the 'Submarine Volcanic Base' cycle, the 'Early Shield Volcanic cycles', the 'Composite Stratovolcano' cycle, the 'Caldera Forming' eruptive cycle and the 'Post-Caldera' eruptive cycle, leading to the currently active hydrothermal system [33].

As a result of its volcanic history, the region hosts spectacular onshore and offshore geological formations that clearly depict the stages of the volcano's evolution, as well as breathtaking geomorphological landmarks such as the iconic caldera, domes, offshore basins and canyons. It is also home to a spectacular active hydrothermal system that has endowed Nisyros with a number of hot springs along its coastline, hot steam emissions, fumaroles and of course the hydrothermal craters at Lakki. Some of the designated geosites of international value of the Geopark are Stefanos Crater, one of the largest and most well-preserved hydrothermal craters in the world featuring degassing fumaroles (Figures 4 and 5a), the Parletia lava neck at Nikia (Figure 5b), the scoria cones at Avlaki, the volcanic cone of Strongyli to the north of Nisyros and the voluminous pumice deposits at Cape Katsouni.

Figure 4. Stefanos is the largest hydrothermal crater of Nisyros and one of the largest of its kind in the world. Its dimensions are 260 m × 350 m, with a depth of 27 m and an estimated age of more than 600 years.

Apart from research, the region's rich geodiversity attracts the interest of many students and early career scientists, thus giving them the chance to work on an open geological laboratory [50–52]. It is an ideal place for excursions of scientists, students as well as the public in general, because it is the most characteristic and interesting volcanic edifice in Greece and hosts well-preserved geological and geomorphological formations, which not only produce scientific results, but also disseminate information in an easy-to-understand manner [7].

The flora and fauna of Nisyros Geopark present great interest both because of the volcanic nature of the region and also due to its geographic position on the immigration routes of the Asian species towards southern Europe and vice versa. There are two areas that belong to the European network Natura 2000 (areas GR4210032 and GR4210007) and three areas identified as Wildlife Refuges. The documented presence of 450 species of flora and 85 species of avifauna highlight Nisyros as a place worthy of special protection and study. This also includes seven species of reptiles, like Kourkoutavlos (traditional name of the lizard Agame stelio)—huge black and brown lizards hiding under volcanic rocks— as well as the presence of Monachus monachus—Mediterranean seals on its shores. The dense bushy vegetation includes the thorny burnet (Sacropoterium spinosum), the lavender (Lavandula stoechas), the hoary rock rose and Gallipoli rose (Cistus criticus and Cistus salvifolius), the thyme (Thimus capitatus), the laurel (Daphne gnidioides) and the Nisyros bellflower (Campanula nisyria), which is a unique endemic plant species.

(a)   (b)

Figure 5. Nisyros is a prime example of an open volcanological laboratory, featuring a number of geological features of great educational and scientific interest: (**a**) Active fumarole within the bottom of Stefanos Crater, with characteristic sulfur deposits (crystals) formed around its vent. Approximate diameter 3–4 cm; (**b**) The Parletia Mound, one of the best examples of a rhyolitic lava neck formed between the two successive eruptions of the Caldera Eruptive Cycle of Nisyros [33]. It is exposed on the eastern caldera wall at an approximate altitude of 300 m a.s.l.

Nisyros Geopark goes beyond presenting a unique volcanic history and the beauty of its nature. It is found at the crossroads between western and oriental civilizations, and the fascinating cultural and historical heritage of the island is courtesy of its geographical position and its circular and mountainous volcanic morphology, making it a rather inaccessible island, where values are more easily preserved [53]. It also exposes the splendor of art and civilization, expressed through prehistoric and historic locations and monuments. Prehistoric witnesses are present from Early Neolithic times, in the form of relicts from the Cycladic and Minoan civilizations, while historical records start from the Hellenistic epoch and continue uninterrupted through Roman, Byzantine, Venetian and Ottoman eras, into modern times [54–56]. Each of these periods has enriched the region with ancient caverns and spas, magnificent castles like Palaiokastro, the most well-known fortress of Nisyros (Figure 6), churches and monasteries with hagiographic frescoes (like the Monastery of Panagia Spiliani). In this respect, and considering its small size, Nisyros aUGGp differs from all Dodecanese and Aegean islands and stands out as a strong candidate for UNESCO's Global Geoparks project.

Figure 6. Palaiokastro, the acropolis of the ancient Greek town of Nisyros, stands today as the best-preserved fortress of the Classical Period. Built by huge volcanic blocks, it is a representative example of the everlasting connection between this volcanic landscape and its inhabitants throughout millennia.

In charge of the operation of Nisyros Geopark is the Municipality of Nisyros in cooperation with the Municipal Public Benefit Enterprise of Nisyros (DIKEN). All decisions regarding the financial management and daily operation of the Geopark are also made by the Municipality of Nisyros in cooperation with DIKEN. As a result, local participation in decision-making is guaranteed. Moreover, some of the employees of DIKEN, who may be considered as local decision-makers, participate in the administrative scheme of the Geopark's management body; hence, having an important role in the decision-making process. The Geopark's management body president is the Mayor of Nisyros.

3. Materials and Methods

For the communication, promotion and networking needs of the Geopark concerning both the public and institutional bodies, an integrated communication strategy was formulated. The practices followed were: (a) the initial design of the Geopark's corporate identity; (b) the design and development of the institution's website; (c) the design of a series of printed information materials and specially designed signage (environmental graphics) for the needs of the Geopark; (d) the design and development of a multimedia application specially designed for smartphones for both Android and iOS environments. The main objective of the communication strategy followed was for the distinctive features of the above actions to reflect the recognition of the institution as a geo-environmental monument of unique and global value in a modern way. For this purpose, an extensive study and research was conducted in the first phase regarding the design of corporate identities in related institutions and Geoparks, as well as the specific characteristics of architectural information and design trends utilized by their respective websites. Based on the research conducted in this first stage, the Geopark's corporate identity was designed and developed, as well as a series of specially designed icons aimed at conceptually representing the main activities and characteristics of the Geopark.

In a second stage, after extensive research, the UX/UI (User Experience/User Interface) design solutions for the needs of the website were formulated through a human-centered design approach, focused on responsive design and covering both its use through desktop computer systems, as well as through smartphones and tablets. For the needs of the website, with specialized functionality regarding publicity and communication, a content management platform system (CMS) was formulated, which supports the posting and publication of relevant information material, announcements and other information. The website, as a communication and reporting hub for the Geopark's activities, was designed with the following information architecture: Initial Page (Landing Page), Nisyros Page, Nisyros Geopark Page, Geosites Page and Interactive Map Page (Figure 7—see 'Results' for a complete description).

For the needs of the dissemination of information on the Geopark's activities, a leaflet with a modern design and easy-to-understand content was produced, aiming at providing sufficient information and comfort of use by visitors and locals alike. The brochure, in the form of a foldable poster, provides rich informational material backed by scientific data on the geology of the Geopark, as well as historical facts concerning Nisyros, enriched with a specially designed topographical map of the Geopark area illustrating the paths and points of interest, featuring diagrams and photographic material. The leaflet provides a geological explanation of the points of interest (visitor observation points), useful information on potential hazards, appropriate clothing by visitors and other safety tips, as well as contact information by using a QR hyperlink directing to the geopark's website.

In combination with the informational leaflet, a digital guide was designed and implemented for Nisyros Geopark in order to improve the services of the institution. The user of the application can discover the morphological characteristics of the area and learn about the history of the development of the Geopark through an interdisciplinary framework of up-to-date material.

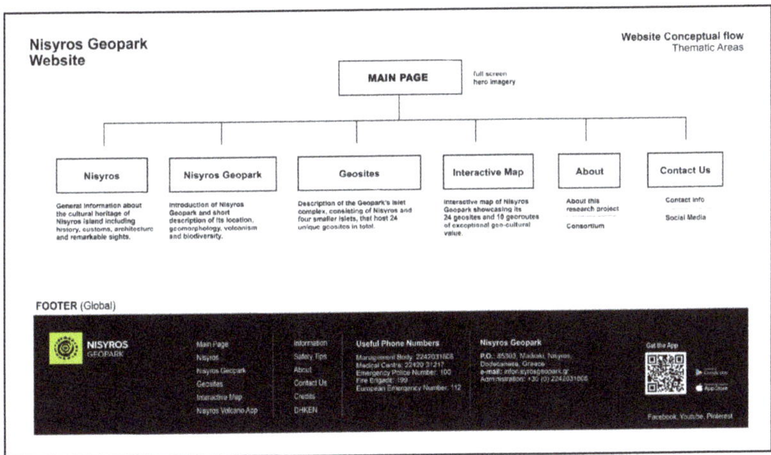

Figure 7. Conceptual flow of the official Nisyros Geopark website, showing the site's available thematic areas.

4. Results

4.1. Website

The modern, responsive and fully interactive official website of Nisyros Geopark www.nisyrosgeopark.gr (accessed on 25 January 2023) constitutes one of the main digital communication tools of high quality and design, providing users with all of the initial information they need to be introduced to its various aspects and help them plan their visit (Figures 8 and 9). The website is currently available in English, while its translations into Greek and other significantly spoken European and Asian languages are under progress and/or consideration. The Home section is the first level of communication (landing page) with the web visitor to the Nisyros Geopark. It showcases the main components of the website's structure, which include the header with the linked subpages to further content, the main body depicting brief information about the Geopark as well as the footer that displays useful information for visitors. Upon official recognition of Nisyros as a UGGp, this level will be the most important in terms of information architecture, since it will essentially consist of live (daily updated) feeds/actions of the Geopark through dedicated sub-areas where the Geopark's "Latest News", the Geopark's actions (Events) and various announcements will be hosted. These areas will provide rich informational material in a chronological order that will follow suit with the Geopark's work/actions and progress through time.

The "Nisyros" section brings the visitor into contact with the island of Nisyros through the following subsections: (a) Mythology, (b) History, (c) Nisyrian People, (d) Customs and Traditions, (e) Nisyrian Villages, (f) Museums and (g) Sights. It starts with the mythological context about the creation of the region (Giantomachy, in 'Mythology'), followed by the presentation of the long history of Nisyros with rich informational material ('History'). The 'Nisyrian People' section provides important information about the modern history, folklore, customs and traditions of the island, with the aim of highlighting the unique intellectual wealth of Nisyros, as well as the work of important Nisyrians. The 'Customs and Traditions' section presents information material on the customs and traditions of Nisyros, the remarkable architecture of the island, the folk culture and the various cultural events of the island. The remaining three sections provide brief descriptions on the island's four main villages, the different museums available for exploration as well as other important geological, natural and cultural sites around the Geopark.

Figure 8. Example of the user-friendly interface of the official Nisyros Geopark website. High resolution, widescreen images accompanied by short, explanatory text regarding a panoramic geosite toward Stefanos Crater and its surroundings.

Figure 9. Example of a starting page of one of the sections of the Nisyros Geopark website (Geosites). Specific attention was given to the presentation of high-quality, full screen imagery of different thematic areas of the Geopark.

The section "Nisyros Geopark" introduces the visitor to the concept of the Geopark, which is directly linked to the geocultural environment of the island complex of Nisyros and the surrounding islets. Its subsections are: (a) Location, (b) Islets Complex, (c) Geomorphology, (d) Volcanism and (e) Biodiversity. The subsection 'Location' includes the clear geographical positioning of Nisyros Geopark in Greece, while the subsection 'Islets Complex' provides a brief description of the physical and anthropogeographical characteristics of both Nisyros and the individual islets that make up the Geopark. The subsection 'Geomorphology' lists the physiographic and morphotectonic features that make up the volcanic environment of the Geopark. In the subsection "Volcanism", the visitor can get in touch with the rich geological history of the area. Finally, the subsection 'Biodiversity' provides useful information on the flora and fauna of the island complex of the Geopark, which thrive due to the fertile soils of volcanic origin and feature scarcity and interesting geographical spread.

The "Geosites" section highlights the 24 designated geosites of the Geopark, providing all the important scientific information accompanied by visual material to the public. It is linked to the last section, "Interactive Map", that presents the 10 proposed georoutes of the geopark while focusing on the interactive experience of the users through the interactive map that displays all the points of interest concerning the geological heritage. It consists of the sub-sections: (a) Geosites, (b) Georoutes and (c) Travels and tours. The subsections 'Geosites' and 'Georoutes' provide visitors with the opportunity to navigate the walking routes of the Geopark, with emphasis mainly on the island of Nisyros, through easy-to-use and user-friendly interactive maps with text, photos, video and sound. The subsection 'Travels and tours' will further diversify the website's content in the future, by including interactive material concerning the other islets of the Geopark, so that visitors will have the opportunity to tour this remote region of the Geopark, currently inaccessible to direct visits.

4.2. Nisyros Volcano App

The Geopark's first educational mobile application, named 'Nisyros Volcano App', is a free, offline and easy to use virtual guide focusing on the area of Lakki within the central caldera of Nisyros Volcano (Figure 10). This area was primarily selected due to its international geological significance, as it is located within the youngest volcanic caldera of the South Aegean Volcanic Arc and is a representative example of an onshore active hydrothermal field [31]. It comprises the eastern part of the caldera's bottom, having a very smooth relief and featuring a number of active hydrothermal craters and lots of fumaroles. One of the craters, Stefanos, is the main touristic highlight of Nisyros Geopark and is one of the largest of its kind in the world [44].

The app features an easy-to-navigate interface (main page) immediately accessed after the opening page once users hit 'Enter'. All the information provided by the app consists of plain language text descriptions, high resolution photos, figures and videos, interactive 2D and 3D maps as well as 360° panoramas. It is organized into three main informational levels, each one of which represents a set of subunits on the interface.Users can easily swipe from one level to another, to find different subunits corresponding to the type of information they want to access (Figure 11). On the upper right side of the main page is a hamburger menu providing instant access to all the subunits of the app, while two buttons were more recently added to provide access to the smartphone's camera, giving the ability to scan the different QR codes used at the Geopark's services. As with the website, the app is currently available in English, while its translations into Greek and other significantly spoken European and Asian languages are under progress and/or consideration.

The first informational level consists of six subunits. The first subunit is named 'About Nisyros Island' and provides a first, brief but efficient introduction to the geological and cultural environment of the island. It presents the geographic and geodynamic position of the island, describes the main types of geological and geomorphological features that make it unique and links its volcanic history to the rich ancient Greek mythology, which had already attempted to geo-interpret the active volcano's behavior. Preceding the text is a simplified interactive map of Nisyros that features five important points of interest, which are the four settlements of the island: Mandraki, the capital and main port, Pali, Emborios and Nikia, as well as the Lakki Hydrothermal Field. Users can touch each one of those points to access a page that provides additional information regarding not only the settlements themselves, but also other significant archeological, cultural or even spiritual points of interest nearby. The descriptions are accompanied by a dedicated image gallery, through which users can admire high resolution photos of a given point of interest, complete with plain language captions.

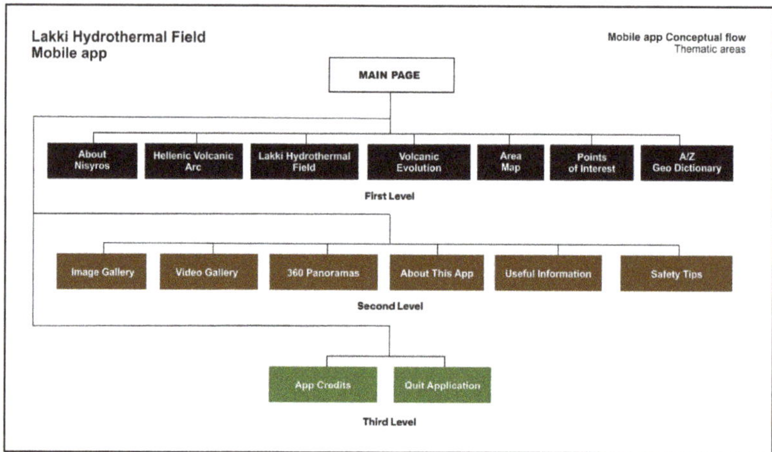

Figure 10. Conceptual flow of the 'Nisyros Volcano App', showing the different informational levels.

Figure 11. 'Nisyros Volcano App's' main user interface design, with the icons representing a different subunit of the three total provided informational levels.

The second subunit is named 'Hellenic Volcanic Arc' and its information provides the first step in understanding the general geodynamic regime in which volcanism at Nisyros occurred. It gives a plain language explanation of the complex geotectonic processes that have shaped the southeastern region of the Aegean Sea for the past millions of years and gives additional details about the other volcanic centers that comprise the Hellenic Volcanic Arc, along with a simple, static map showing their location. By placing the Geopark's geological highlights in a broader context that also includes other well-known islands in the Aegean, like Santorini and Milos, its outstanding geological heritage can be made easier to understand, and thus, noticed and admired. It is important to note that the informative text here, and also in other subunits of the app, is presented as both a short and a long version. The short version gives an efficient description of the most important information on each topic, while the long version takes this a step further, by providing additional details for visitors who have more time in the area or who want to explore more later at home. The short version is not cut in half, with the remaining text in the longer version, but instead is a complete paragraph itself. To access the longer version, users must touch the 'more' button.

The third subunit is named 'Lakki Hydrothermal Field' and specifies the geological processes that power one of the most impressive such fields for the past 24,000 years. It describes what a hydrothermal crater is and how it can form, while providing a comprehensive graphic timeline of the history of the hydrothermal craters over the years. The goal of this subunit is not only to scientifically describe the hydrothermal field, but also help users realize the important difference between the type of eruption that formed the impressive craters and the volcanic eruptions of Nisyros in general. The timeline also aims to give an age context of the craters, thus proving the undoubtable bond that connects volcanism with the local communities of Nisyros, who both struggled and were benefited throughout the island's history.

The fourth subunit is named 'Volcanic Evolution' and is one of the most important throughout the app. It describes the six eruptive cycles that gradually built Nisyros over the past 160,000 years, from mild underwater volcanic activity to a composite stratovolcano that produced voluminous eruptions and eventually formed the edifice seen today [33]. The successive eruptive cycles of Nisyros are individually presented at the end of the subunit, in a form of a timeline that consists of the title of each cycle, a short text description and a high-quality Digital Elevation Model of Nisyros island, highlighting the areas covered by each volcanic cycle's eruptive products. This way visitors are provided with a first means of geo-interpretation for the entire island, as the information can be used to start describing the different volcanic products of Nisyros, regardless of the app's main focus on Lakki hydrothermal field.

The fifth subunit is the Lakki Hydrothermal Field 'Area Map', and the core of the mobile application. It is a virtual guide that gives the opportunity of exploring the three designated points of interest within Lakki, by providing a comprehensive 2D and 3D map of the area. At the beginning, an informative text gives general directions on the accessibility and safety of Lakki, and presents the three points of interest (Stefanos Crater, Lofos Hydrothermal Area and Polyvotis Craters). By exploring the map, users will then find a 2D map of the Lakki area, that includes the location of the craters along with their official names, the main asphalt road that leads to Lakki, the three points of interest and the walking trails that connect them. While in the area, the user's location is also visible in real-time, without the need for an internet connection. This further improves the accessibility to the points of interest, as it can be combined with the informative panels and directions directly found on site. A complete legend of the map can be activated or hidden by touching the button to the right of the 2D mode (Figure 12). The button to the right of the Legend activated 3D mode. Again, visible are the main asphalt road, part of the broader 'Caldera—Lakki' route (one of the ten official georoutes of Nisyros Geopark), the walking trail and the points of interest. This time, the identification of the geosites is easier due to the use of a satellite image background for the map. The final button to the right activates the elevation of the walking trail. It is a useful addition for the app, because it can help plan an excursion at the area by providing the changes in elevation, giving visitors an expected level of difficulty for the georoute. All the points of interest both on the 2D and the 3D maps are interactive and connect to the next subunit.

The sixth and final subunit of the first informative level is the 'Points of Interest' table. It provides detailed information about the geology of the three designated points of interest (geosites) within Lakki, which are Stefanos, the larger hydrothermal crater at the bottom of which visitors can walk, Lofos Hydrothermal area and the Polyvotis Craters. It complements the descriptions with a collection of impressive photos from each of the geosites.

The second informative level consists of another six subunits (seven to twelve). The seventh subunit is the 'Geo Dictionary'. It is an A–Z dictionary that explains complex but important terms used throughout the informative texts of the mobile application, including not only geological, but also other terms as well. The goal of the dictionary is to ensure that users will make the most out of the educational purposes of the app, even when it is impossible to avoid the use of scientific words that a non-specialist audience may not be familiar with. For this reason, the Geo Dictionary can also be accessed by touching underlined terms found within the text of other subunits throughout the app. Doing so will instantly bring the user to the term's description within the dictionary. It also simplifies additional geologic terms that may be encountered during a visit at the Geopark or in general. The enrichment of the application by adding this dictionary was considered a necessity, primarily due to the lack of such easy-to-use explanatory means by other Geoparks.

Figure 12. Screenshot examples of the contents of Nisyros Volcano App. Here, the 'Volcanic Evolution', 'Area Map' and 'About Nisyros' subunits are presented.

The eighth subunit is the 'Photo Gallery' of the app. It includes the best of the high-resolution images used throughout the app, both from the Lakki area and the rest of Nisyros Geopark. The layout of the gallery is such that users are aware of all the photos they have already seen (a green check mark appears on the upper right of the photo). They can also select their favorites, by activating the heart icon on the upper right of the photo, and then accordingly short all the images, on the 'Short by' button. Upon selection of an image, the rest appears at the bottom of the interface, providing instant access. All photos are accompanied by a title and a short explanatory caption (Figure 13).

The ninth subunit is the 'Video Gallery' of the mobile application. It includes four high resolution aerial videos filmed by the Geopark's cooperation colleagues by using photography UAVs. The videos present stunning high-altitude views of the hydrothermal field, aerial views of Stefanos and the degassing fumaroles found at its interior, as well as a flight within Megalos Polyvotis crater. Users can play, pause and select a different time marker at any given moment when viewing the videos.

Another spectacular addition to the mobile application is the tenth subunit, the '360 panoramas'. It includes three high resolution UAV panoramas that can be interactively explored at all directions, giving the idea of flying above Nisyros. Views are provided for Lakki, the Polyvotis craters and the caldera of Nisyros, above the Nikia settlement.

Complying with the regulations of UNESCO regarding safety and geoconservation, the eleventh subunit of the app presents the 'Safety Tips' of Nisyros Geopark. It is a comprehensive catalog of 12 points that include rules and suggestions for the safety of visitors, as the area is an active volcano, and also for the protection of geological and natural heritage.

Figure 13. Screenshot examples of the contents of Nisyros Volcano App. Here, the second informational level's total subunits, the 'Image Gallery' and 'Video Gallery' subunits are presented.

Finally, the twelfth subunit 'About this app' provides a brief explanation of the purposes of the mobile application, its connection to the official website of Nisyros Geopark and its availability.

The third and last informative level of the app consists of the remaining three subunits. 'Useful Info' describes the operational framework of Nisyros Geopark and provides useful links, phone numbers, social media and websites for the Municipality of Nisyros, Nisyros DIKEN and Nisyros Geopark. 'App credits' includes credits to the scientific and technical team that gathered the informative material, the photos and videos, wrote the texts and developed the mobile application (Figure 14). It also includes selected references and papers that helped with the scientific aspect of the texts. Finally, the 'Exit Application' subunit allows for instant deactivation of the app.

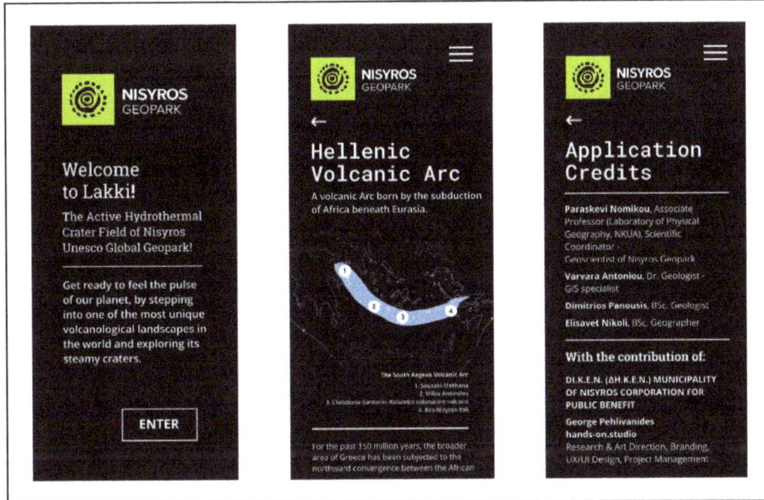

Figure 14. Screenshot examples of the contents of Nisyros Volcano App. Here, the opening page, the 'Hellenic Volcanic Arc' and 'App Credits' subunits are presented.

4.3. Leaflet

The purpose of the new informative leaflet of Nisyros Geopark is to further complement the services provided by the 'Nisyros Volcano App', once again regarding the Lakki hydrothermal area. It includes brief sections of the main information that is also presented within the app, while focusing on the area's 2D and 3D maps. Thus, an alternative, more traditional way of navigating through Lakki is provided, allowing for full accessibility of the region for locals and tourists that do not have the means or simply do not want to download and install the mobile application.

Each foldable level of the leaflet presents, step by step, the geological frame and history of Nisyros, starting with the general geodynamic setting and eventually focusing on the three points of interest at Lakki. The first foldable level includes a brief introduction to the Lakki hydrothermal field of Nisyros, along with the Geopark's logo and the official website on the front side. On the rear side, it features 11 simple but significant pieces of safety information, as determined by the scientific team that studied the volcanic landscape and visited all the included touristic spots within Lakki during all seasons. It is very important for these 'Safety Tips' to be the easiest to access; thus, they are included both within the mobile application and the very first foldable level of the leaflet.

The second foldable level focuses on the South Aegean Volcanic Arc (or the Hellenic Volcanic Arc), presenting a brief description of its geodynamic setting, history and the location of Nisyros Geopark within it. It also provides an over simplified map of southern Greece, highlighting the location of the other volcanic fields that belong to the arc, as well as the boundary along which the subduction of the African plate beneath the Eurasian plate takes place.

The third foldable level presents the six successive eruptive cycles of the volcano. It includes a brief description regarding the scientific methods that contributed to deciphering the volcanic evolution of Nisyros, and then showcases six detailed geologic maps with the land cover of each cycle's volcanic products, gradually forming the modern-day landscape of the island. It is therefore made clear to the reader that Nisyros is an entirely volcanic island and traces of its violent past can be found all along its region. It is also the first time that the most up to date volcanic evolution maps of Nisyros are used on an informative leaflet. This level provides a fine example of how the leaflet can be used along with the digital products; a visitor can gain instant insight into the volcanic cycles of Nisyros and then proceed to find out more within the mobile application, and even further at the Geopark's website.

The fourth foldable level presents the Lakki area's 3D map, an inclined satellite view with the main road to the hydrothermal craters highlighted (a road that is also part of the broader 'Caldera—Lakki' route, one of the ten official georoutes of Nisyros Geopark). It is a static, paper version of the map presented within the mobile application. Alongside the map, brief information and high-resolution aerial imagery of the points of interest are provided. The level also includes the elevation profile that illustrates the topography along the section of the 'Caldera-Lakki' route and the paths, while highlighting the location of the three points of interest.

Finally, the fifth foldable level is the full, unfolded extent of the leaflet with total paper dimensions of A2 (42 × 59.4 cm—Figure 15). It presents the Lakki area's 2D map, based on a detailed digital elevation model. It includes the main road to the hydrothermal craters and other secondary parts of the road network, the location of the main such craters within Lakki, the three points of interest as well as the trail that connects them. Detailed altitude information is also provided (easy to discern contour lines) as well as the names and the peaks of the lava domes that fill the western part of the caldera and dominate the views above Lakki. An index map to the left of the unfolded poster showcases the entire area with the five distinctive islands that compose the landscape of Nisyros Geopark, while the Geopark's website and mobile application are accessible via the provided QR code.

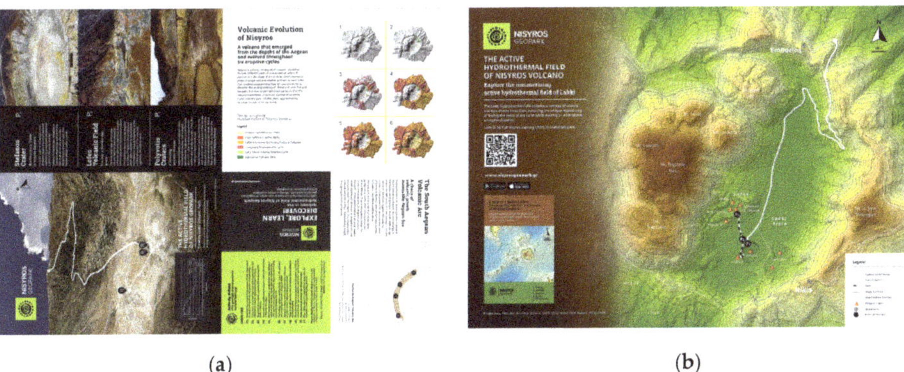

(a)   (b)

Figure 15. Prototype images of the full extent of the foldable poster (fifth foldable level): (**a**) Back view presenting all the successive fold levels described above; (**b**) Front view representing the topographical area of Lakki, within the caldera of Nisyros, along with the georoute, the trail connecting the points of interest and the location of the most famous hydrothermal craters.

4.4. Guidebook

To successfully approach the closer (local communities) and wider (visitors) social environments of the Geopark, its scientific team decided to create a socially appealing, informative guidebook with the aim of scientific education and the rise of environmental consciousness. The guidebook condenses all available important information, regarding every aspect of Nisyros Geopark related to geology, archaeology, culture, activities, educational events, biodiversity and natural disasters.

The guidebook is titled 'The Nisyros Geopark Guidebook' and is a small, paper book of A5 dimension, with a durable cover and a total of 64 pages, written in English language with the aim of future translations in Greek and other widely used European and Asian languages.

The information presented by the guidebook is rendered through short, synoptic texts written in plain language. In this way, scientific content becomes easier and more pleasing to understand by audiences of all ages. Texts are accompanied by maps, diagrams, brief tables and rich photographic material, while instant access to the digital products is thoroughly provided via QR codes and links to the website and the Geopark's social media accounts. The final product provides a traditional and artistic opportunity for easy reading, transport and use by visitors and locals alike, providing information that can be decisive even for planning a trip to Nisyros.

At the beginning is the special thanks section and the link to the Geopark's social media. Then, readers encounter a general touristic map of Nisyros featuring the settlements, the location of the geosites and the georoutes. The map covers both pages of that particular section, to allow for a convenient use. Following are two pages that host the map's legend and a geotouristic index map of the entire Nisyros Geopark, also showcasing its general geographic position (Figure 16).

The main content of the guidebook then starts, with general information on the geodynamic context that gave birth to volcanism in the area, the mythological background established for millennia by the island's first inhabitants as well as a brief description of the history and culture of Nisyros, from past centuries to present.

What follows is the detailed presentation and description of the 24 official geosites of Nisyros Geopark. For each geosite, representative photographic material is provided along with highlighted geological, geoenvironmental and other useful information. All of the above are accompanied by special QR codes that link readers to each geosite's dedicated page on the Geopark's website, after being scanned by a smart mobile device (Figure 17). The section of the geosites descriptions is followed by the presentation of cultural and archaeological points of interest (museums, archaeological spaces). Then follow the main safety and function rules and regulations of Nisyros Geopark (Code of Conduct—Safety Tips). The appearance of the safety tips here, as well as in the mobile application and the leaflet, is necessary for the even management of the Geopark and, mainly, for the personal safety of all its visitors.

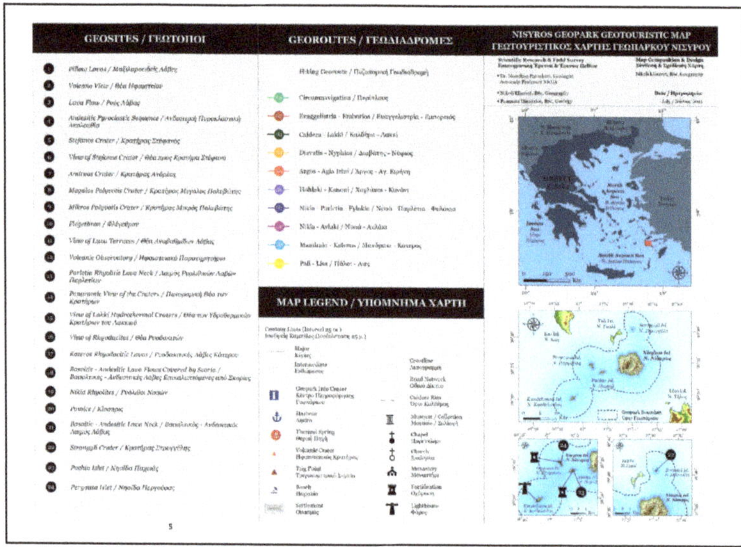

Figure 16. View of the introductory map's legend and the broader maps that portray the location of Nisyros Geopark, at the beginning of the guidebook, in English and Greek.

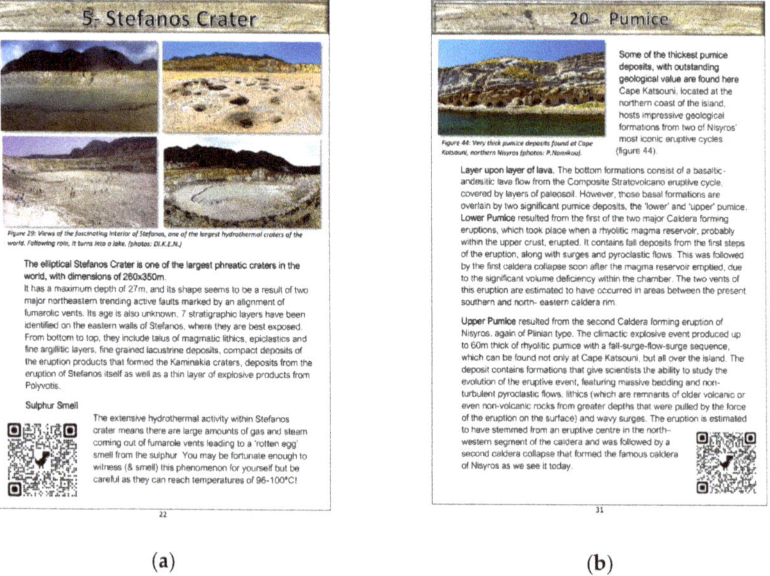

(a)   (b)

Figure 17. Examples of the content of the guidebook, regarding the detailed presentation of the geosites. Each description is accompanied by images and a geologically styled QR code that links to the website: (**a**) Description of Stefanos Crater (Geosite 5); (**b**) Description of Pumice outcrops at northern Nisyros (Geosite 20).

The following section of the guidebook offers a detailed presentation of the 10 georoutes of the geopark, having a similar structure to the geosites section. Again, it is accompanied by maps, photographs and also more useful information regarding the length of each georoute, the approximate traverse time, the difficulty level and the type of points of interest to be encountered along the way. This section also features QR codes that provide access to the website, where users can also search and use the digital interactive map of the Geopark, along with the guidebook. Following the georoutes description is a section dedicated to the Geopark's unique and rich biodiversity.

The final section of the guidebook consists of the Geoscientific dictionary, a concept realized by the Geopark's scientific team that is also present within the mobile application (Figure 18). Its goal is the simple and efficient clarification of unknown and sometimes complex scientific terms that are used throughout the guidebook. This is the same as in the mobile application, but now accessible even through this traditional means of information dissemination, readers can search for a hard-to-understand term and gain more relative insight. What is an extra addition to the guidebook, not found in the mobile application, is an easy-to-use Nisyros rock type identification guide, following the dictionary. It can be used by people of all ages and has a primarily educational character, allowing for an easy identification of the basic igneous rock types found at the Geopark under the concept of a game.

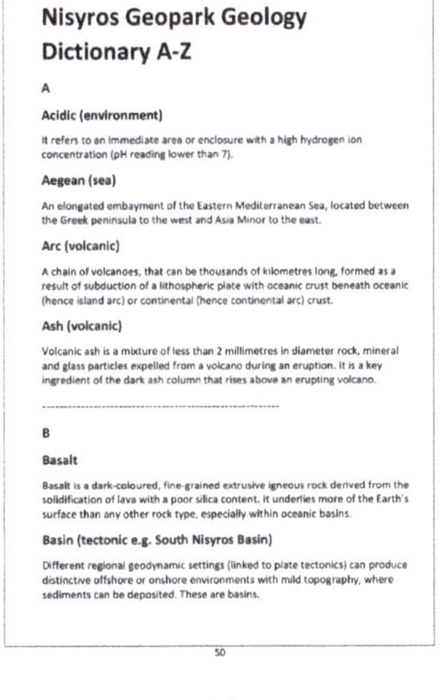

(**a**) (**b**)

Figure 18. Examples of the content of the guidebook: (**a**) The first page of the dedicated section to the A-Z geological dictionary, also present within the 'Nisyros Volcano App'; (**b**) The Rock Guide at the end of the guidebook.

4.5. Panels—Signs

A total of three geotouristic informative panoramas (panels) have been installed at the Lakki hydrothermal field, to the area's three points of interest. The panoramas, which share the same design pattern, include high-definition representative photos of each of the points of interest, a short text description of their main geological highlight, their location within the elevation model of the route and a schematic timeline of events related to the activity of the hydrothermal system. They also include a QR code that links to the website, so that visitors can instantly gain access to more information about these important geosites. The panoramas are installed on an inclined, table-like mount of durable wood and placed on areas that do not prevent complete and unobscured observation of each geosite. A large geotouristic map of Nisyros island has also been installed at the main port of Mandraki, welcoming visitors and providing the first useful directions to the Geopark's geosites, georoutes and other points of interest.

Apart from Lakki, twelve additional informative signs have been installed to the rest of the geosites, ten with dimensions of 60 × 40 cm and two with dimensions of 70 × 90 cm. Each of the signs consists of two different sections regarding the information provided for each of the geosites, written in both English and Greek. The first section is always about the geology of the geosites and the second is about other topics related to the geosite or its direct natural environment, such as biodiversity and cultural or religious significance. High resolution photos along with detailed captions are provided on every sign, showcasing other important places of interest directly related to the geosites. At the right side of each sign, a geotouristic map section is present showcasing the location of the current and neighboring geosites as well as georoutes, archaeological sites, churches, monasteries and other interesting places in the proximity. The final section of each sign contains the most significant rules from the Geopark's Code of Conduct, regarding the respect and protection of the natural environment. It also includes useful information on the management body and a QR code linking to each geosite's dedicated page on the Nisyros Geopark website (Figure 19).

Figure 19. Examples of informative signs and panels at each geosite. (**Upper left**) Prototype layout of informative sign with dimensions of 60 × 40 cm, regarding the pillow lavas at Mandraki, Geosite 1; (**Upper right**) Informative panel installed at Megalos and Mikros Polyvotis craters, Geosites 8 and 9, respectively, and Point of Interest 3 at Lakki; (**Lower left**) Prototype layout of the informative panel at Stefanos, Geosite 5 and Point of Interest 1 of Lakki; (**Lower right**) Composite informative sign with dimensions 70 × 90 cm, regarding view of rhyodacites and the islets Pachia and Pergoussa, Geosites 16, 23 and 24 respectively.

In order to ensure the safety of the Geopark's local communities and visitors, the installation of special safety signage regarding areas that are prone to landslides was considered a priority, not only to the proximity of the geosites, but also to other locations along the island's road network, where the risk of a landslide is higher. Landslides at Nisyros are common and this is primarily due to the combination of the island's steep morphology with the successive alternation of volcanic lithologies, each with a different degree of erosion. Especially during the autumn and winter seasons, extreme and sudden weather phenomena can produce significant landslide events.

5. Discussion

Nisyros aUGGp, an island geopark in the southeastern Aegean Sea, features outstanding geological, natural and cultural characteristics, making it an ideal candidate for UNESCO's Global Geoparks list. Located at the southeastern edge of the South Aegean Volcanic Arc, it hosts dramatic landscapes shaped after the many volcanic eruptions of the past 160,000 years, each one with its own character, that left their scars both on the onshore and the offshore areas of the geopark. A fantastic trail network, along with 24 geosites, await both visitors, in order to give them the chance to easily explore the area and align with nature, and locals, to help them re-appreciate the place they call home. The offshore area of the geopark is constantly under the spotlight of scientific research of global impact, that has shed light on impressive underwater volcanic structures like craters, lava domes and fractures, as well as a pre-historic caldera, called Avyssos (the Abyss, for its depth of almost 700 m), from where the largest volcanic event of the Eastern Mediterranean occurred 161,000 years ago [30–32]. The offshore area features a number of basins, as well as a regional horst, Kondeliousa, a direct result of the complex geodynamic processes that shape the geopark's broader region and the Dodecanese. These structures are bounded by significant faults that are still active up to this day, are well monitored and their study yields important details on the evolution of the area [49].

The cultural heritage of the area is also of great significance because, despite its small size, it has been constantly habituated for millennia. Directly linked to the myth of Gigantomachy and the fight between Polyvotis giants and Poseidon, the place has not only earned its place among the famous and globally known Hellenic mythology, but also inherited parts of the cultural identity of other civilizations who conquered it and controlled it. Impressive castles, like Palaiokastro—one of the most well-preserved Greek fortresses worldwide—the Mandraki castle, remnants of ancient habituations and a great number of churches and chapels from the Byzantine years have ultimately encapsulated the region's history and given rise to the traditions and tangible and intangible heritage of locals today.

Tourist infrastructure development, educational activities, partnerships with neighboring UGGps, improvements in already existing structures and a management body that fully integrates local populations in the decision-making processes of the geopark, help make Nisyros Geopark a strong candidate. The educational and touristic material now includes publications, informative leaflets, signs and maps, while the new, modern and user-friendly website, along with the mobile application and several other educational tools, like a Story Map [7], complement the experience. As a candidate, Nisyros Geopark will join Greece's seven other officially recognized UNESCO Global Geoparks in the Hellenic Geoparks Forum, which consists of the Geoparks of Lesvos, Psiloritis, Chelmos-Vouraikos, Vikos-Aoos, Sitia, Grevena-Kozani and Kefalonia-Ithaka. Out of those, only Lesvos Geopark features volcanic geoheritage, although its geological characteristics relate to extinct volcanism. Thus, Nisyros will be the second volcanic Geopark and Greece's first active volcano to be included in the list once it achieves the UNESCO Global Geopark status.

Connection with the geopark is by boat from neighboring Kos, where people can get to either by boat from Piraeus port or by plane. The road network of the island, which is properly developed and combined with the trail network, helps visitors to visit every corner of the island. There are also marine activities that involve sailing around the island and visiting the surrounding islets.

Inspired by the actions and initiatives of many other UGGps over the past few years, especially during the COVID-19 pandemic [2], the management body of Nisyros aUGGp in collaboration with the scientific team and a team of graphic design, website and mobile application construction specialists, set up the production of a modern, up to date official website for the Geopark and a user friendly mobile application regarding the touristic highlight of Nisyros, the hydrothermal craters field, as a high priority. In this way, the aspiring Geopark is already able to join the dynamic world of digital science information and dissemination, and provide its residents and visitors with the opportunity to explore its best geological, biological and cultural features in an accessible, friendly and easy to use manner, thus representing the mature level of communication of the Geopark with the public. These digital products also enhance the visibility of the Geopark, complement the touristic services provided by its management body and easily assist the more traditional means of information communication material to the public, which are the signs, the leaflet and the Nisyros Geopark guidebook, as they all connect to the digital products through links and QR codes. Thus, not only does the Geopark allow for a classic approach to the exploration of the geosites and georoutes, with a handy and innovative map and a book, but also provides instantly accessible digital means of information gathering, communication and geological interpretation, through the website and the mobile application.

Although still at an early stage, the new products have already started being used by many tourists who arrived at Nisyros during the summer tourism season of 2022. During the Geopark's evaluation field trip at the end of October 2022, it was noticed that previously less explored areas and geosites of the Geopark now receive more attention due to the new informative panels installed. Furthermore, the new foldable leaflet gradually replaced the older ones provided by the ticket office at the hydrothermal field of Lakki, as well as the volcanological museum. Areas and routes that are presented within the leaflet's main map of Lakki, such as the path toward the Polyvotis craters (POI 3), have increasingly started to see groups of tourists who were previously almost unaware of their existence and importance. Hence, a significant contribution to the further and better understanding, and eventually respecting, of the natural and cultural environment of the Geopark can already be observed.

A further step for the future, however, is the quantitative evaluation of the use and impact of the Geopark's products presented here, especially the website and the mobile application, along with the production of new ones.

The Geopark's eventual inclusion in the Global Geoparks Network, once it is officially recognized as a UGGP, will contribute to making this special corner of Greece further known, despite its touristic mecca neighbors in the Cyclades and the Dodecanese. With its outstanding beauty, unique geodiversity and cultural heritage, as well as high scientific value, Nisyros aUGGp has it all. Its dedicated administrative and scientific teams work around the clock to further improve, innovate and motivate everyone involved in the project, to align with the mission and the goals of the Global Geoparks Network (GGN— https://globalgeoparksnetwork.org/ accessed on 29 December 2022). That said, the links between geological heritage and all other aspects of the area's natural and cultural heritage clearly demonstrate that geodiversity is the foundation of all ecosystems and the basis of human interaction with the landscape [12].

6. Conclusions

Nisyros Geopark is an open geological, biological and cultural laboratory, not only giving scientists from different fields of expertise, but also people who visit, the opportunity to experience an active volcano up-close, to discover the different volcanic products that construct its geological history, to identify the five eruptive cycles and even walk inside one of the biggest hydrothermal craters on the planet and listen to the rumble of the living planet Earth. As a result, this paper presented the Geopark's official website, its first virtual guide (mobile application), a new informative leaflet, a number of installed informative panels and signs, as well as a guidebook, as the first efforts made towards the goal of achieving UGGp status. The significance of preserving its thrilling heritage while managing to promote and disseminate the science behind it, has undoubtedly dictated the need to create a series of products which, through the combination of modern and traditional means, will bring Nisyros, from now on, closer to everyone.

Author Contributions: Conceptualization, P.N. (Paraskevi Nomikou), D.E. and P.N. (Panagiotis Nastos); methodology, P.N. (Paraskevi Nomikou), D.P., E.N. and G.P.; software, G.P., E.N., V.A., M.A. and A.B.; validation, P.N. (Paraskevi Nomikou), V.A., D.P., E.N., G.P., D.E. and P.N. (Panagiotis Nastos); investigation, D.P., E.N., G.P. and E.C.-J.; resources, P.N. (Paraskevi Nomikou), V.A. and E.N.; data curation, P.N. (Paraskevi Nomikou), D.P., E.N. and V.A.; writing—original draft preparation, D.P., E.N. and G.P.; writing—review and editing, P.N. (Paraskevi Nomikou), V.A., D.P., E.N. and G.P.; visualization, D.P., E.N. and G.P.; supervision, P.N. (Paraskevi Nomikou); project administration, P.N. (Paraskevi Nomikou), D.E. and P.N. (Panagiotis Nastos); funding acquisition, P.N. (Paraskevi Nomikou) and E.N. All authors have read and agreed to the published version of the manuscript.

Funding: The creation of the mobile application "Nisyros Volcano App", as well as the informative leaflet and the three geotouristic panoramas highlighting the hydrothermal field of Nisyros Geopark, were conducted in the framework of the Reasearch Project "Creation of Visitor's Leaflet and Virtual Tour Guide for the geo-route of the Hydrothermal Craters of Nisyros Volcano" which was funded by DIKEN (Municipal Public Benefit Enterprise of Nisyros). The series of the informative geosites' signs were designed and installed under the auspices of the MUNICIPALITY OF NISYROS in the framework of the Research Project "Highlighting the geosites of Nisyros using digital applications for the inclusion of the island in the Global Geoparks Network". The creation of the official website of Nisyros Geopark was also funded by DIKEN, while The Nisyros Geopark Guidebook received no external funding.

Data Availability Statement: The data presented in this study are available on request from the corresponding author.

Acknowledgments: The authors would like to acknowledge the Mayor of Nisyros Christofis Koronaios and the people who work at the Municipality of Nisyros and DIKEN, for funding the accommodation and rental cars during data collection field trips to Nisyros, as well as providing significant information that contributed to the production of the presented products. George Pehlivanides (hands-on.studio) for Research and art direction, branding, UX/UI Design and project management. Koukotzilas Ioannis (monoscopic studio) for the website development of Nisyros Geopark and Vatsikouras Dimitris (anifactum studio) for the Icon set design—iconography. 'The Nisyrian Studies Society', for providing important information on cultural, tangible and intangible heritage of Nisyros throughout the ages. Evaggelia Metaxa and RAY, for undertaking the high-quality prints of the informative panels and signs. George Torizis, carpenter, and his team, for the installation of all the informative signs on the field and the construction of their respective mounts. 'Petastra Rent a Car and Motorbike', for providing the authors team with cars for all the vital excursions at Nisyros to gather data, supervise and make progress on the products. Finally, the people of Nisyros, whom hospitality, local knowledge and experience of their home island furtherly improved the content of the products and provided fruitful motivation towards the realization of the goal of Nisyros Geopark to become recognized as a UNESCO Global Geopark.

Conflicts of Interest: The authors declare no conflict of interest.

References

1. Dowling, R.K. Global Geotourism—An Emerging Form of Sustainable Tourism. *Czech J. Tour.* **2014**, *2*, 59–79. [CrossRef]
2. Fassoulas, C.; Nikolakakis, E.; Staridas, S. Digital Tools to Serve Geotourism and Sustainable Development at Psiloritis UNESCO Global Geopark in COVID Times and Beyond. *Geosciences* **2022**, *12*, 78. [CrossRef]
3. Stolz, J.; Megerle, H.E. Geotrails as a Medium for Education and Geotourism: Recommendations for Quality Improvement Based on the Results of a Research Project in the Swabian Alb UNESCO Global Geopark. *Land* **2022**, *11*, 1422. [CrossRef]
4. Farsani, N.T.; Coelho, C.; Costa, C. Geotourism and geoparks as novel strategies for socio-economic development in rural areas. *Int. J. Tour. Res.* **2011**, *13*, 68–81. [CrossRef]
5. Zouros, N. Global Geoparks Network and The New Unesco Global Geoparks Programme. *Bull. Geol. Soc. Greece* **2017**, *50*, 284–292. [CrossRef]
6. Gordon, J.E. Geoheritage, Geotourism and the Cultural Landscape: Enhancing the Visitor Experience and Promoting Geoconservation. *Geosciences* **2018**, *8*, 136. [CrossRef]
7. Antoniou, V.; Nomikou, P.; Panousis, D.; Zafeirakopoulou, E. Nisyros Volcanic Island: A Geosite through a Tailored GIS Story. *Geosciences* **2021**, *11*, 132. [CrossRef]
8. Catana, M.M.; Brilha, J.B. The Role of UNESCO Global Geoparks in Promoting Geosciences Education for Sustainability. *Geoheritage* **2020**, *12*, 1. [CrossRef]

9. Giardino, M.; Justice, S.; Olsbo, R.; Balzarini, P.; Magagna, A.; Viani, C.; Selvaggio, I.; Kiuttu, M.; Kauhanen, J.; Laukkanen, M.; et al. ERASMUS+ Strategic Partnerships between UNESCO Global Geoparks, Schools, and Research Institutions: A Window of Opportunity for Geoheritage Enhancement and Geoscience Education. *Heritage* **2022**, *5*, 677–701. [CrossRef]
10. Xu, K.; Wu, W. Geoparks and Geotourism in China: A Sustainable Approach to Geoheritage Conservation and Local Development—A Review. *Land* **2022**, *11*, 1493. [CrossRef]
11. UNESCO (United Nations Educational, Scientific and Cultural Organization). *UNESCO Global Geoparks. Celebrating Earth Heritage, Sustaining Local Communities*; UNESCO: Paris, France, 2016; pp. 3–13.
12. UNESCO. Geoparks Fundamental Features. Available online: https://en.unesco.org/global-geoparks/focus#fundamental (accessed on 21 December 2022).
13. Martini, B.G.; Zouros, N.; Zhang, J.; Jin, X.; Komoo, I.; Border, M.; Watanabe, M.; Frey, M.L.; Rangnes, K.; Van, T.T.; et al. UNESCO Global Geoparks in the "World after": A multiplegoals roadmap proposal for future discussion. *Episodes* **2021**, *45*, 29–35. [CrossRef] [PubMed]
14. UNESCO. Statutes of the International Geoscience and Geoparks Programme (IGCP). Available online: https://unesdoc.unesco.org/ark:/48223/pf0000234539.locale=en (accessed on 10 December 2022).
15. Zouros, N.; Rangnes, K. The European Geoparks Network: Operation and Procedures. *Schriftenr. Dt. Ges. Geowiss.* **2016**, *88*, 31–36. [CrossRef]
16. UNESCO. List of UNESCO Global Geoparks. Available online: https://en.unesco.org/global-geoparks/list#list (accessed on 21 December 2022).
17. Rosado-González, E.M.; Sá, A.A.; Palacio-Prieto, J.L. UNESCO Global Geoparks in Latin America and the Caribbean, and Their Contribution to Agenda 2030 Sustainable Development Goals. *Geoheritage* **2020**, *12*, 1–15. [CrossRef]
18. United Nations. *Transforming Our World: The 2030 Agenda for Sustainable Development*; United Nations: New York, NY, USA; Available online: https://sdgs.un.org/2030agenda (accessed on 21 December 2022).
19. Gill, J.C. Geology and the Sustainable Development Goals. *Episodes* **2017**, *40*, 70–76. [CrossRef]
20. Cayla, N. An Overview of New Technologies Applied to the Management of Geoheritage. *Geoheritage* **2014**, *6*, 91–102. [CrossRef]
21. Aldighieri, B.; Testa, B.; Bertini, A. 3D Exploration of the San Lucano Valley: Virtual Geo-routes for Everyone Who Would Like to Understand the Landscape of the Dolomites. *Geoheritage* **2016**, *8*, 77–90. [CrossRef]
22. Kim, H.-S.; Lim, C. Developing a geologic 3D panoramic virtual geological field trip for Mudeung UNESCO global geopark, South Korea. *Episodes* **2019**, *42*, 235–244. [CrossRef]
23. Perotti, L.; Bollati, I.M.; Viani, C.; Zanoletti, E.; Caironi, V.; Pelfini, M.; Giardino, M. Fieldtrips and Virtual Tours as Geotourism Resources: Examples from the Sesia Val Grande UNESCO Global Geopark (NW Italy). *Resources* **2020**, *9*, 63. [CrossRef]
24. Williams, M.; McHenry, M. The Increasing Need for Geographical Information Technology (GIT) tools in Geoconservation and Geotourism. *Geoconservation Res.* **2020**, *3*, 17–32. [CrossRef]
25. Gambino, F.; Borghi, A.; D'Atri, A.; Gallo, L.M.; Ghiraldi, L.; Giardino, M.; Martire, L.; Palomba, M.; Perotti, L.; Macadam, J. TOURinSTONES: A Free Mobile Application for Promoting Geological Heritage in the City of Torino (NW Italy). *Geoheritage* **2019**, *11*, 3–17. [CrossRef]
26. Papanikolaou, D.; Nomikou, P. Tectonic structure and volcanic centers at the eastern edge of the aegean volcanic arc around Nisyros island. *Bull. Geol. Soc. Greece* **2001**, *34*, 289–296. [CrossRef]
27. Nomikou, P.; Papanikolaou, D.; Alexandri, M.; Sakellariou, D.; Rousakis, G. Submarine volcanoes along the Aegean volcanic arc. *Tectonophysics* **2013**, *597–598*, 123–146. [CrossRef]
28. Allen, S.R. Volcanology of the Kos Plateau Tuff, Greece: The Product of An Explosive Eruption in An Archipelago. Ph.D. Thesis, Monash University, Melbourne, Australia, 1998.
29. Allen, S.R.; Stadlbauer, E.; Keller, J. Stratigraphy of the Kos Plateau Tuff: Product of a major Quaternary explosive rhyolitic eruption in the eastern Aegean, Greece. *Int. J. Earth Sci.* **1999**, *88*, 132–156. [CrossRef]
30. Allen, S. Reconstruction of a major caldera-forming eruption from pyroclastic deposit characteristics: Kos Plateau Tuff, eastern Aegean Sea. *J. Volcanol. Geotherm. Res.* **2001**, *105*, 141–162. [CrossRef]
31. Nomikou, P. Contribution to The Geodynamics of the Dodecanese: The Submarine Area of Kos-Nisyros Islands. Ph.D. Thesis, National and Kapodistrian University of Athens, Greece, 2004. (In Greek).
32. Pe-Piper, G.; Piper, D.J.; Perissoratis, C. Neotectonics and the Kos Plateau Tuff eruption of 161 ka, South Aegean arc. *J. Volcanol. Geotherm. Res.* **2005**, *139*, 315–338. [CrossRef]
33. Dietrich, V.J. Geology of Nisyros Volcano. In *Nisyros Volcano*, 1st ed; Dietrich, V.J., Lagios, E., Eds.; Springer International Publishing: Cham, Switzerland, 2018; pp. 57–102. [CrossRef]
34. Pe-Piper, G.; Piper, D.J.W. The Igneous Rocks of Greece. The Anatomy of an Orogen. *Geol. Mag.* **2003**, *140*, 357. [CrossRef]
35. Nomikou, P.; Papanikolaou, D. Extension of active fault zones on Nisyros volcano across the Yali-Nisyros Channel based on onshore and offshore data. *Mar. Geophys. Res.* **2011**, *32*, 181–192. [CrossRef]
36. Tibaldi, A.; Pasquarè, F.; Papanikolaou, D.; Nomikou, P. Tectonics of Nisyros Island, Greece, by field and offshore data, and analogue modelling. *J. Struct. Geol.* **2008**, *30*, 1489–1506. [CrossRef]
37. Nomikou, P.; Papanikolaou, D.; Dietrich, V.J. Geodynamics and Volcanism in the Kos-Yali-Nisyros Volcanic Field. In *Nisyros Volcano*, 1st ed; Dietrich, V.J., Lagios, E., Eds.; Springer International Publishing: Cham, Switzerland, 2018; pp. 13–55. [CrossRef]
38. Di Paola, G.M. Volcanology and petrology of Nisyros Island (Dodecanese, Greece). *Bull. Volcanol.* **1974**, *38*, 944–987. [CrossRef]

39. Innocenti, F.; Manetti, P.; Peccerillo, A.; Poli, G. South Aegean volcanic arc: Geochemical variations and geotectonic implications. *Bull. Volcanol.* **1981**, *44*, 377–391. [CrossRef]
40. Makris, J.; Stobbe, C. Physical properties and state of the crust and upper mantle of the Eastern Mediterranean Sea deduced from geophysical data. *Mar. Geol.* **1984**, *55*, 345–361. [CrossRef]
41. Papanikolaou, D.; Lekkas, E.; Sakellariou, D. Geological structure and evolution of the Nisyros volcano. *Bull. Geol. Soc. Greece* **1991**, *25*, 405–419.
42. Volentik, A.; Vanderkluysen, L.; Principe, C.; Hunziker, J.C. Stratigraphy of Nisyros volcano (Greece). In *The Petrology and Geochemistry of Lavas and Tephras of Nisyros Volcano (Greece)*; Hunziker, J.C., Marini, L., Eds.; Mémoires de Géologie: Lausanne, Switzerland, 2005; Volume 44, pp. 26–66. Available online: https://www.researchgate.net/publication/236611984_Stratigraphy_of_Nisyros_Volcano_Greece (accessed on 10 December 2022).
43. Royden, L.H.; Papanikolaou, D.J. Slab segmentation and late Cenozoic disruption of the Hellenic arc. *Geochem. Geophys. Geosystems* **2011**, *12*, Q03010. [CrossRef]
44. Dietrich, V.; Lagios, E. *Nisyros Volcano*, 1st ed.; Springer International Publishing: Cham, Switzerland, 2018. [CrossRef]
45. Papadimitriou, P.; Karakonstantis, A.; Kapetanidis, V.; Bozionelos, G.; Kaviris, G.; Voulgaris, N. Seismicity and Tomographic Imaging of the Broader Nisyros Region (Greece). In *Nisyros Volcano*, 1st ed.; Dietrich, V.J., Lagios, E., Eds.; Springer International Publishing: Cham, Switzerland, 2018; pp. 245–271. [CrossRef]
46. Papoulia, J.; Makris, J.; Koulakov, I.; Fasoulaka, C.; Drakopoulou, P. Microseismicity and Crustal Deformation of the Dodecanese Volcanic Area, Southeastern Aegean Sea Using an Onshore/Offshore Seismic Array. In *Nisyros Volcano*, 1st ed.; Dietrich, V.J., Lagios, E., Eds.; Springer International Publishing: Cham, Switzerland, 2018; pp. 273–284. [CrossRef]
47. Parcharidis, I.; Lagios, E.; Sakkas, V. Differential Interferometry as a Tool of An Early Warning System in Reducing The Volcano Risk: The Case of Nisyros Volcano. *Bull. Geol. Soc. Greece* **2018**, *36*, 913–918. [CrossRef]
48. Tompolidi, A.-M.; Parcharidis, I.; Sykioti, O. Investigation of Sentinel-1 capabilities to detect hydrothermal alteration based on multitemporal interferometric coherence: The case of Nisyros volcano (Greece). *Procedia Comput. Sci.* **2021**, *181*, 1027–1033. [CrossRef]
49. Nomikou, P.; Krassakis, P.; Kazana, S.; Papanikolaou, D.; Koukouzas, N. The Volcanic Relief within the Kos-Nisyros-Tilos Tectonic Graben at the Eastern Edge of the Aegean Volcanic Arc, Greece and Geohazard Implications. *Geosciences* **2021**, *11*, 231. [CrossRef]
50. Chalkiadakis, N. Volcanic Risk Management at The Nisyros Volcano. Bachelor's Thesis, National and Kapodistrian University of Athens, Greece. Available online: https://pergamos.lib.uoa.gr/uoa/dl/object/2946922 (accessed on 10 December 2022).
51. Chartofili, A. Digital Games in Education: Development and Evaluation of an Interactive Game for Teaching and Enhancing the Location Knowing of Nisyros for Secondary and High School Students. Master's Thesis, University of the Aegean, Greece. Available online: http://hdl.handle.net/11610/18832 (accessed on 10 December 2022).
52. Chatzifoti, E. Nisyros Volcano: Volcanic Risk Management and Risk. Bachelor's Thesis, National and Kapodistrian University of Athens, Greece. Available online: https://pergamos.lib.uoa.gr/uoa/dl/object/1317943 (accessed on 10 December 2022).
53. Dietrich, V.J. Epilogue Nisyros Island, The Inaccessible Outpost Between Orient and Occident, Home of a Restless Giant. In *Nisyros Volcano*, 1st ed.; Dietrich, V.J., Lagios, E., Eds.; Springer International Publishing: Cham, Switzerland, 2018; pp. 321–336. [CrossRef]
54. Liritzis, I.; Oikonomou, A. An updated overview of archaeological sciences research in insular Greek Aegean islands. *Mediterr. Archaeol. Archaeom.* **2021**, *21*, 1–27. [CrossRef]
55. Liritzis, I.; Michael, C.; Galloway, R.B. A significant aegean volcanic eruption during the second millennium B.C. revealed by thermoluminescence dating. *Geoarchaeology* **1996**, *11*, 361–371. [CrossRef]
56. Galloway, R.B.; Liritzis, Y. Provenance of Aegean volcanic tephras by high resolution gamma-ray spectrometry. *Nucl. Geophys.* **1992**, *6*, 405.

Article

Geological Uniqueness and Potential Geotouristic Appeal of Murge and Premurge, the First Territory in Puglia (Southern Italy) Aspiring to Become a UNESCO Global Geopark

Marcello Tropeano [1], Massimo A. Caldara [1], Vincenzo De Santis [1], Vincenzo Festa [1], Mario Parise [1], Luisa Sabato [1,*], Luigi Spalluto [1], Ruggero Francescangeli [1], Vincenzo Iurilli [1], Giuseppe A. Mastronuzzi [1], Marco Petruzzelli [1], Filippo Bellini [1], Marianna Cicala [1], Elio Lippolis [1], Fabio M. Petti [2], Matteo Antonelli [3], Stefano Cardia [1], Jacopo Conti [3], Rafael La Perna [1], Maria Marino [1], Antonella Marsico [1], Enrico Sacco [3], Antonello Fiore [4], Oronzo Simone [4], Salvatore Valletta [4], Umberto S. D'Ettorre [1], Vincenzo De Giorgio [1], Isabella S. Liso [1] and Eliana Stigliano [1]

1 Dipartimento di Scienze della Terra e Geoambientali, Università degli Studi di Bari Aldo Moro, 70100 Bari, Italy
2 Muse—Museo delle Scienze di Trento, 38400 Trento, Italy
3 Dipartimento di Scienze della Terra, Sapienza Università di Roma, 00100 Roma, Italy
4 SIGEA, Società Italiana di Geologia Ambientale, 00100 Roma, Italy
* Correspondence: luisa.sabato@uniba.it

Citation: Tropeano, M.; Caldara, M.A.; De Santis, V.; Festa, V.; Parise, M.; Sabato, L.; Spalluto, L.; Francescangeli, R.; Iurilli, V.; Mastronuzzi, G.A.; et al. Geological Uniqueness and Potential Geotouristic Appeal of Murge and Premurge, the First Territory in Puglia (Southern Italy) Aspiring to Become a UNESCO Global Geopark. *Geosciences* **2023**, *13*, 131. https://doi.org/10.3390/geosciences13050131

Academic Editors: Hara Drinia, Panagiotis Voudouris, Assimina Antonarakou and Jesus Martinez-Frias

Received: 27 March 2023
Revised: 19 April 2023
Accepted: 21 April 2023
Published: 30 April 2023

Abstract: At the end of November 2021, a large area of Puglia (an administrative region in Southern Italy) was officially nominated as new aUGGp (aspiring UNESCO Global Geopark) by the Italian National Commission of UNESCO. This area comprises the northwestern part of the Murge territory, where a Cretaceous sector of the Apulia Carbonate Platform crops out, and part of the adjacent Premurge territory, where the southwestward lateral continuation of the same platform (being flexed toward the Southern Apennines Chain) is covered by thin Plio-Quaternary foredeep deposits. The worldwide geological uniqueness of the aspiring Geopark (Murge aUGGp) is that the area is the only in situ remnant of the Adria Plate, the old continental plate almost entirely squeezed between the Africa and Eurasia Plates. In such a context, the Murge area (part of the Apulia Foreland) is a virtually undeformed sector of Adria, while other territories of the plate are and/or were involved in the subduction/collision processes. In the aspiring Geopark, the crust of Adria is still rooted to its mantle, and the Cretaceous evolution of the continent is widely recorded in the Murge area thanks to the shallow-water carbonate succession of one of the largest peri-Tethys carbonate platforms (the Apulia Carbonate Platform). The aspiring Geopark also comprises the Premurge area, which represents the outer Southern Apennines foredeep, whose Plio-Quaternary evolution is spectacularly exposed thanks to an "anomalous" regional middle-late Quaternary uplift. Despite the presence of numerous geological singularities of international importance, it would be important, from a geotourist point of view, to propose a regional framework of the geology of the aUGGp before introducing visitors to the significance of the individual geosites, whose importance could be amplified if included in the geoevolutionary context of the Murge aUGGp.

Keywords: Southern Italy; Apulia Foreland; Murge aspiring UNESCO Global Geopark (aUGGp); Premurge; Adria; geotourism; Apulia Carbonate Platform; Bradanic Trough

1. Introduction

Starting from 2002, the executive of SIGEA-Puglia (the regional board of a national voluntary association of Italy that attempts to make citizens and public administrations sensitive towards geology) has tried to direct public attention towards a knowledge, census, and cataloguing of regional geosites of Puglia, the easternmost administrative region in Southern Italy.

At the same time, the same small group of geologists has focused attention on the possibility of proposing a geopark in a selected part of the Murge area, the central sector of the Puglia Region (Figure 1). The activity of this group led to a presentation of the main regional geosites [1], to the organization of the VII International Geoheritage Symposium [2], and to the production of the first inventory of geosites of Puglia required by the Regional Administration [3].

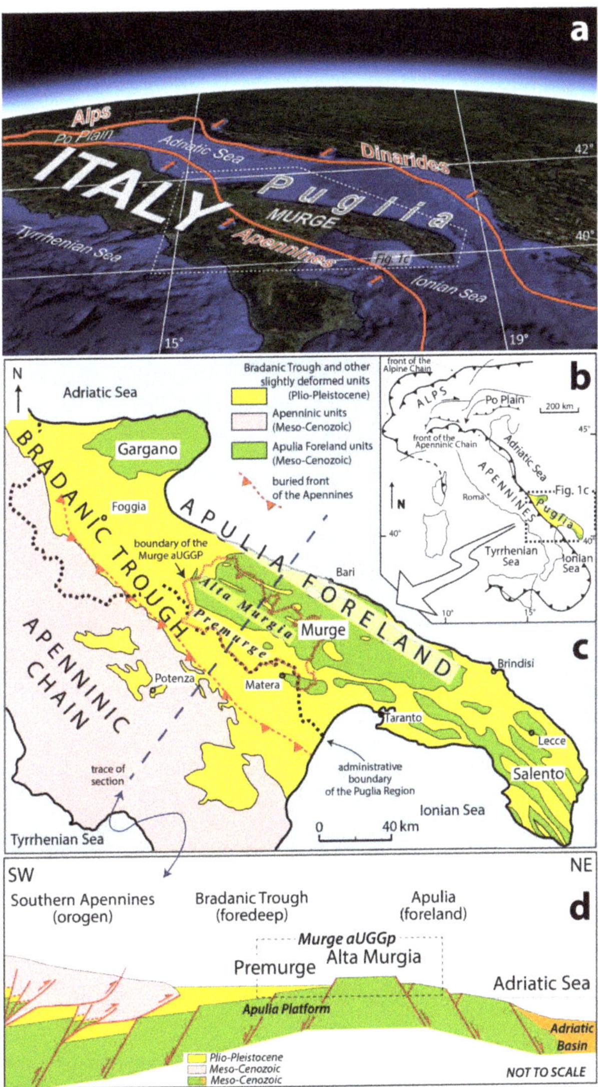

Figure 1. (**a**) Position of the Puglia Region along the Italian peninsula, with location of Figure 1c (dotted line); the Murge area corresponds to the central part of the region. Base map from Google Earth. (**b**) Simplified geological scheme of Italy ([4], mod.) with location of Figure 1c (dotted line). (**c**) Simplified geological map of Southern Italy ([5], mod.) with location of the Murge aspiring UNESCO Global Geopark (aUGGp) and trace of the geological section reported below. (**d**) Cross section showing relationships between Alta Murgia and Premurge in the Murge aspiring Geopark ([6], mod.). Geographic coordinates in the WGS84 system.

At the beginning of 2019, the new management of the Alta Murgia National Park took charge to candidate its territory to become a Geopark, delegating to the "Dipartimento di Scienze della Terra e Geoambientali, Università degli Studi di Bari Aldo Moro" (Earth and Geoenvironmental Sciences Department, Aldo Moro University of Bari, Italy) the task of highlighting the geology of the area. At the end of November 2021, a large area of Puglia was officially proposed as a new aUGGp (aspiring UNESCO Global Geopark) by the Italian National Commission of UNESCO. The aspiring Geopark comprises not only the National Park area, i.e., the northwestern Murge area, where a Cretaceous sector of the Apulia Carbonate Platform crops out (see later in the text), but also the adjacent Premurge area, where the southwestward lateral continuation of the same platform, being flexed toward the Southern Apennines Chain, is covered by thin Plio-Quaternary deposits (Figure 1c,d).

The aim of this work is to briefly report the synthesis carried out by the task force of Bari University to describe the geological uniqueness that led the area to be nominated as an aUGGp. This should be the introductory theme for geotourists visiting the aspiring Geopark.

2. Preliminary Geographic Information

To avoid confusion in the text, it is important to explain the meaning of two pairs of widely, and often alternately, used geographical terms: Puglia/Apulia and Murge/Murgia.

2.1. Puglia/Apulia

As noted before, Puglia is an administrative region in Southern Italy (Figure 1c), the boundaries of which do not exactly match those of Apulia, the old and more internationally known area between the "spur" (the Gargano promontory) and the "heel" (the Salento Peninsula) of Italy (Figure 2).

Figure 2. The shape of Italy recalls that of a boot, and the Puglia Region extends between the spur and the heel. This picture comes from [7]. Compare with Figure 1b.

From a geological standpoint, researchers mainly assign the term Apulia to the area corresponding to the Apulia Foreland, sometimes including the Bradanic Trough (Figure 1c). In this broad sense, Apulia corresponds to the exposed area located east of the front of the Southern Apennines, and this is how we use Apulia in the present work. Apulia is also the name that is sometimes used to indicate the "small" continental plate located between the Africa and Eurasia ones (see later in the text). This small plate is better known as Adria, and this is the term we use in the present work.

2.2. Murge/Murgia

The word "murgia" (pl. "murge"), with the first letter in lowercase, is a term no longer used, and it indicates in Southern Italy a rocky hill, often with a steep slope bounding it from an adjacent area. The term probably comes from the Latin word "murex", meaning rock (cliff); it became a relatively used toponym that, as with other words and capitalizations of the first letter, has assumed a geographical meaning and indicates specific localities or territories. The plural term with the first letter in uppercase and always preceded by the Italian article "le" (le Murge) indicates the wide central area of the Apulia Foreland characterized by the presence of a series of plateaus made up of karstified carbonate rocks (Figure 3). The topographically highest plateau is named "Murge Alte" or "Alta Murgia" (literally: high part of Murge), and it is split into two sectors by a morphotectonic depression (the Gioia del Colle saddle). The northwestern sector of this highest plateau roughly corresponds to the Alta Murgia National Park. Eventually, Premurge literally means "before Murge" (Figure 3).

2.3. The Proposed Geopark Name

For the name of the aUGGp, in order to suggest an easy term to indicate and remember the geopark, we proposed to use only the geographic toponym "Murge", even if the selected area was that of the Alta Murgia National Park plus part of the Premurge area. This led us to play with words proposing the crasis "MurGEopark", which seems to indicate that it is "the last piece of Adria, the (almost) lost continent" (see the reason below), with the idea that plate tectonics are seen as a puzzle, and thereby igniting the imagination regarding the lost continent.

3. Geological Evolution of Apulia: The Leitmotiv of the Murge aUGGp

To understand the geology of Apulia, to which the Murge and Premurge areas belong, it is necessary to consider the geodynamic scenario in which the rocks that form Southern Italy developed and/or were involved over time.

This implies that we need to consider the plate tectonics theory [8–10], a scientific paradigm formulated to explain mountain building and many other phenomena distributed at the planetary scale. The idea that this territory of the Puglia Region is a candidate for being a Geopark starts from this paradigm because the area can be interpreted as a part of the little exposed remnant (the Apulia Foreland) of a plate that has almost completely disappeared by subduction [11,12] (Figure 4), the latter process being one of the main tenets of the theory [13,14]. Since the late 19th century, long before the formulation of the plate tectonics theory, it had been assumed that the Mediterranean region now occupied by the Italian peninsula and the Adriatic Sea was previously occupied by Adria, an old continent [15]. The complex distribution of the mountain chains around the Mediterranean Sea led to Adria being considered as a wider continental area, i.e., a promontory of the African continent [16] originally occupying the area east of the present day Baetic Cordillera, south of the present-day Alpine Chain, and the present-day Carpathian-Pannonian region up to large parts of the present-day Balkans and Anatolia [17].

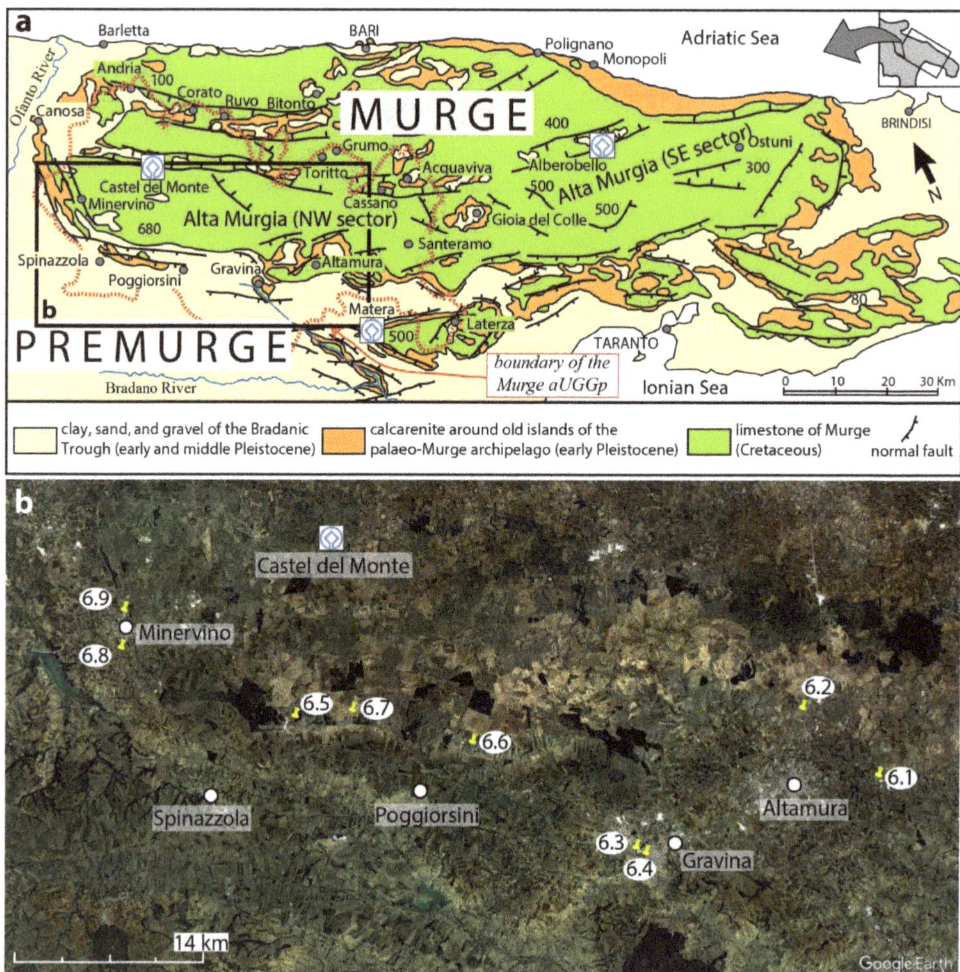

Figure 3. (**a**) Structural sketch of the Murge area (see inset for the location). Note the location of the three UNESCO World Heritage sites (Castel del Monte, Matera and Alberobello) quoted in chapter 5. The approximate boundary of the Murge aspiring UNESCO Global Geopark (Murge aUGGp) is reported. After [18,19]. (**b**) Location of the geosites (numbers 6.1–6.9) described in chapter 6. Geographic coordinates are reported in the same chapter. Base map from Google Earth.

This "Greater Adria" was firstly included in plate tectonics scenery by [20], and it was considered a relatively small continental plate split from Africa (the parent plate) that never totally disconnected from the latter [17,21]. This small plate was also called Apulia Plate [22], or Greater Apulia, which, according to a more complex palaeogeographic interpretation, resulted from two different microcontinents welded together during the Late Permian-Early/Middle Triassic to the northern sector of that part of the Pangea Supercontinent, which, in turn, later became the Africa Plate [23,24]. In the framework of the breakup of Pangea, from the middle Jurassic [25], the Alpine Tethys Ocean progressively separated the Eurasia Plate from the Africa-Adria ones, given that Adria connected to Africa through a narrow bridge ([17,21,26], among many others) (Figures 5 and 6a), and is comparable to the present-day terrestrial geographic relationship between Arabia and Africa plates.

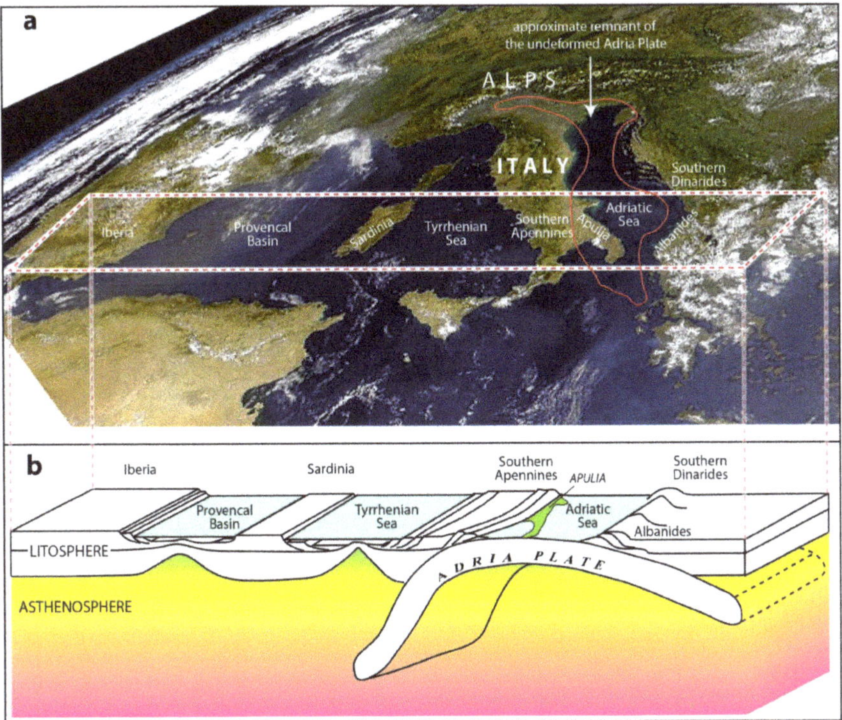

Figure 4. (**a**) Western Mediterranean Sea and surrounding regions, showing the approximate present-day shape of Adria and the portion of Earth surface, to which the 3D sketch refers (Figure 4b). NASA-Shuttle view of the central-western Mediterranean ([27], mod.); (**b**) 3D geodynamic model showing the double subduction of the Adria Plate (from [28], mod.).

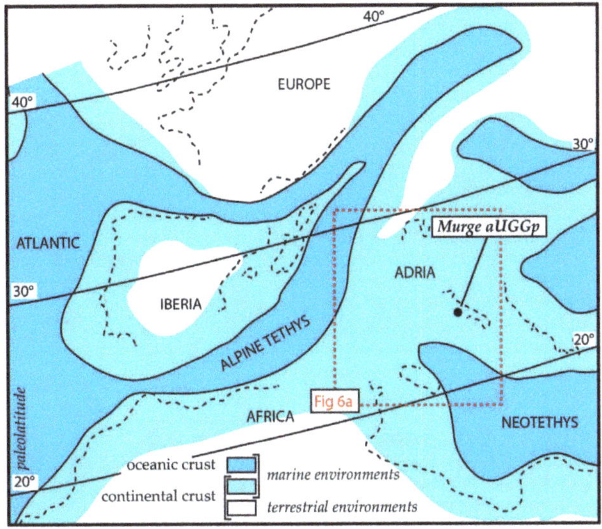

Figure 5. Palaeogeographic map of Adria at the beginning of the Late Cretaceous. Deeply simplified after [26], among many others.

During the Mesozoic, the portion of the Adria Plate later involved in the Southern Apennines orogenic system represented a passive margin [29,30] characterized by an alternation of deep basins and carbonate platforms (peri-Adriatic platforms) [31–33] (Figure 6a). In this context, slow subsidence rates compensated by shallow-water carbonate sedimentation favored the progradation and aggradation of these platforms [34–36], which, although flanked by deep-sea basins, were periodically connected to each other by continental bridges that allowed the migration of dinosaurs [17,20,21] (Figure 6a). As far as Puglia is concerned, it is important to underline the presence of one [37–39] of these platforms, known as the Apulia Carbonate Platform, whose vestiges can be recognized in the large carbonate rock masses of Gargano, Murge, and Salento [29] (Figure 1c).

Even if the Adria Plate originated as a consequence of the breakup of Pangea and a subsequent continental drift, the current configuration of Southern Italy is due to the following process of plate convergence and mountain building [21,40,41]. In particular, Apulia is today affected by a double subduction, producing the Apennine orogenic system to the west of the Adria Plate and the Dinaric/Hellenic orogenic system east of the same plate [28,42–44] (Figures 4 and 6). This dynamism can be appreciated only keeping in mind the deep-time concept, since the present-day geological framework should be considered a fixed frame of a long geodynamic movie, whose known stages for the development of Apulia date back to the Permian-Triassic (Puglia 1 well [29]), well before the beginning of the Apennines orogenic processes, which were taking place since the Oligocene (at least), and which still affect Southern Italy [45,46].

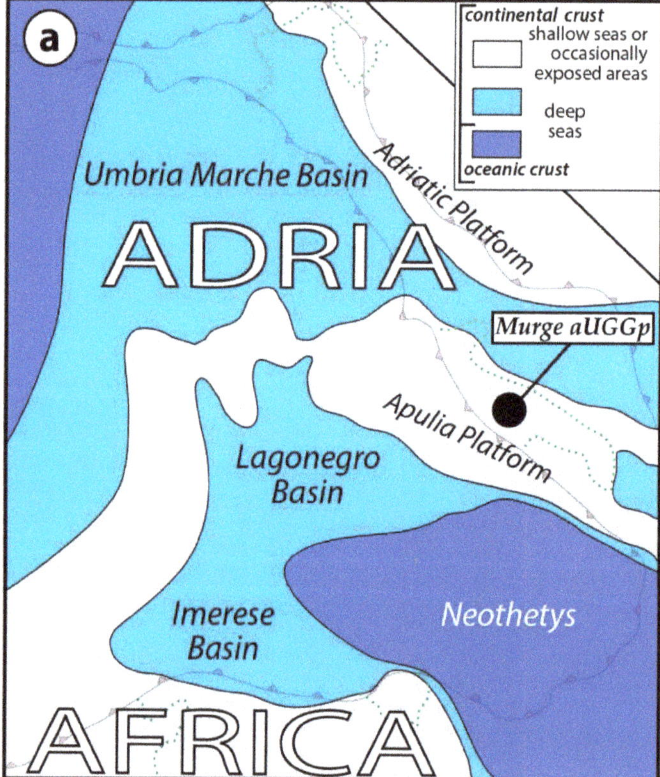

Figure 6. *Cont.*

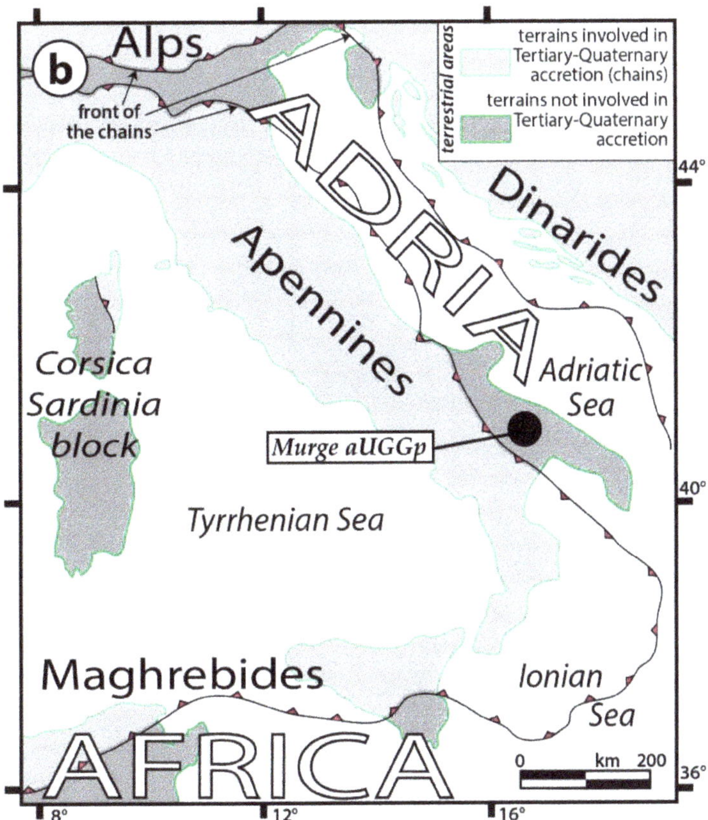

Figure 6. (**a**) Cenomanian paleogeography, according to [47]. Note the position of Murge within the Apulia Carbonate Platform. (**b**) Present-day structural sketch of Italy. Note the small size of the remnant of Adria Plate, surrounded by active chains. After [48], mod. Geographic coordinates in the WGS84 system.

It is worth recalling that the three main domains of an orogenic system are: (i.) the foreland, i.e., the area not yet reached by mountain deformations; (ii.) the foredeep (or, the "foreland basin", which is the more stringent definition used if what is being referred to is the Apennines foredeep), i.e., the flexed area at the foot of the mountain chain that receives sediment eroded from the reliefs; and (iii.) the chain, i.e., the mountainous area whose rocks are tectonically superimposed on each other and strongly deformed [9,49–52]. In such a dynamic context, the chain migrates towards the foredeep, progressively incorporating it into the mountain reliefs; at the same time, the foredeep migrates towards the flexing foreland, which becomes the bedrock of the foredeep itself; and the foreland "observes" this migration, waiting to be involved in the flexure in its area facing the chain (migrating foredeep in [53–55]). As regards the Apennines orogeny, it involved the western sector of the Adria Plate; as a consequence, the Italian peninsula shows the present-day three domains of the orogenic system (chain, foredeep, and foreland) corresponding, from west to east, respectively, to the Apennines Chain, the Apennines foredeep (i.e., from north to south, the Po Plain, the central-northern Adriatic Sea, and the Bradanic Trough, comprising the Premurge area), and the Apulia Foreland (i.e., Gargano, Murge, and Salento) (Figure 1b,c) [56–59].

During the Apennines orogeny, the upper crust (mainly the sedimentary cover) of the Mesozoic palaeogeographic domains of the western side of the Adria Plate was delaminated from its basement; several thrust sheets were tectonically stacked along the western portion of the Apulia Carbonate Platform, progressively forming the Southern Apennines Chain [46] (Figures 1 and 6b). The eastern portion of the Apulia Carbonate Platform, the sector still not involved in thrusting, corresponds to Apulia (i.e., the Apulia Foreland and the bedrock of the Bradanic Trough), where the crust of Adria is still rooted to its mantle [60,61]. During the formation of the Apennines, the Apulia Foreland itself was arched (e.g., [43], and references therein) and divided into blocks, which underwent relative lowering and uplift. The most raised areas correspond to the Gargano (a promontory reaching altitudes of about 1000 m), to the Murge (a plateau system that reaches altitudes of about 680 m), and to the Salento (formed by hilly reliefs, the Serre Salentine, with maximum elevations of about 200 m) (Figure 1). Normal faults affecting the Apulia Foreland formed two morphostructural staircases, respectively, dipping towards the east and into the Adriatic Sea, and towards the west as far as below the Southern Apennines [62,63] (Figure 1). This last morphostructural staircase corresponds to the bedrock of the Bradanic Trough, representing the most recent Southern Apennines foredeep [64].

4. Main Geological Features of the Murge aUGGp

The Murge aUGGp comprises the Alta Murgia area, where a Cretaceous sector of the Apulia Carbonate Platform crops out, and the adjacent Premurge area, where the southwestward lateral prosecution of the same platform, being flexed toward the Southern Apennines Chain, is covered by Plio-Quaternary foredeep deposits (Figure 1).

The Murge anatomy reflects part of the long geological history of Puglia, documented by field evidence that dates back to about 140 million years ago, i.e., the early Cretaceous [29,65]. At that time, the region that was going to be the Apulia Foreland was a wide shallow-marine inter-tropical area basically comparable to the present-day Bahamas [36]. There, on the Apulia Carbonate Platform, carbonate muds, which gradually were becoming limestones, deposited over millions of years up to about 66 million years ago (end of the Cretaceous), when, as mentioned, the platform underwent subaerial exposure. Later, in the area now corresponding to the Murge, these limestones were affected by extensional tectonics and a horst and graben structure, roughly NW-SE striking, leading the area to be characterized by topographic highs and lows [19,48,63,66]. Subsequently, during late Pliocene (about 3 million years ago), this region was affected by subsidence, induced by the eastward migration of the Apenninic orogenic system, which caused the return of the sea on limestones that had been exposed for a long time, previously experiencing several phases of karstification [67,68]. Occupying a region made up of highs and lows, this relative sea-level rise created a wide archipelago whose islands corresponded to the structural highs of the previously described faults system [59,69,70] (Figure 7).

Cretaceous limestones that today characterize the Murge karst landscape were the bedrock of both the islands and the structural depressions (straits and seaways) among islands (Figure 7a). A more important and deeper seaway, the Bradanic Trough, connected the central-northern Adriatic Sea to the Ionian Sea between the migrating Apennines and the Murge archipelago (Figure 7a). The slow subsidence of the whole Murge region caused the progressive submersion of the archipelago and coarse-grained coastal deposits, formed by a mix of skeletal carbonate fragments and detritus eroded from the exposed limestones, accumulated on island flanks [69,71,72] (shallow-marine carbonate sands in Figure 7a,b). Since palaeoenvironments were controlled by morphotectonic features of the Cretaceous bedrock, carbonate systems with different geometries and facies distribution developed (see examples in [71–74]).

After diagenesis, these carbonate sediments would become the easily-dug porous carbonate soft-rocks representing the bedrock on which rupestrian towns in the Murge and Premurge areas (e.g., Gravina in Puglia, Matera, Laterza, and Ginosa) developed [75,76] (Figure 8).

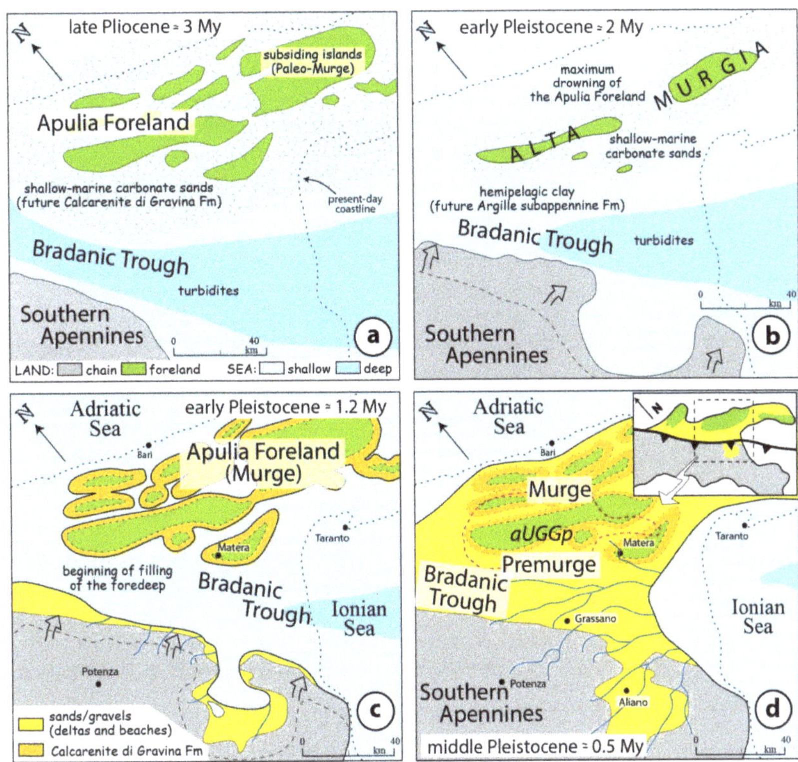

Figure 7. The late Pliocene to middle Pleistocene evolution (**a**–**d**) of the Southern Apennines foredeep system. From [5], mod. (Compare each picture with Figures 1c and 3.)

Figure 8. The thin Bradanic Trough (foredeep) succession along the Gravina (canyon) of Gravina in Puglia town, at the toe of Murge. After having eroded the Quaternary foredeep succession, the stream cut the bedrock, made up of Cretaceous limestones. Note the angular unconformity between tilted Cretaceous strata and the overlying sub-horizontal younger sedimentary units. Photo by D. Belfiore [77].

About 1.5 million years ago, the maximum relative rise of the sea on the islands of the palaeo-Murge archipelago was reached; this phenomenon left subaerially exposed only the highest reliefs of the Murge, corresponding to Alta Murgia (Figure 7b). At the same time, the sediments transported by rivers crossing the Apennines began to feed the Bradanic Trough seaway, which was progressively reached by offshore clays and filled with coastal sands and gravels [78–81] (Figure 7c). These same sediments also reached areas of the old archipelago, filling the narrow straits between the old islands; the latter were almost completely buried by sediment, with the exception of the most elevated ones [59,70] (Figure 7d). At the passage between the early and middle Pleistocene, the Apulia Foreland began to undergo a still active tectonic uplift [82–84], which would have progressively led the Murge to exceed 600 m of altitude. The beginning of uplift was the geological "moment" when the hydrographic network that now characterizes the Murge area established [85]. At the same time, the drainage network began to dissect the original flat top of the Premurge area, today at more than 450 m of altitude. Here, the rivers running on sands and gravels (uppermost deposits of the foredeep infill) progressively cut these topmost sediments, reaching the underlying clays. Locally, the drainage network also reached deeper and more ancient carbonate rocks, creating canyons locally called "gravine" (the plural of "gravina") [86–88] and disclosing how the Premurge bedrock corresponds to outcropping rocks of the Murge (Figure 8).

The "anomalous" middle-upper Quaternary uplift of Murge and its flanks was attributed to the presence of a thick lithosphere in correspondence with the present-day foreland that led to the buckling of the plate [83,84]. As a consequence: (i) old rocks of Adria crop out in the Alta Murgia, the latter being the hinge of the largest lithospheric antiform of the world (i.e., [43]); (ii) stratigraphic and structural features of the outer side of a subsiding foredeep are exposed today in the Premurge area [79].

5. The Geotouristic Appeal of the Murge aUGGp

The 2011 Arouca Declaration [89] defines geotourism "as tourism which sustains and enhances the identity of a territory, taking into consideration its geology, environment, culture, aesthetics, heritage and the well-being of its residents." There is a close connection between geotourism and geoparks since, according to the official UNESCO site [90], "UNESCO Global Geoparks are single, unified geographical areas where sites and landscapes of international geological significance are managed with a holistic concept of protection, education and sustainable development. [. . .] While a UNESCO Global Geopark must demonstrate geological heritage of international significance, the purpose of a UNESCO Global Geopark is to explore, develop and celebrate the links between that geological heritage and all other aspects of the area's natural, cultural and intangible heritages".

In accordance with [91], geology should also represent a cultural and social discipline and the first activity to pursue this objective is the scientific research on geosites understood as "one of the components of a given territory, on a par and together with fauna and flora or with cultural heritages such as works of art, monuments . . . etc." Therefore, well-studied (and correctly proposed) geosites would represent key points to suggest ecotouristic trips in which geology could be one of the main cultural interests for visitors (geotourism). A large number of geosites of international and national relevance characterizes. the Murge aUGGp but, at the moment, many individual geosites of the Murge aUGGp, even if listed for their geological importance, are either not open to the public or are proposed to tourists based upon other subjects far from (or ignoring) the geological ones. As suggested in this paper, the most interesting approach to propose geotourism activities in the area would be to preface any visit or tour with a geological regional view and to select some sites trying to follow the geological history of the area. Keeping in mind this aspect, each geotour or geosite could represent the opportunity to discover a geological world not strictly linked to a local peculiarity. In this regard, it is important to reiterate that:

-   The area represents a virtually undeformed sector of Adria, a continental plate located between Africa and Eurasia Plates and almost totally involved in subduction/collision processes [21,92].
-   The crust of Adria is still rooted to its mantle, as suggested by the analysis of several seismic lines crossing Apulia ([60] and quoted references), and the Cretaceous evolution of the plate is spectacularly recorded in the Murge area due to the outcropping shallow-water carbonate succession of one of the biggest peri-Adriatic carbonate platforms (the Apulia Carbonate Platform) [32].
-   An "anomalous" regional middle-late Quaternary foreland uplift led to spectacular exposition from the bedrock of a complete foredeep succession, pointing out that the latter can commonly be observed only after being deformed and involved in the mountain chain or visualized from subsurface data (i.e., seismic lines), especially in marine settings [42,43,54,93].

Geosites were described following a geochronologic criterium. Accordingly, the main geosites recognized within the Murge aUGGp have been divided into eight main categories, each characterized by at least one geosite of international value. Since several sites offer more than one significant geologic feature, the identified geosites can fall in one or more of the following categories: Spatial and/or Panoramic geosites; Apulia Carbonate Platform geosites; Bradanic Trough geosites; Quaternary Uplift-related geosites; Karst geosites; Tectonics-related geosites; Hydrogeology-related and Water-related geosites; and Man and Geology geosites. The presence of three UNESCO World Heritage Sites (Figure 3), one within the boundary of the aUGGp (Castel del Monte) and two in its surroundings (Trulli di Alberobello and Sassi di Matera), represents further tourist attractions that add value to the Murge aUGGp. Specifically, the Sassi di Matera rupestrian districts, after being the European Capital of Culture in 2019, saw an outstanding increase in tourism that reverberated in the adjacent Murge area, and that could also be oriented towards geotouristic targets [5,75,76].

6. Some Examples of Geosites of the Murge aUGGp

In the following sections, some geosites of potential geotouristic interest, and falling into one or more of the eight categories mentioned above, will be proposed. The concise geological description of each site is just sufficient to represent the scientific reasons underlying their importance. These geosites rest in the western sector of the Murge aUGGp, at the turn of Murge and Premurge, and are described moving from SE to NW (see quoted localities in Figure 3). Moreover, regarding the choice of how to link in an ideal geotouristic route, these sites are left to the sensitivity of both the guides and the visitors, and to logistical constraints as well.

6.1. Apulia Carbonate Platform Geosites: The Disused Quarry "Cava Pontrelli" of Altamura

40°48′24.92″ N; 16°37′14.48″ E

(a)   *The dinosaur tracksite*

In May 1999, in the vicinity of Altamura, on the bottom of a disused quarry ("Cava Pontrelli"), one of the world's largest dinosaur track sites was discovered (Figure 9) [94].

Figure 9. (**a**) The bottom of a disused quarry ("Cava Pontrelli") is one of the world's largest dinosaurs track sites. It records the passage of hadrosaurs and ankylosaurs on an original muddy surface during late Cretaceous. Note the impressive number of tracks, most of them not always referred to a trackway. (**b**) A detail of the surface showing, among many tracks, three coloured dinosaur trackways [94].

Up until this discovery, the Apulia Carbonate Platform was considered as a Bahamian-type isolated platform, but the presence of several dinosaur footprints in the carbonate succession led to a deep palaeogeographic revision, suggesting the presence of bridges between Periadriatic Platforms and the main continents of that time (Africa and Eurasia) to justify the occurrence of these terrestrial vertebrates [37–39,95]. Studied quadrupedal trackways suggest the passage of hadrosaurs and ankylosaurs on an original muddy surface during the late Cretaceous (early Campanian) [94,96,97].

(b) *Orbitally-controlled shallowing-upward peritidal sequences*

The carbonate succession cropping out along the walls of "Cava Pontrelli" is made up of about 50 m thick peritidal and shallow subtidal facies associations showing a shallowing-upward cyclic arrangement [98]. The occurrence of cyclic variations in the stratigraphic record is a widespread feature, and its study (cyclostratigraphy) in Mesozoic carbonate platform successions led to the improvement of the accuracy and resolution of geochrono-logic timetables [99,100]. This kind of study has been proposed in some selected portions of the Apulia Carbonate Platform succession [101–103] and is in progress in the lower Campanian section of "Cava Pontrelli" [96]. It is important to highlight that this easily accessible abandoned quarry with high walls could be used to discuss past climate change, cyclicity, and sea-level variations, comparing past and present-day climate change, and showing how environments change over time.

At the moment, the visit is possible exclusively through contacts with the Alta-mura Municipality.

6.2. Karst Geosites/Man and Geology Geosites: The "Grotta di Lamalunga" (Lamalunga Cave), in the Vicinity of Altamura

40°51′55.20″ N; 16°34′54.13″ E

The "Grotta di Lamalunga" (Figure 10) is one of the many underground karst features in the Altamura territory. With its entrance located on the right side of a typical karst valley (lama), the cave gained international recognition after 1993, when a complete Neanderthal skeleton was discovered by cavers in a small chamber of the karst system [104,105].

Figure 10. (**a**) Entrance of the "Grotta di Lamalunga" closed by a metal dome. (**b**) Group of geotourists with 3D glasses in the Lamalunga Visitor Center. (**c**) The skull of the Neanderthal skeleton discovered in 1993 [104].

Dating the calcite coatings above the bones provided an age comprised between 128 and 170 ka, whereas analysis of DNA revealed that this human belonged to Homo neanderthalensis [106]. In addition to the remarkable importance of the "Altamura man", the Lamalunga Cave hosts within the system many palaeontological remains and fossils yet to be studied, which cover several passages of the karst systems and are totally embedded in calcite crusts. The site is therefore of high relevance for analysis of the palaeoclimatic and environmental conditions for this sector of the Murge [107].

For information and a 3D virtual tour of the cave, it is possible to contact the "Centro Visite Lamalunga" (Lamalunga Visitor Center) (Figure 10).

6.3. Spatial and/or Panoramic Geosites/Bradanic Trough Geosites/Quaternary Uplift-Related Geosites: The Panoramic Point of Gravina in Puglia

40°49′32.46″ N; 16°24′27.68″ E

In addition to being a meaningful portion of the small remnant of the Adria Plate, Murge and Premurge represent one of the few worldwide examples in which different evolutionary steps of the transition from foreland to external foredeep (foreland ramp) are well exposed. As mentioned above, in some places of the Premurge area, it is possible to observe both the bedrock of the foreland basin, i.e., the same limestones of Alta Murgia, and the whole foredeep sedimentary wedge pinching-out the foreland (Figure 11; see also Figure 1d). This relatively condensed lower Pleistocene succession is basically characterized by shallow-marine deposits comprising mainly, from the bottom to the top, bioclastic temperate-water (heterozoan) coastal-carbonates (Calcarenite di Gravina Fm), silty clay shelfal hemipelagites (Argille subappennine Fm), and coastal/alluvial sandy and gravelly deposits (Monte San Marco Fm) (Figure 11b,c) [59,79–81,108].

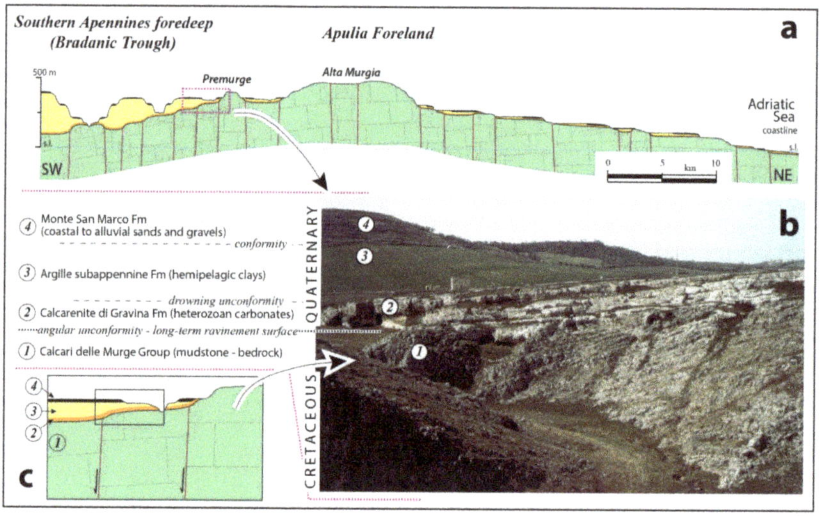

Figure 11. (**a**) Geological cross-section of Murge (Apulia Foreland) and Premurge (Bradanic Trough), showing the relationships between Cretaceous and Quaternary deposits. Compare with Figure 1d. (**b**) Photo from the panoramic point in the vicinity of the stadium parking of Gravina in Puglia town. Note the relationships between the faulted Cretaceous bedrock and the Quaternary foredeep succession. (**c**) Geological cross-section showing (inset) the virtual position of the outcrop of (**b**) and how to appreciate the foredeep sedimentary wedge pinching-out the foreland. The whole figure from [42,69,74,109] mod.

One of the best places to observe the whole foredeep stratigraphy and to understand its location within the geological framework of the entire region is from the town of Gravina in Puglia and its surroundings (Figures 8 and 11b). At these panoramic points, a significant synthesis of the whole history of the outer foredeep can be appreciated, from subsidence (the succession filling the basin) to uplift (the deepening of the canyon).

Some panels in the town explain the geology of the area (Figure 12). Other panoramic points are from the stadium parking [42,109].

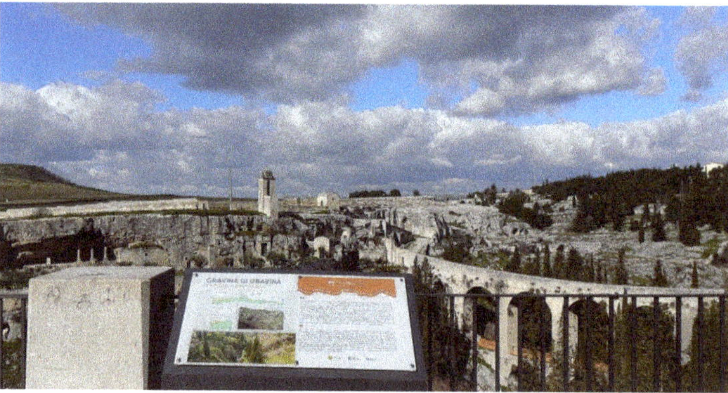

Figure 12. The Aqueduct Sant'Angelo—Fontana della Stella (corresponding to the bridge at Gravina in Puglia town). Note the explanatory panel in front of the panoramic point. Photo courtesy of Giusy Schiuma.

6.4. Hydrogeology- and Water-Related Geosites: The Sant'Angelo—Fontana della Stella Aqueduct, an Example of Tapping and Distributing Water since Middle Age

40°49′11.50″ N; 16°24′49.58″ E

Presence of water in karst areas has always been a serious problem for the local communities: the peculiar aspects of karst hydrogeology [110–112], with most of the water being rapidly drained underground, to create the complex subterranean network of conduits and caves, caused past inhabitants to face seriously the issue of looking for water, and preserving it during the summer season [113–115]. For these reasons, evidence of historical management of water resources is of extreme importance and represents a remarkable heritage to preserve.

At Gravina in Puglia, the Aqueduct Sant'Angelo—Fontana della Stella is a wonderful example of the ability to build underground hydric works, able to function for centuries (Figures 8 and 12).

As documented by historical sources, construction of the aqueduct started in 1743 [111,116], even though possibilities of a likely older origin have been postulated. With an overall length of some 3.5 km, the aqueduct is one of the best preserved underground man-made works for the collection and transport of water resources in Southern Italy [117]. A system of underground galleries (average height 1.75 m, width 0.77 m), connected to the surface by inspection wells (to clean periodically and to manage the hydraulic work), allowed the waters to flow toward the town. The subterranean system ends up at the right valleyside of the Gravina canyon; to pass the deep valley, and let the water reach the final destination, a 90 m long bridge-canal was built across the canyon. The Sant'Angelo—Fontana della Stella aqueduct is a very important heritage [118,119], since it is one of the most significant ancient subterranean water-system in the Puglia Region (Figure 8).

A panel displayed on the abutment of the bridge provides some information on the aqueduct (Figure 12).

6.5. Spatial and/or Panoramic Geosites/Tectonics-Related Geosites: The Alta Murgia (or Murge Alte) Scarp

40°58′35.75″ N; 16°13′45.32″ E

As described before, the Apulia Foreland represents the edge of a wide WNW-ESE trending antiform whose flanks correspond to down faulted blocks of the Apulia Carbonate Platform. The Murge and Premurge areas are also characterized by the extensional activity of these faults, generally attributed to Pliocene and early Pleistocene times [66,120]. One of the most important of these faults is the NW-SE striking fault of "Valle del Bradano" [121], corresponding to the scarp that separates the Murge from the Premurge (i.e., from the Bradanic Trough) (Figure 13). The steep and straight Alta Murgia scarp (ca. 35 km long and up to 200 m high, from ca. 450 m up to ca. 650 m a.s.l.) includes the most impressive morphotectonic evidence of early Pleistocene tectonics. The faulted bedrock (Cretaceous limestones of the Apulia Platform) is exposed in the higher free face of the slope and covered, downwards, by Pleistocene deposits. These deposits are slope carbonate breccias organized in steep and short fans often coalescing and forming a long string bordering the northwestern part of the Murge area, and they record the interaction of active tectonics with climate change [122,123]. Only some quarry walls show the presence of the normal-fault plain (Figure 14).

Figure 13. The Alta Murgia scarp, bounding the Murge from the Bradanic Trough (Premurge), corresponds to a receded fault plane (dashed red line). Landscape seen from the Premurge area.

Figure 14. (**a**) The Alta Murgia scarp cut in an abandoned quarry and seen from the top of the Murge. Note the regularized sloping surface and the beginning of the Premurge area at its toe. (**b**) Clinostratified reddish breccia deposits (on the left of the quarry wall) pinching out against the fault plain bounding the white sub-horizontal Cretaceous beds (on the right of the quarry wall). (**c**) Detail of the contact between the Pleistocene breccia deposits (on the left) and the Cretaceous bedrock (on the right) (dashed white line).

The scarp is the scenery of almost all touristic itineraries straddling Murge and Pre-murge areas. Breccia deposits crop out along road cuts or not accessible (active or abandoned) quarries. Explanatory panels have not yet been realized.

6.6. Apulia Carbonate Platform Geosites/Tectonics-Related Geosites: The Oligocene Calcare a Planorbis Formation

40°55′43.33″ N; 16°18′43.24″ E

The Oligocene Calcare a Planorbis Fm crops out at the top of the Alta Murgia scarp (Jazzo Madama locality), unconformably resting on Cretaceous carbonates (Figure 15); from a panoramic point of view close to Poggiorsini, it is possible to appreciate that the formation developed on two different and not coeval depocenters [124] (Figure 15). This stratigraphic architecture can be attributed to the synsedimentary development of an active strike-slip basin [125]. The geosite has an international value both for its tectonic meaning in the area (the only record of tectonics between Cretaceous and Pliocene in the Apulia Foreland) and for the possibility that correlates its continental paleontological content to the distant Paratethys. As suggested by [125], the site could offer a link with a similar setting in the Apennines, where a geosite with lacustrine succession linked to strike slip tectonics [126] is proposed in a touristic path devoted to promoting geology, following the writings and the paintings of Carlo Levi (a famous Italian artist, who suffered a political exile during the fascist era) [5].

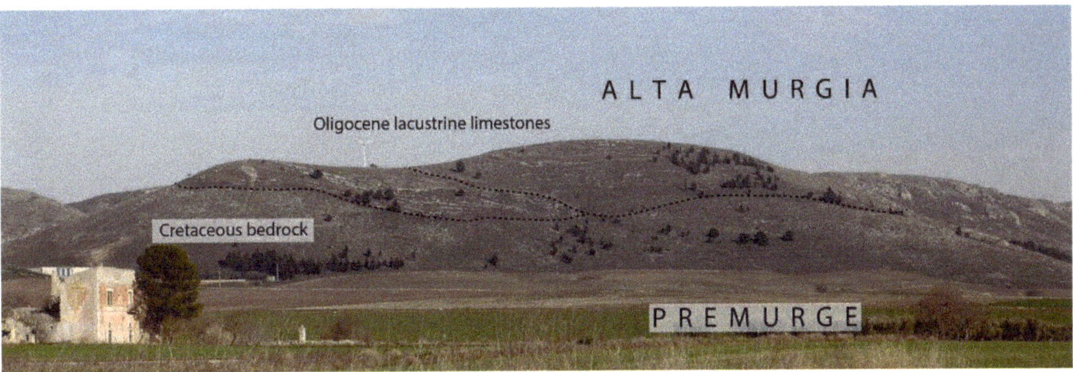

Figure 15. The Oligocene lacustrine succession perched at the top of the Alta Murgia scarp. After [125].

The lacustrine succession can be observed from panoramic points along the road linking Gravina in Puglia and Minervino. Explanatory panels have not yet been made.

6.7. Apulia Carbonate Platform Geosites/Tectonics-Related Geosites: The intra-Cretaceous Unconformity

40°59′15.85″ N; 16°10′55.92″ E

The oldest tectonic phase recognized in the Murge area is Turonian in age and is highlighted by a regional unconformity in the Cretaceous succession and by the presence of bauxites along the unconformity. The abandoned bauxite mines (open pits) of Murgetta Rossa (Figure 16), in the vicinity of Spinazzola village, represent a touristic attraction, too, since the variety of uncommon intense colours of the outcropping rocks (reddish, brownish, yellowish, greenish), due to the abundant presence of oxide and phyllosilicate minerals, immediately excites and intrigues any visitor ("Murgetta Rossa" literally means "small red rocky area").

Figure 16. One of the abandoned bauxite mines of "Murgetta Rossa".

Outcropping residual rocks are hosted in palaeokarst (canyon-like) features [67] and the relationships between faulting and residual rock formations can be easily observed in the field. This peculiarity makes the site an ideal stratigraphic, mineralogical, and tectonic case study of international attractiveness in different fields of the Earth Science [127–133].

The site is equipped with panels for visits, but information is basically devoted to its aesthetic value, to biodiversity, and to industrial archeology.

6.8. Bradanic Trough Geosites: A Gilbert-Type Delta in Carbonate Succession

41° 4′22.98″ N; 16° 3′39.48″ E

During the early Pleistocene, as a consequence of the migration of the Southern Apennines orogenic system, the Apulia Foreland underwent a relatively rapid increase in regional subsidence, and regional transgression resulted on the flexed foreland, i.e., the paleo-Premurge (Figure 7). The return of the sea on this karst area led to the deposition of a thin (no more than a few tens of metre-thick) mantle of carbonates (the Calcarenite di Gravina Fm) on the Cretaceous bedrock (Figures 11 and 17).

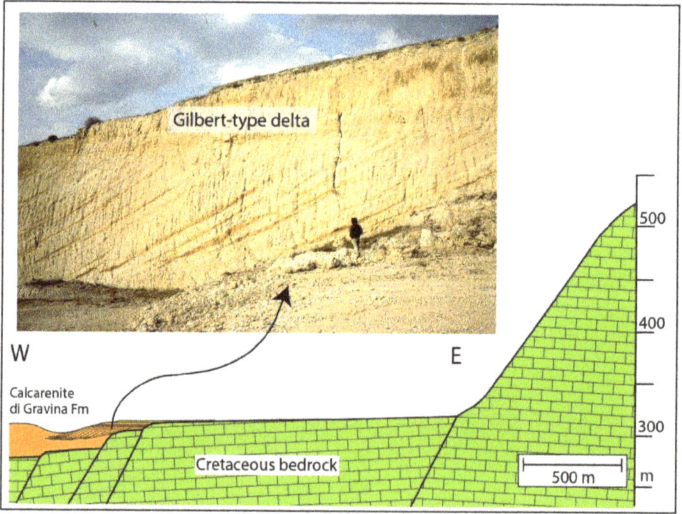

Figure 17. The Minervino delta and its location at the toe of the Murge Alte scarp mod [134].

A lower Pleistocene Gilbert-type delta, encased in offshore facies, outcrops along the western margin of Alta Murgia, west of Minervino town, and is well visible along the walls of an abandoned quarry, with a total thickness of about 12 m. This kind of delta shows a characteristic tripartite geometry with an upper part characterized by sub-horizontal strata, and a middle part composed of inclined beds (up to 30°/35°) passing to a lower part with sub-horizontal strata (Figure 17). Basically, this kind of delta develops at the foot of uplifting mountain regions directly facing the sea. The uniqueness of the Minervino delta comes from the fact that it developed on a flexuring and a subsiding karst setting, affected by tensional tectonics rather than at the foot of an uplifting mountain chain affected by compressional tectonics; further, it is composed only of carbonate extraclasts, i.e., rounded fragments of Cretaceous limestones [69,109,134]. Only one other gravelly carbonate delta has been described in international literature: it was observed in Croatia and developed during the Eocene at the base of an uplifting carbonate thrust [135].

The geosite is not equipped for visits, even if it is easy to access and can be reached with a short walk through the olive fields.

6.9. Karst Geosites/Man and Geology Geosites: The "Grotta di San Michele" (San Michele Cave) at Minervino

41°5′39.25″ N; 16°4′34.75″ E

An important aspect of the relationship between man and geology in the Murge is represented by religious and worship issues. As in many other parts of the world, karst caves have become worship sites for different cults. One of the best examples is the Michaelic worship; several towns in the Murge area host such sites with rupestrian churches, such as at Gravina in Puglia, Altamura, and Andria. However, its more remarkable expression is represented by a natural limestone cave at Minervino Murge, which was later transformed into a church dedicated to the Archangel Michael ("Grotta di San Michele") (Figure 18).

Figure 18. The interior of San Michele Cave during recent studies.

Recent studies were performed to define the stability assessment in the rock mass of the cave [136,137], since this rupestrian place is visited by Italian and foreign tourists. Nowadays, a picturesque procession takes place on the occasion of the patronal feast in honor of Saint Michael, on 29 September, whilst on 8 May a pilgrimage starts from the town and ends in the church with a religious ceremony. It brings to mind a famous path connecting different places of worship dedicated to the Archangel (in Puglia, in Italy, and in the world), all closely linked to geological features. In Puglia, many karst caves are dedicated to the Archangel Michael, the most important being the oldest shrine in Western Europe at Monte Sant'Angelo, in the Gargano Promontory, which became a UNESCO World Heritage Site in 2011.

The San Michele Cave is opened to public and can be visited, with information provided by searching the web with the words, "Grotta di San Michele, Minervino".

7. Conclusions

Due to its geodiversity, the Murge aUGGp represents a good opportunity to disseminate geological knowledge to a wide and diversified audience. It also represents a study area that still contains a great variety of geological items to be discovered and/or described. Several geological topics can be followed both by researchers and geotourists, but each of them should be traced back to the geological history of the Puglia region, which is in turn closely related to the evolution of the Adria Plate, a piece of the plate tectonics puzzle that played a pivotal role in the Thetys and later in the development of the Mediterranean realm. For this reason, the local geological singularities should not distract us from the unifying geological reasons that led to its candidacy for being a geopark.

8. A General Remark

In our experience, most of the protected areas in Italy suffer from a lack of geological knowledge and/or related dissemination and, even when the geological reason is the predominant for the establishment of a protected area, this boils down to the production of a report often lost in the back of a drawer, with further activities almost exclusively devoted to biological aspects of the area (biodiversity largely prevailing over geodiversity). Furthermore, the link between the geological heritage and all other aspects of the territory is often attributed only to the landscape, meant as a scenography with only an aesthetic value. Taking into account that in our territories it could be difficult to find landscapes not influenced or modified by human activity, this means that the dissemination interests (among many others) mainly concern prehistory, history, architecture, agriculture, rurality, and gastronomy, all proposed without any concrete link with geology.

Our paper represents an introduction to the scientific grounds that led to the nomination of the Murge and Premurge areas as an aspiring UNESCO Global Geopark due to the realization of a scientific dossier. The dossier was produced by a working group of the "Dipartimento di Scienze della Terra e Geoambientali" (Department of Earth and Geoenvironmental Sciences) of the "Aldo Moro" University of Bari (Italy), corresponding to the first authors of the present paper (M.A.C., V.D.S., V.F., M.P., L.S. (Luisa Sabato), L.S. (Luigi Spalluto), R.F., V.I. and G.A.M., coordinated by M.T.). The dossier and the present paper derive both from thirty years of experience and study in the Murge and Premurge areas made by the working group and from other data collected and interpreted thanks to the collaboration with all the other authors. After these efforts, we hope that visitors and geotouristic operators realize that a holistic approach is not just about eating a sandwich in front of a silent geosite.

Author Contributions: Conceptualization, M.T. and L.S. (Luisa Sabato); investigation, all the Authors; writing—original draft preparation, M.T., L.S. (Luisa Sabato), M.A.C., V.D.S., V.F., M.P. (Mario Parise) and L.S. (Luigi Spalluto); writing—review and editing, M.T., L.S. (Luisa Sabato) and M.C.; visualization, M.T. and L.S. (Luisa Sabato); supervision, M.T. and L.S. (Luisa Sabato); project administration, M.T. and G.A.M.; funding acquisition, M.T., V.F., M.P. (Mario Parise) and L.S. (Luisa Sabato). All authors have read and agreed to the published version of the manuscript.

Funding: The "Dipartimento di Scienze della Terra e Geoambientali" (Department of Earth and Geoenvironmental Sciences) of the "Aldo Moro" University of Bari (Italy) was involved by the Alta Murgia National Park to produce the geological section of the UNESCO Global Geopark application dossier though the "Convenzione di Ricerca: Il patrimonio geologico delle Murge per la candidatura GEOPARCO UNESCO del Parco Nazionale dell'Alta Murgia nel 2020". In addition to this financial commitment, the Department has invested energy and time in obtaining funding to activate: 3 PhD fellowships, a university taught course for external users, and several research projects, all of them regarding studies and dissemination of geology in the area. These activities are dedicated to culturally and scientifically support (over time) the idea that, as regards geology, the area can aspire to become a UNESCO Global Geopark. Research, still in progress, is now financially supported by the University of Bari through: The HORIZON EUROPE SEEDS fund "S63—Patrimonio naturalistico e turismo culturale ed escursionistico in aree protette (pa.na.c.e.a.) Divulgazione dei concetti di "sviluppo sostenibile" e di "conservazione e gestione della geo/biodiversità" nel Parco Nazionale Alta Murgia, aspiring UNESCO Global Geopark" (to M. Tropeano). Codice Identificativo Progetto S63, CUP H91I21001670006; PhD grants to: (i.) E. Lippolis: "Studio di percorsi natura di carattere didattico/geoturistico nell'ambito del candidato Geoparco delle Murge (aUGGp): la stratigrafia delle successioni carbonatiche affioranti nei comuni delle aree interne delle Murge (Poggiorsini, Spinazzola e Minervino Murge) come base di divulgazione geo-scientifica" ["Study of didactic/geotouristic nature-trails in the context of the candidate Geopark of Murge (aUGGp): the stratigraphy of the carbonate successions outcropping in the municipalities of the internal areas of Murge (Poggiorsini, Spinazzola and Minervino Murge) as a basis for geo-scientific dissemination"] aimed at the correct transfer of geological topics to a non-expert audience. PhD Grant concessed by the "Agenzia per lo Sviluppo Territoriale" ("D.M. 725, 22-06-2021—Dottorato Comunale"); (ii.) F. Bellini: "Valorizzazione del 'capitale naturale geologico' a fini turistici (geoturismo) del Parco Nazionale Alta Murgia (candidato Geoparco UNESCO) e realizzazione di piattaforme e siti digitali dedicati ai visitatori e alla comunità degli smartworkers" ["Enhancement of the 'geological natural capital' for tourism purposes (geotourism) of the Alta Murgia National Park (UNESCO Geopark candidate) and creation of platforms and digital sites dedicated to visitors and the community of smartworkers"], in collaboration with the Everywhere TEW Company, aimed at identify geotouristic routes linking geosites of broad interest; and (iii.) U.S. D'Ettorre: "Fenomeni di desertificazione nelle Murge pugliesi ed analisi quali-quantitativa delle risorse idriche" ["Desertification phenomena in the Apulian Murge, and quali-quantitative analysis of hydric resources"]. Research funds to the project "GeoSciences: un'infrastruttura di ricerca per la Rete Italiana dei Servizi Geologici—GeoSciences IR" (codice identificativo domanda: IR0000037); CUP: I53C22000800006. Piano Nazionale di Ripresa e Resilienza, PNRR, Missione 4, Componente 2, Investimento 3.1, "Fondo per la realizzazione di un sistema integrato di infrastrutture di ricerca e innovazione" finanziato dall'Unione Europea—Next Generation EU (to V. Festa). research funds to the project "Interventions for exploration of karst phenomena", Apulia Region, Environmental Division, 2019–2021 (to M. Parise).

Institutional Review Board Statement: Not applicable.

Informed Consent Statement: Not applicable.

Data Availability Statement: Not applicable.

Acknowledgments: The authors acknowledge the Academic Editor and anonymous reviewers for their comments and suggestions that helped us to improve the manuscript. The corresponding author L. Sabato also thanks the Editor and Assistants.

Conflicts of Interest: The authors declare no conflict of interest.

References

1. Fiore, A.; Valletta, S. *Il Patrimonio Geologico Della Puglia. Territorio e Geositi*; Fiore, A., Valletta, S., Eds.; SIGEA: Avigliana, Italy, 2010.
2. Bentivenga, M.; Geremia, F.; Fiore, A.; Valletta, S. Geoheritage: Protecting and Sharing. In Proceedings of the 7th Int. Symp. ProGEO on the Conservation of the Geological Heritage "Geoheritage: Protecting and Sharing", Bari, Italy, 24–28 September 2012.
3. Mastronuzzi, G.; Valletta, S.; Damiani, A.; Fiore, A.; Francescangeli, R.; Giandonato, P.B.; Iurilli, V.; Sabato, L. *Geositi Della Puglia. Ricognizione e Verifica Dei Geositi e Delle Emergenze Geologiche Della Regione Puglia*; Mastronuzzi, G., Valletta, S., Damiani, A., Fiore, A., Francescangeli, R., Giandonato, P.B., Iurilli, V., Sabato, L., Eds.; Graphic Concept Lab.: Bari, Italy, 2015.
4. Doglioni, C. Foredeeps versus Subduction Zones. *Geology* **1994**, *22*, 271. [CrossRef]

5. Sabato, L.; Tropeano, M.; Festa, V.; Longhitano, S.G.; Dell'Olio, M. Following Writings and Paintings by Carlo Levi to Promote Geology within the "Matera-Basilicata 2019, European Capital of Culture" Events (Matera, Grassano, Aliano—Southern Italy). *Geoheritage* **2019**, *11*, 329–346. [CrossRef]

6. Ghielmi, M.; Tropeano, M.; Carubelli, P.; Pugliese, A.; Sabato, L. Evoluzione Tettono-Stratigrafica Nel Pliocene e Quaternario, Con Particolare Riferimento All'avanfossa Appenninica. In *Basilicata*; Tropeano, M., Sabato, L., Schiattarella, M., Eds.; Società Geologica Italiana: Bologna, Italy, 2020; pp. 51–61.

7. Zackabier. Europe According to Creative People—What Europe's Countries Look Like. Available online: https://www.youtube.com/watch?v=2L-ZYcqqGds (accessed on 15 March 2023).

8. Wilson, J.T. Hypothesis of Earth's Behaviour. *Nature* **1963**, *198*, 925–929. [CrossRef]

9. Dewey, J.F.; Horsfield, B. Plate Tectonics, Orogeny and Continental Growth. *Nature* **1970**, *225*, 521–525. [CrossRef]

10. Dewey, J.F.; Bird, J.M. Mountain Belts and the New Global Tectonics. *J. Geophys. Res.* **1970**, *75*, 2625–2647. [CrossRef]

11. Cuffaro, M.; Riguzzi, F.; Scrocca, D.; Antonioli, F.; Carminati, E.; Livani, M.; Doglioni, C. On the Geodynamics of the Northern Adriatic Plate. *Rend. Lincei* **2010**, *21*, 253–279. [CrossRef]

12. Carminati, E.; Petricca, P.; Doglioni, C. Mediterranean Tectonics. In *Encyclopedia of Geology (Second Edition)*; Alderton, D., Scott, A.E., Eds.; Academic Press: Cambridge, MA, USA, 2021; Volume 4, pp. 408–419.

13. Benioff, H. Orogenesis and Deep Crustal Structure—Additional Evidence from Seismology. *Bull. Geol. Soc. Am.* **1954**, *65*, 385–400. [CrossRef]

14. White, D.A.; Roeder, D.H.; Nelson, T.H.; Crowell, J.C. Subduction. *Geol. Soc. Am. Bull.* **1970**, *81*, 3431–3432. [CrossRef]

15. Suess, E. *Das Antlitz Der Erde (The Face of the Earth)*; F. Tempsky: Prague, Czech Republic, 1883.

16. Argand, É. La Tectonique de l'Asie. In Proceedings of the Proceedings of the XIIIth International Geological Congress, Brussels, Belgium, 10 August 1924; pp. 171–372.

17. Channell, J.; D'Argenio, B.; Horvath, F. Adria, the African Promontory, in Mesozoic Mediterranean Palaeogeography. *Earth Sci. Rev.* **1979**, *15*, 213–292. [CrossRef]

18. Pieri, P.; Festa, V.; Moretti, M.; Tropeano, M. Quaternary Tectonic Activity of the Murge Area Apulian Foreland—Southern Italy. *Ann. Geofis.* **1997**, *40*, 1395–1404. [CrossRef]

19. Tropeano, M.; Pieri, P.; Moretti, M.; Festa, V.; Calcagnile, G.; Del Gaudio, V.; Pierri, P. Tettonica Quaternaria Ed Elementi Di Sismotettonica Nell'area Delle Murge (Avampaese Apulo). *Il Quat.* **1997**, *10*, 543–548.

20. Dewey, J.F.; Pitman, W.C.; Ryan, W.B.F.; Bonnin, J. Plate Tectonics and the Evolution of the Alpine System. *Geol. Soc. Am. Bull.* **1973**, *84*, 3137–3180. [CrossRef]

21. Van Hinsbergen, D.J.J.; Torsvik, T.H.; Schmid, S.M.; Maţenco, L.C.; Maffione, M.; Vissers, R.L.M.; Gürer, D.; Spakman, W. Orogenic Architecture of the Mediterranean Region and Kinematic Reconstruction of Its Tectonic Evolution since the Triassic. *Gondwana Res.* **2020**, *81*, 79–229. [CrossRef]

22. Lort, J.M. The Tectonics of the Eastern Mediterranean: A Geophysical Review. *Rev. Geophys.* **1971**, *9*, 189–216. [CrossRef]

23. Stampfli, G.M.; Mosar, J. The Making and Becoming of Apulia. *Mem. Sci. Geol.* **1999**, *51*, 141–154.

24. Stampfli, G.M. Plate Tectonics of the Apulia-Adria Microcontinents. In *CROP PROJECT: Deep Seismic Exploration of the Central Mediterranean and Italy*; Finetti, I.R., Ed.; Elsevier: Amsterdam, The Netherlands, 2005; pp. 747–766. ISBN 9780080457604.

25. Bernoulli, D.; Jenkyns, H.C. Alpine, Mediterranean, and Central Atlantic Mesozoic Facies in Relation to the Early Evolution of the Tethys. In *Modern and Ancient Geosynclinal Sedimentation*; Sp. Publ. 19; Dott, R.H., Shaver, R.H., Eds.; SEPM (Society of Economic Paleontologists and Mineralogists): Tulsa, OK, USA, 1974; pp. 129–160.

26. Stampfli, G.M.; Hochard, C. Plate Tectonics of the Alpine Realm. *Geol. Soc. Lond. Spec. Publ.* **2009**, *327*, 89–111. [CrossRef]

27. Catalano, R. Appunti Di Geologia Regionale. Available online: http://www.siripro.it/dipgeopa.asp?structure=education&where=regionale&cap=09&subc=5&lang=it (accessed on 15 March 2023).

28. Doglioni, C.; Moretti, I.; Roure, F. Basal Lithospheric Detachment, Eastward Mantle Flow and Mediterranean Geodynamics: A Discussion. *J. Geodyn.* **1991**, *13*, 47–65. [CrossRef]

29. Ricchetti, G.; Ciaranfi, N.; Luperto Sinni, E.; Mongelli, F.; Pieri, P. Geodinamica Ed Evoluzione Sedimentaria e Tettonica Dell'avampaese Apulo. *Mem. Soc. Geol. Ital.* **1988**, *41*, 57–82.

30. Bernoulli, D. Mesozoic-Tertiary Carbonate Platforms, Slopes and Basins of the External Apennines and Sicily. In *Anatomy of an Orogen: The Apennines and Adjacent Mediterranean Basins*; Springer: Dordrecht, The Netherlands, 2001; pp. 307–325.

31. Mostardini, F.; Merlini, S. Appennino Centro Meridionale. Sezioni Geologiche e Proposta Di Modello Strutturale. *Boll. Soc. Geol. Ital.* **1986**, *7*, 117–202.

32. Zappaterra, E. Carbonate Paleogeographic Sequences of the Periadriatic Region. *Boll. Soc. Geol. Ital.* **1990**, *109*, 5–20.

33. Zappaterra, E. Source-Rock Distribution Model of the Periadriatic Region. *Am. Assoc. Pet. Geol. Bull.* **1994**, *78*, 333–354. [CrossRef]

34. D'Argenio, B. Le Piattaforme Carbonatiche Periadriatiche. Una Rassegna Di Problemi Nel Quadro Geodinamico Mesozoico Dell'area Periadriatica. *Mem. Soc. Geol. Ital.* **1974**, *13*, 137–161.

35. Ricchetti, G. Nuovi Dati Stratigrafici Sul Cretaceo Delle Murge Emersi Da Indagini Nel Sottosuolo. *Boll. Soc. Geol. Ital.* **1975**, *94*, 1083–1108.

36. Eberli, G.P.; Anselmetti, F.S.; Betzler, C.; Konijnenburg, J.-H.; Van Bernoulli, D. Carbonate Platform to Basin Transitions on Seismic Data and in Outcrops: Great Bahama Bank and the Maiella Platform Margin, Italy. In *Seismic Imaging of Carbonate Reservoirs and Systems*; Eberli, G.P., Masaferro, J.L., Sarg, J.F., Eds.; American Association of Petroleum Geologists: Tulsa, OK, USA, 2004; pp. 207–250. ISBN 0891813624.
37. Bosellini, A. Dinosaurs "Re-Write" the Geodynamics of the Eastern Mediterranean and the Paleogeography of the Apulia Platform. *Earth Sci. Rev.* **2002**, *59*, 211–234. [CrossRef]
38. Zarcone, G.; Petti, F.M.; Cillari, A.; Di Stefano, P.; Guzzetta, D.; Nicosia, U. A Possible Bridge between Adria and Africa: New Palaeobiogeographic and Stratigraphic Constraints on the Mesozoic Palaeogeography of the Central Mediterranean Area. *Earth Sci. Rev.* **2010**, *103*, 154–162. [CrossRef]
39. Randazzo, V.; Di Stefano, P.; Schlagintweit, F.; Todaro, S.; Cacciatore, M.S.; Zarcone, G. The Migration Path of Gondwanian Dinosaurs toward Adria: New Insights from the Cretaceous of NW Sicily (Italy). *Cretac. Res.* **2021**, *126*, 104919. [CrossRef]
40. Boccaletti, M.; Elter, P.; Guazzone, G. Plate Tectonic Models for the Development of the Western Alps and Northern Apennines. *Nat. Phys. Sci.* **1971**, *234*, 108–111. [CrossRef]
41. Hsü, K.J. Origin of the Alps and Western Mediterranean. *Nature* **1971**, *233*, 44–48. [CrossRef]
42. Cicala, M. Relationships between Flexure of the Apulia Foreland and Sedimentation: Onshore and Offshore Examples. Ph. D. Thesis, University of Bari "Aldo Moro", Bari, Italy, 2023.
43. Cicala, M.; Festa, V.; Sabato, L.; Tropeano, M.; Doglioni, C. Interference between Apennines and Hellenides Foreland Basins around the Apulian Swell (Italy and Greece). *Mar. Pet. Geol.* **2021**, *133*, 105300. [CrossRef]
44. De Alteriis, G. Different Foreland Basins in Italy: Examples from the Central and Southern Adriatic Sea. *Tectonophysics* **1995**, *252*, 349–373. [CrossRef]
45. Gueguen, E.; Doglioni, C.; Fernandez, M. On the Post-25 Ma Geodynamic Evolution of the Western Mediterranean. *Tectonophysics* **1998**, *298*, 259–269. [CrossRef]
46. Patacca, E.; Scandone, P. Geology of the Southern Apennines. *Boll. Soc. Geol. Ital.* **2007**, *7*, 75–119.
47. Patacca, E.; Scandone, P. Il Contributo Degli Studi Stratigrafici Di Superficie e Sottosuolo Alla Conoscenza Dell'Appennino Lucano. In Proceedings of the Atti del Congresso Ricerca, Sviluppo e Utilizzo delle Fonti Fossili: Il Ruolo del Geologo, Potenza, Italy, 30 November–2 December 2013; pp. 97–153.
48. Festa, V. Cretaceous Structural Features of the Murge Area (Apulian Foreland, Southern Italy). *Ecol. Geol. Helv.* **2003**, *96*, 11–22.
49. Aubouin, J. *Geosynclines*; Elsevier: Amsterdam, The Netherlands, 1965.
50. Dewey, J.F. Continental Margins: A Model for Conversion of Atlantic Type to Andean Type. *Earth Planet Sci. Lett.* **1969**, *6*, 189–197. [CrossRef]
51. Dickinson, W.R. Plate Tectonics and Sedimentation. In *Tectonics and Sedimentation*; Sp. Publ. 22; SEPM Society for Sedimentary Geology: Tulsa, OK, USA, 1974.
52. Holmes, A.; Duff, P.M.D. *Holmes' Principles of Physical Geology*; Duff, P.M.D., Ed.; Stanley Thornes Publishers Ltd.: Cheltenham, UK, 1993; ISBN 0-7487-4381-2.
53. Bally, A.W.; Gordy, P.L.; Steward, G.A. Structure, Seismic Data, and Orogenic Evolution of the Southern Canadian Rocky Mountains. *Bull. Can. Pet. Geol.* **1966**, *14*, 337–381.
54. Allen, P.A.; Homewood, P. *Foreland Basins*; Allen, P.A., Homewood, P., Eds.; Blackwell Publishing Ltd.: Oxford, UK, 1986; ISBN 9781444303810.
55. Roberts, D.G.; Bally, A.W. *Regional Geology and Tectonics: Principles of Geologic Analysis*; Roberts, D.G., Bally, A.W., Eds.; Elsevier: Amsterdam, The Netherlands, 2012; ISBN 9780444530424.
56. D'Argenio, B.; Pescatore, T.; Scandone, P. Schema Geologico Dell'Appennino Meridionale (Campania e Lucania). In Proceedings of the Moderne vedute sulla geologia dell'Appennino, Roma, Italy, 16–18 February 1973; pp. 49–72.
57. Boccaletti, M.; Calamita, F.; Deiana, G.; Gelati, R.; Massari, F.; Moratti, G.; Ricci Lucchi, F. Migrating Foredeep-Thrust Belt Systems in the Northern Apennines and Southern Alps. *Palaeogeogr. Palaeoclimatol. Palaeoecol.* **1990**, *77*, 3–14. [CrossRef]
58. Boccaletti, M.; Ciaranfi, N.; Cosentino, D.; Deiana, G.; Gelati, R.; Lentini, F.; Massari, F.; Moratti, G.; Pescatore, T.; Ricci Lucchi, F.; et al. Palinspastic Restoration and Paleogeographic Reconstruction of the Peri-Tyrrhenian Area during the Neogene. *Palaeogeogr. Palaeoclimatol. Palaeoecol.* **1990**, *77*, 41–50. [CrossRef]
59. Tropeano, M.; Sabato, L.; Pieri, P. Filling and Cannibalization of a Foredeep: The Bradanic Trough, Southern Italy. *Geol. Soc. Lon. Spec. Publ.* **2002**, *191*, 55–79. [CrossRef]
60. Finetti, I.R.; Del Ben, A. Crustal Tectono-Stratigraphic Setting of the Adriatic Sea. In *CROP PROJECT: Deep Seismic Exploration of the Central Mediterranean and Italy*; Finetti, I.R., Ed.; Elsevier Science: Amsterdam, The Netherlands, 2005; pp. 519–547.
61. Scrocca, D.; Carminati, E.; Doglioni, C. Deep Structure of the Southern Apennines, Italy: Thin-Skinned or Thick-Skinned? *Tectonics* **2005**, *24*, 1–20. [CrossRef]
62. Ricchetti, G. Contributo Alla Conoscenza Strutturale Della Fossa Bradanica e Delle Murge. *Boll. Soc. Geol. Ital.* **1980**, *99*, 421–430.
63. Pieri, P. Principali Caratteri Geologici e Morfologici Delle Murge. *Murgia Sotter.* **1980**, *2*, 13–19.
64. Casnedi, R. La Fossa Bradanica: Origine, Sedimentazione e Migrazione. *Mem. Soc. Geol. Ital.* **1988**, *41*, 439–448.
65. Spalluto, L.; Pieri, P.; Ricchetti, G. Le Facies Carbonatiche Di Piattaforma Interna Del Promontorio Del Gargano: Implicazioni Paleoambientali e Correlazioni Con La Coeva Successione Delle Murge. (Italia Meridionale, Puglia). *Boll. Soc. Geol. Ital.* **2005**, *124*, 675–690.

66. Iannone, A.; Pieri, P. Caratteri Neotettonici Delle Murge. *Geol. Appl. Idrogeol.* **1982**, *18*, 147–159.

67. Sauro, U. A Polygonal Karst in Alte Murge (Puglia, Southern Italy). *Z. für Geomorphol.* **1991**, *35*, 207–223. [CrossRef]

68. Parise, M. Surface and Subsurface Karst Geomorphology in the Murge (Apulia, Southern Italy). *Acta Carsologica* **2011**, *40*, 79–93. [CrossRef]

69. Tropeano, M.; Sabato, L. Response of Plio-Pleistocene Mixed Bioclastic-Lithoclastic Temperate-Water Carbonate Systems to Forced Regressions: The Calcarenite Di Gravina Formation, Puglia, SE Italy. *Geol. Soc. Lon. Spec. Publ.* **2000**, *172*, 217–243. [CrossRef]

70. Tropeano, M.; Sabato, L.; Pieri, P. The Quaternary «Post-Turbidite» Sedimentation in the South-Apennines Foredeep (Bradanic Trough-Southern Italy). *Boll. Soc. Geol. Ital.* **2002**, *1*, 449–454.

71. Pomar, L.; Tropeano, M. The Calcarenite Di Gravina Formation in Matera (Southern Italy): New Insights for Coarse-Grained, Large-Scale, Cross-Bedded Bodies Encased in Offshore Deposits. *Am. Assoc. Pet. Geol. Bull.* **2001**, *85*, 661–689. [CrossRef]

72. Mateu-Vicens, G.; Pomar, L.; Tropeano, M. Architectural Complexity of a Carbonate Transgressive Systems Tract Induced by Basement Physiography. *Sedimentology* **2008**, *55*, 1815–1848. [CrossRef]

73. Tropeano, M.; Spalluto, L.; Meloni, D.; Moretti, M.; Sabato, L. 'Isolated Base-of-slope Aprons': An Oxymoron for Shallow-marine Fan-shaped, Temperate-water, Carbonate Bodies along the South-east Salento Escarpment (Pleistocene, Apulia, Southern Italy). *Sedimentology* **2022**, *69*, 345–371. [CrossRef]

74. Longhitano, S.G.; Tropeano, M.; Chiarella, D.; Festa, V.; Mateu-Vicens, G.; Pomar, L.; Sabato, L.; Spalluto, L. The Sedimentary Response of Mixed Lithoclastic-Bioclastic Lower- Pleistocene Shallow-Marine Systems to Tides and Waves in the South Apennine Foredeep (Basilicata, Southern Italy) Tidalites Field Trips Special Volume—Tidalites, Matera, Field Trip T3. *Geol. Field Trips* **2021**, *13*, 1–54. [CrossRef]

75. Tropeano, M.; Sabato, L.; Festa, V.; Capolongo, D.; Casciano, C.I.; Chiarella, D.; Gallicchio, S.; Longhitano, S.G.; Moretti, M.; Petruzzelli, M.; et al. "Sassi", the Old Town of Matera (Southern Italy): First Aid for Geotourists in the "European Capital of Culture 2019". *Alp. Mediterr. Quat.* **2018**, *31*, 133–145. [CrossRef]

76. Chiarella, D.; Festa, V.; Sabato, L.; Tropeano, M. The City of Matera and Its 'Sassi' (Italy): An Opportunity to Broadcast Geology in the European Capital of Culture 2019. *Geol. Today* **2019**, *35*, 174–178. [CrossRef]

77. Belfiore, D. Ponte Di Gravina.Jpg. Available online: https://commons.wikimedia.org/wiki/File:Ponte_di_Gravina.jpg (accessed on 15 March 2023).

78. Sabato, L. Quadro Stratigrafico-Deposizionale Dei Depositi Regressivi Nell'area Di Irsina (Fossa Bradanica). *Geol. Romana* **1996**, *32*, 219–230.

79. Pieri, P.; Sabato, L.; Tropeano, M. Significato Geodinamico Dei Caratteri Deposizionali e Strutturali Della Fossa Bradanica Nel Pleistocene. *Mem. Soc. Geol. Ital.* **1996**, *51*, 501–515.

80. Sabato, L.; Tropeano, M.; Pieri, P. Problemi Di Cartografia Geologica Relativa Ai Depositi Quaternari Del F° 471 "Irsina". Il Conglomerato Di Irsina: Mito o Realtà? *Il Quat.* **2004**, *17*, 391–404.

81. Sabato, L.; Tropeano, M.; Cilumbriello, A.; Gallicchio, S.; Pieri, P. Stratigraphic Architecure of Bradanic Trough Deposits (Southern Italy): Geological News and Their Potential Use in Applied Fields. *Rend. Online Soc. Geol. Ital.* **2011**, *12*, 83–86.

82. Ciaranfi, N.; Maggiore, M.; Pieri, P.; Rapisardi, L.; Ricchetti, G.; Walsh, N. *Considerazioni Sulla Neotettonica Della Fossa Bradanica*; CNR: Bari, Italy, 1979.

83. Doglioni, C.; Mongelli, F.; Pieri, P. The Puglia Uplift (SE Italy): An Anomaly in the Foreland of the Apenninic Subduction Due to Buckling of a Thick Continental Lithosphere. *Tectonics* **1994**, *13*, 1309–1321. [CrossRef]

84. Doglioni, C.; Tropeano, M.; Mongelli, F.; Pieri, P. Middle-Late Pleistocene Uplift of Puglia: An Anomaly in the Apenninic Foreland. *Mem. Soc. Geol. Ital.* **1996**, *51*, 101–117.

85. Gioia, D.; Sabato, L.; Spalluto, L.; Tropeano, M. Fluvial Landforms in Relation to the Geological Setting in the "Murge Basse" Karst of Apulia (Bari Metropolitan Area, Southern Italy). *J. Maps* **2011**, *7*, 148–155. [CrossRef]

86. Boenzi, F. Gravine. In *Italia—Atlante dei tipi geografici dell'IGM*; Istituto Geografico Militare: Firenze, Italy, 2004; pp. 164–166.

87. Donnaloia, M.; Giachetta, E.; Capolongo, D.; Pennetta, L. Evolution of Fluviokarst Canyons in Response to the Quaternary Tectonic Uplift of the Apulia Carbonate Platform (Southern Italy): Insights from Morphometric Analysis of Drainage Basins. *Geomorphology* **2019**, *336*, 18–30. [CrossRef]

88. Parise, M.; Federico, A.; Delle Rose, M.; Sammarco, M. Karst Terminology in Apulia (Southern Italy). *Acta Carsologica* **2003**, *32*, 65–82. [CrossRef]

89. Arouca Declaration on Geotourism November 12, 2011 Portugal. Available online: https://www.europeangeoparks.org/?p=223 (accessed on 15 March 2023).

90. UNESCO Global Geoparks (UGGp). Available online: https://en.unesco.org/global-geoparks (accessed on 15 March 2023).

91. Panizza, M.; Piacente, S. Geomorphosites: A Bridge Betwenn Scientific Research, Cultural Integration and Artistic Suggestion. *Alp. Mediterr. Quat.* **2005**, *18*, 3–10.

92. Van Hinsbergen, D.J.J.; Mensink, M.; Langereis, C.G.; Maffione, M.; Spalluto, L.; Tropeano, M.; Sabato, L. Did Adria Rotate Relative to Africa? *Solid Earth* **2014**, *5*, 611–629. [CrossRef]

93. Butler, R.W.H. Relationships between the Apennine Thrust Belt, Foredeep and Foreland Revealed by Marine Seismic Data, Offshore Calabria. *Boll. Soc. Geol. Ital.* **2009**, *128*, 269–278. [CrossRef]

94. Petti, F.M.; Antonelli, M.; Sacco, E.; Conti, J.; Petruzzelli, M.; Spalluto, L.; Cardia, S.; Festa, V.; La Perna, R.; Marino, M.; et al. Geothematic Map of the Altamura Dinosaur Tracksite (Early Campanian, Apulia, Southern Italy). *Geol. Field Trips* **2022**, *14*, 1–19. [CrossRef]

95. Nicosia, U.; Petti, F.M.; Perugini, G.; Porchetti, S.D.; Sacchi, E.; Conti, M.A.; Mariotti, N.; Zarattini, A. Dinosaur Tracks as Paleogeographic Constraints: New Scenarios for the Cretaceous Geography of the Periadriatic Region. *Ichnos* **2007**, *14*, 69–90. [CrossRef]

96. Interventi Urgenti Di Protezione e Conservazione Delle Impronte Di Dinosauro e Della Paleo Superficie Alla Cava Pontrelli in Altamura (BA). Relazione Finale—Marzo 2020—Nell'ambito Del Bando Promosso Dal MiBACT—CIG: 736256767D CUP: F72. C16000040001; 2020; Bari, Italy. Available online: https://www.dinosauridialtamura.it/lavori-in-cava/ (accessed on 15 March 2023).

97. Petti, F.M.; Antonelli, M.; Citton, P.; Mariotti, N.; Petruzzelli, M.; Pignatti, J.; D'Orazi Porchetti, S.; Romano, M.; Sacchi, E.; Sacco, E.; et al. Cretaceous Tetrapod Tracks from Italy: A Treasure Trove of Exceptional Biodiversity. *J. Mediterr. Earth Sci.* **2020** *12*, 167–191.

98. Iannone, A. Facies Analysis of Upper Cretaceous Peritidal Limestones Characterised by the Presence of Dinosaur Tracks (Altamura, Southern Italy). *Mem. Sci. Geol.* **2003**, *55*, 1–12.

99. Strasser, A.; Hilgen, F.J.; Heckel, P.H. Cyclostratigraphy Concepts, Definitions, and Applications. *Newsl. Stratigr.* **2006**, *42*, 75–114. [CrossRef]

100. Strasser, A. Hiatuses and Condensation: An Estimation of Time Lost on a Shallow Carbonate Platform. *Depos. Rec.* **2015**, *1*, 91–117. [CrossRef]

101. Spalluto, L. Sedimentology and High-Resolution Sequence Stratigraphy of an Early Cretaceous Shallow-Water Carbonate Succession Cropping out in the Western Gargano Promontory (Puglia, Southern Italy). *GeoActa* **2008**, *1*, 173–191.

102. Spalluto, L. Facies Evolution and Sequence Chronostratigraphy of a "Mid"-Cretaceous Shallow Water Carbonate Succession of the Apulia Carbonate Platform from the Northern Murge Area (Apulia, Southern Italy). *Facies* **2012**, *58*, 17–36. [CrossRef]

103. Spalluto, L.; Caffau, M. Stratigraphy of the Mid-Cretaceous Shallow-Water Limestones of the Apulia Carbonate Platform (Murge, Apulia, Southern Italy). *Ital. J. Geosci.* **2010**, 335–352. [CrossRef]

104. Vacca, E.; Pesce Delfino, V. Three-Dimensional Topographic Survey of the Human Remains in Lamalunga Cave (Altamura, Bari, Southern Italy). *Coll. Antropol.* **2004**, *28*, 113–119. [PubMed]

105. Pesce Delfino, V.; Vacca, E. Report of an Archaic Human Skeleton Discovered at Altamura (Bari), in the "Lamalunga" District. *Hum. Evol.* **1994**, *9*, 1–9. [CrossRef]

106. Lari, M.; Di Vincenzo, F.; Borsato, A.; Ghirotto, S.; Micheli, M.; Balsamo, C.; Collina, C.; De Bellis, G.; Frisia, S.; Giacobini, G.; et al. The Neanderthal in the Karst: First Dating, Morphometric, and Paleogenetic Data on the Fossil Skeleton from Altamura (Italy). *J. Hum. Evol.* **2015**, *82*, 88–94. [CrossRef]

107. Mastronuzzi, G.; Mirella, M.; Piscitelli, A.; Parise, M.; Scardino, G. Patrimonio Culturale e Geositi Dell'area Murgiana. In Proceedings of the Paesaggi di Pietra, Ostuni, Italy, 18 October 2019; pp. 43–50.

108. Pieri, P.; Sabato, L.; Tropeano, M. Evoluzione Tettonico-Sedimentaria Della Fossa Bradanica a Sud Dell'Ofanto Nel Pleistocene. In Proceedings of the Guida alle escursioni. 77° Congresso Nazionale della Società Geologica Italiana, Quaderni della Biblioteca Provinciale di Matera, Bari, Italy, 18–20 September 1994; pp. 35–54.

109. Spalluto, L.; Cicala, M.; Festa, V.; Sabato, L.; Tropeano, M. From the Apulia Foreland to the Bradanic Trough (Puglia, Southern Italy): A One-Day Geological Field Trip "Jumping" from Cretaceous to Pleistocene in the Murge Area. *Geol. Field Trips Maps* **2023**, *Submitted*.

110. Bakalowicz, M. Karst Groundwater: A Challenge for New Resources. *Hydrogeol. J.* **2005**, *13*, 148–160. [CrossRef]

111. Parise, M.; Galeazzi, C.; Germani, C.; Bixio, R.; Del Prete, S.; Sammarco, M. The Map of Ancient Underground Aqueducts in Italy: Updating of the Project, and Future Perspectives. In Proceedings of the International Congress in Artificial Cavities "Hypogea 2015", Rome, Italy, 11–15 March 2015; pp. 235–243.

112. Stevanovic, Z. *Karst Aquifers—Characterization and Engineering*; Stevanović, Z., Ed.; Professional Practice in Earth Sciences; Springer International Publishing: Cham, Switzerland, 2015; ISBN 978-3-319-12849-8.

113. Laureano, P. *The Water Atlas: Traditional Knowledge to Combat Desertification*; Bollati Boringhieri: Turin, Italy, 2001; ISBN 84-933371-6-1.

114. Parise, M.; Sammarco, M. The Historical Use of Water Resources in Karst. *Environ. Earth Sci.* **2015**, *74*, 143–152. [CrossRef]

115. Valipour, M.; Ahmed, A.T.; Antoniou, G.P.; Sala, R.; Parise, M.; Salgot, M.; Sanaan Bensi, N.; Angelakis, A.N. Sustainability of Underground Hydro-Technologies: From Ancient to Modern Times and toward the Future. *Sustainability* **2020**, *12*, 8983. [CrossRef]

116. Parise, M.; Bixio, R.; Burri, E.; Caloi, V.; Del Prete, S.; Galeazzi, C.; Germani, C.; Guglia, P.; Meneghini, M.; Sammarco, M. The Map of Ancient Underground Aqueducts: A Nation-Wide Project by the Italian Speleological Society. In Proceedings of the 15th International Congress of Speleology, Kerrville, TX, USA, 16–26 July 2009; pp. 2027–2032.

117. Bixio, R.; Parise, M.; Saj, S.; Traverso, M. *Opera Ipogea*; A.G.E. s.r.l.: Urbino, Italy, 2007; pp. 105–112.

118. Fiore, A.; Martimucci, V.; Parise, M.; Sammarco, M. Artificial Caves of Apulia, a Poorly Exploited Cultural Heritage. In Proceedings of the 7th International Symposium ProGEO on the Conservation of the Geological Heritage "Geoheritage: Protecting and Sharing", Bari, Italy, 24–28 September 2012; pp. 58–59.

119. Martimucci, V.; Parise, M.; Antonicelli, A.; Antonucci, M.P.; Carbonara, M.; Chieco, M.; Pace, F. The Regional Registers of Natural and Artificial Caves of Apulia (Southern Italy): Recent Developments from a Joint Project Regione Puglia—Federazione Speleologica Pugliese. In Proceedings of the 7th Int. Symp. ProGEO on the Conservation of the Geological Heritage "Geoheritage: Protecting and Sharing", Bari, Italy, 24–26 September 2012; pp. 43–50.
120. Ciaranfi, N.; Ghisetti, F.; Guida, M.; Iaccarino, G.; Lambiase, S.; Pieri, P.; Rapisardi, L.; Ricchetti, G.; Torre, M.; Tortorici, L.; et al. *Carta Neotettonica Dell'Italia Meridionale*; CNR: Bari, Italy, 1983.
121. Martinis, B. *Sulla Tettonica Delle Murge Nord-Occidentali*; Rendiconti della Classe di Scienze fisiche, matematiche e naturali; Accademia Nazionale dei Lincei: Rome, Italy, 1961; p. 31.
122. Caldara, M.; Ciaranfi, N. Le Brecce Polifasiche Quaternarie Delle Murge Settentrionali. *Mem. Soc. Geol. Ital.* **1988**, *41*, 685–695.
123. Stigliano, E. Geological Features of an Aspiring Geopark (AUGGp): The Murgeopark. Master's Thesis, University of Bari "Aldo Moro", Bari, Italy, 2021.
124. Iannone, A. Sequenze Lacustri Carbonatiche Nelle Murge Nord-Occidentali. *Mem. Soc. Geol. Ital.* **1996**, *51*, 217–225.
125. Lippolis, E.; Sabato, L.; Spalluto, L.; Tropeano, M. Geotourism around Poggiorsini: Unexpected Geological Elements for a Sustainable Tourism in Internal Areas of Murge (Puglia, Southern Italy). *Rend. Online Soc. Geol. Ital.* **2023**, *59*, 152–158. [CrossRef]
126. Onofrio, V.; Tropeano, M.; Festa, V.; Moretti, M.; Sabato, L. Quaternary Transpression and Lacustrine Sedimentation in the San Lorenzo Area (Sant'Arcangelo Basin, Italy). *Sediment. Geol.* **2009**, *222*, 78–88. [CrossRef]
127. D'Argenio, B.; Mindszenty, A. Bauxites and Related Paleokarst: Tectonic and Climatic Event Markers at Regional Unconformity. *Eclogae Geol. Helv.* **1995**, *88*, 453–499.
128. Mindszenty, A.; D'Argenio, B.; Aiello, G. Lithospheric Bulges Recorded by Regional Unconformities. The Case of Mesozoic-Tertiary Apulia. *Tectonophysics* **1995**, *252*, 137–161. [CrossRef]
129. Mongelli, G. Ce-Anomalies in the Textural Components of Upper Cretaceous Karst Bauxites from the Apulian Carbonate Platform (Southern Italy). *Chem. Geol.* **1997**, *140*, 69–79. [CrossRef]
130. Mongelli, G.; Boni, M.; Buccione, R.; Sinisi, R. Geochemistry of the Apulian Karst Bauxites (Southern Italy): Chemical Fractionation and Parental Affinities. *Ore Geol. Rev.* **2014**, *63*, 9–21. [CrossRef]
131. Mongelli, G.; Buccione, R.; Sinisi, R. Genesis of Autochthonous and Allochthonous Apulian Karst Bauxites (Southern Italy): Climate Constraints. *Sediment. Geol.* **2015**, *325*, 168–176. [CrossRef]
132. Buccione, R.; Mongelli, G.; Sinisi, R.; Boni, M. Relationship between Geometric Parameters and Compositional Data: A New Approach to Karst Bauxites Exploration. *J. Geochem. Explor.* **2016**, *169*, 192–201. [CrossRef]
133. Agosta, F.; Manniello, C.; Cavalcante, F.; Belviso, C.; Prosser, G. Late Cretaceous Transtensional Faulting of the Apulian Platform, Italy. *Mar. Pet. Geol.* **2021**, *127*, 104889. [CrossRef]
134. Sabato, L. Delta Calcareo Terrazzato Nella Calcarenite Di Gravina (Pleistocene Inferiore) (Minervino, Murge Nord-Occidentali). *Mem. Soc. Geol. Ital.* **1996**, *51*, 517–526.
135. Postma, G.; Babic, L.; Zupanic, J.; Roe, S.L. Delta-Front Failure and Associated Bottomset Deformation in a Marine, Gravelly Gilbert-Type Fan Delta. In *Fan Deltas: Sedimentology and Tectonic settings*; Nemec, W., Steel, R., Eds.; Blackie & Son: Glasgow, UK, 1988; pp. 91–102.
136. Cardia, S. STARM (Stability Assessment in Rock Masses)—An Integrated Approach by Combining Rock Mechanics, Remote Sensing and Engineering Geology. Ph. D. Thesis, University of Bari "Aldo Moro", Bari, Italy, 2023.
137. Cardia, S.; Langella, F.; Pagano, M.; Palma, B.; Sabato, L.; Tropeano, M. 3D Digital Analysis for Geo-Structural Monitoring and Virtual Documentation of the Saint Michael Cave in Minervino Murge, Bari (Italy). *Digit. Appl. Archaeol. Cult. Herit.* **2023**, *Submitted*.

Article

"Geo-Archaeo-Routes" on the Island of Lemnos: The *"Nalture"* Experience as a Holistic Geotouristic Approach within the Geoethical Perspective

Maria V. Triantaphyllou [1,*], Nikolaos Firkasis [1], Theodora Tsourou [1], Emmanuel Vassilakis [1], Evangelos Spyrou [1], Olga Koukousioura [2], Argyro Oikonomou [1] and Athanasios Skentos [1]

[1] Department of Historical Geology and Palaeontology, Faculty of Geology and Geo-Environment, National and Kapodistrian University of Athens, Panepistimioupolis, 15784 Athens, Greece; nfirkasis@gmail.com (N.F.); ttsourou@geol.uoa.gr (T.T.); evasilak@geol.uoa.gr (E.V.); evspyrou@geol.uoa.gr (E.S.); aoikon@geol.uoa.gr (A.O.); thanasis.skentos@aecom.com (A.S.)
[2] School of Geology, Aristotle University of Thessaloniki, 54124 Thessaloniki, Greece; okoukous@geo.auth.gr
* Correspondence: mtriant@geol.uoa.gr; Tel.: +30-210-727-4893

Abstract: The geosites of Lemnos represent local touristic products that, beyond their high aesthetic value, display significant scientific links to the geological past as well as prehistory and history, archaeology, mythology and religious heritage of the island. The unique wealth of Lemnos geosites in combination with the abundance of archaeological sites, cultural monuments and museums composes the basis of what we define here as *"Geo-Archaeo-Routes"*: certain routes that can be geographically defined, offered, guided and finally followed by the touristic masses. The outcome of the performed quantitative Lemnos geosite assessment enables decision making, thus providing a toolbox useful for sustainable Geo-Archaeo-tourism development at a local level and forms the basis for designing *"Geo-Archaeo-Routes"*. *"Geo-Archaeo-Routes"* are particularly favorable of environmentally friendly alternative types of tourism, attracting naturalists, hikers, fans of cultural or religious tourism and many others who represent a major part of the touristic needs of the 21st century. The established hiking and road *"Geo-Archaeo-Routes"* on Lemnos Island may represent a distinctive touristic product as they offer a high level of *"nalture"* entertainment, blending "nature with culture" in the framework of a holistic geotouristic approach.

Keywords: *"nalture"*; geotourism; geoethics; *Geo-Archaeo-Routes*; Lemnos Island

Citation: Triantaphyllou, M.V.; Firkasis, N.; Tsourou, T.; Vassilakis, E.; Spyrou, E.; Koukousioura, O.; Oikonomou, A.; Skentos, A. *"Geo-Archaeo-Routes"* on the Island of Lemnos: The *"Nalture"* Experience as a Holistic Geotouristic Approach within the Geoethical Perspective. *Geosciences* **2023**, *13*, 143. https://doi.org/10.3390/geosciences13050143

Academic Editors: Jesus Martinez-Frias and Deodato Tapete

Received: 4 April 2023
Revised: 2 May 2023
Accepted: 8 May 2023
Published: 12 May 2023

1. Introduction

Nowadays, the tourist visit is not a simple period for physical rest but is much more an attempt at spiritual release and elimination of daily stress. Thus, this time interval is often planned in the context of visiting areas with natural beauty profoundly linked to geological processes that have left their traces in the morphology of the Earth's surface (e.g., [1]).

The tectonic, paleogeographical and geomorphological evolution of the Greek landmass during the last 10 million years not only curved the morphological relief, but also resulted in the genesis of geological sites with unique characteristics; e.g., interesting sedimentary structures, rare or characteristic fossils, tectonic structures, significant mineralogical–petrological occurrences, ongoing geomorphological and geological processes, caves, etc. [2], comprising geotopes and geosites. Geotopes are defined as the smallest geographical unit with such prominent geological features [3,4], while geosites combine natural geoscientific monuments with aesthetic, naturalistic, cultural, historical, touristic and educational values (e.g., [4,5]). Apparently, geotopes consist of attractive touristic sites, thus being the heart of geotourism, a recently developed alternative form of tourism [1,6–11] that contributes to the local economy of an area and to its sustainable development [12–14] through environmental management that integrates geodiversity awareness and promotes sustainable economic growth and employment (e.g., [15]).

Geosites can therefore be considered as natural capital that should be preserved to be available for the preferences of the future generations in the sense of sustainable development [16]; namely, the intergenerational welfare that can be maximized by guaranteeing their enjoyability for the future [16,17]. However, a series of geoethical issues may be raised, when geotopes are incorporated in the line of geotouristic development. The massive touristic exploitation and malpractices unavoidably leave negative marks on the natural environment as many anthropogenic impacts on geosites may lead to their irreversible degradation [17]. The responsible management of geosites and their protection can be achieved only by applying the values of geoethics that raise awareness and responsibility on geological heritage conservation, and stress out the important link with geoenvironmental education activities [18,19], towards a holistic geotouristic approach. As pointed out by [20], it is only through a geoethical perspective that geotourism can contribute to our understanding of the Earth as a system, through the relationships that bind the parts to the whole, thus linking people with their land [18,19].

The environmentally friendly anthropogenic imprint on the geoenvironment, associated with archaeological sites and monuments, museums and religious sites, as well as the sustainable production of local goods of agriculture and wine-growing, plays a major role in fostering the cultural education of tourist masses together with the respect for the natural capital of geotopes in a geoethical perspective, which is actually linked to the rational use of nonrenewable resources. Apparently, geotourism activities minimize their environmental impacts and avoid the risk of exceeding the threshold of ecological and social sustainability only when guided by the geoethical principles that, besides increasing awareness for sustainable geoheritage management, also foster the public's understanding of natural hazards, such as climate change, sea level rise and flooding [18,19,21]. With the values of geoethics, the integration between geodiversity and cultural resources is able to develop the needed sense of responsibility for preserving geoheritage, not only for the enjoyability of the present communities but also for assuring its existence for the future generations [15,22].

Greece denotes an exceptional example of a geotope ensemble with solid geotouristic potential [2,7,8,23–26]. In particular [6,7], after evaluating more than 500 geotopes of the Hellenic territory, concluded that Greece holds high geotouristic perspectives that can further support the vital tourism sector and the regional development of the Greek economy via the management of the geotopes as attractive local tourism products.

Lemnos Island in the northeastern Aegean Sea represents an interesting case study for geotouristic development. The island is featured by both rich natural and cultural capital, which is why it has been designated as an Area of Outstanding Natural Beauty [27] and traditional settlements [28]. It displays a smooth landscape, carved within millions of years by the volcanic activity, the water runoff, the sea and the wind, which resulted in a spectacular agro-pastoral environment hosting numerous geosites and some of the largest and most important Mediterranean wetlands [29]. The land of Lemnos exhibits among others, some of the oldest human settlements of hunters and fishermen in the Aegean Sea dating back to the 11th millennium BC [30]. Furthermore, the island is widely known for its distinguished household economy and traditional high-quality local wines, meats, cheeses, fruits, vegetables, herbs and handmade pasta. Concerning the status of geotourism development on the island, there has recently been an important detailed effort to identify, visualize and present on the Web the numerous Lemnos geosites, geomorphosites and cultural sites in order to promote to the public the island's geological and geomorphological heritage [31,32].

The scope of the present study is to highlight specific geological locations together with the natural environment and biodiversity of Lemnos Island. This is achieved in the context of a geotouristic approach that combines the natural capital represented by local geosites with important archaeological sites, as well as local food and wine tasting experiences, comprising a *"nalture"* geotouristic experience. The term *"nalture"* is introduced in this study to describe the geotouristic bind of nature with culture in terms of symbiosis [28], also described by [15,33,34], and not under the traditional dualistic opposition that regards the concept of nature as something separated from human beings (e.g., [35]). We consider the *"nalture"* geotouristic experience to be realized in the framework of environmental ethics (e.g., [36]), concerning humans in their culture but also residing in nature, therefore involving sustainable use of the environmental resources (i.e., geosites).

As a result, alternative dynamic *"Geo-Archaeo-Routes"* are proposed to build and formulate a touristic product in the geoethical perspective of regional sustainable development. The rationale behind establishing *"Geo-Archaeo-Routes"* is that according to [37], "the biophysical landscape itself loads throughout history a series of interwoven human traces", therefore it is more than evident that the geoheritage interacts with the cultural assets, forming geocultural sites [38]. The concept is enriched with the involvement of gastronomical and wine tasting experiences that can evolve the *"Geo-Archaeo-Routes"* touristic product to a focal point for regional touristic development. Thus, *"Geo-Archaeo-Routes"* can represent a distinctive touristic product for numerous tourist groups as they offer a high level of *"nalture"* entertainment, hence enjoying the natural beauty of geosites in relation to the marks of human influence on the geoenvironment.

In the context of the geoethical values, we consider the proposed *"Geo-Archaeo-Routes"* to offer to the public awareness not only a brief description of their elements, but a considerable geoscientific documentation that will enable the recognition of the heritage value of geosites for audiences outside the specialists [38]. Therefore, besides the information provided for the study area and the applied methodology (Sections 2 and 3), the following Section 3 includes not only a description but an extensive documentation of the described geosites and also their assessment based on the scientific literature and evaluation criteria. As a result, nine hiking and road *"Geo-Archaeo-Routes"* are defined for a holistic geotouristic development of Lemnos Island. Finally, in the synthesis Section 5, potential future steps for the realistic realization of the proposed *"Geo-Archaeo-Routes"* are described in relation to the opportunities and limitations concerning the geotouristic development of the island.

2. Study Area

2.1. Physicogeographical and Geomorphological Setting

Lemnos Island located in the North Aegean region (Figure 1) is the eighth largest island of Greece with an area of 475.6 km^2. The highlights of the Lemnos landscape comprise the presence of coastal and inland sand dunes, interesting geological formations, extensive coastal wetlands and agropastoral land [39]. The island displays smooth morphology, being almost flat with the highest elevation of Mount Skopia (Vigla) at 470 m a.s.l., located at the northwestern part of the island. The island's terrain is mostly volcanic with low relief formations and medium inclines. Throughout the coastal zone there are low hills, 250–350 m high, except for the eastern part of the island as well as the bay of Moudros where there are extended plains. The faults of the island are well exposed in the central and western part forming narrow and shallow basins. Thus, the western part is steeper and hilly, also displaying semi-mountainous parts, as thick pyroclastic deposits cover the underlying sedimentary sequences, producing characteristic morphological cliffs due to erosional processes. The central and eastern part presents a flat relief and fertile soils, dominated by a lowland farmland mosaic, around soft hills. The hydrographic drainage network consists of streams of seasonal flow, having a very poor drainage that, combined with the small amount of rainfall, is not favorable for the creation of prominent alluvial fans. The most important coastal landforms are the sand dunes in the area of the Aliki lagoon, and in the northern part of the island, in the area of Katalakos and Gomati beach. Tombolo formations have been identified at the NE part of Lemnos, at the Fakos peninsula as well as at the bay of Plaka in the NW part of the island. Lemnos is characterized by an extended and rich-in-sandy-beaches coastaline of 259.3 km, with the Gulf of Moudros being the most prominent feature of the coastline. The slope of the coasts in Lemnos varies.

There are coasts with a small slope (0°–30°) found mainly in the NW part, with a medium slope (30°–40°) in the northern and southeastern parts and with a large slope (>40°) that appear in the western and part of the northern area of the island as well as on the southern coastline of the Fakos peninsula. The landforms of the hinterland are characterized by tafoni and volcanic structures [40].

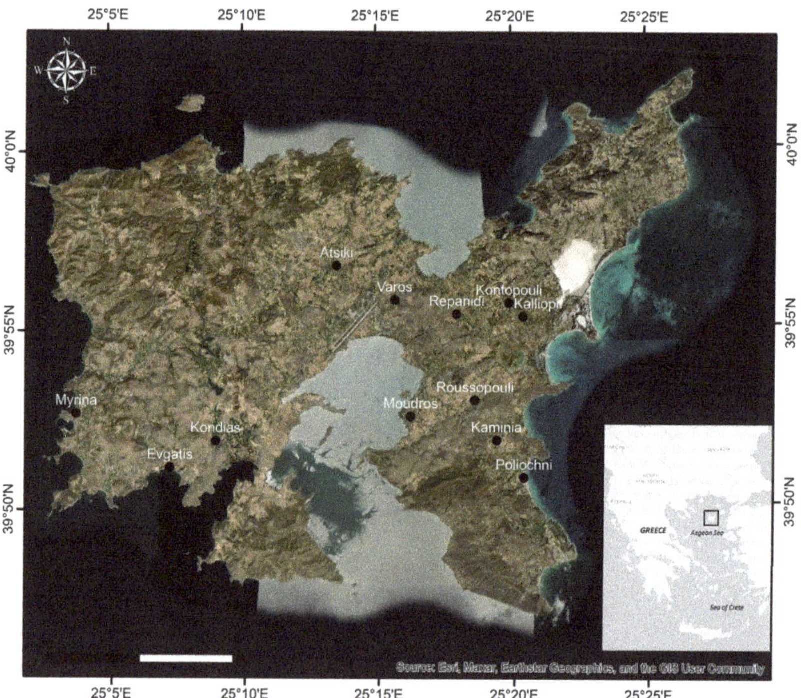

Figure 1. Map of the study area and location of the considered Aegean island of Lemnos.

The climate is temperate with mild winters, prevailing blowing northeastern winds and a dry season lasting from March to October, while the annual precipitation is about 500 mm [41,42]. In the winter, the average monthly temperature is below 10 °C, while the average annual maximum temperature is 27.3 °C. Fog is present throughout the year with a higher frequency in between September and April. Interestingly, the sunshine level in the North Aegean is of the highest in Greece, reaching 2734 h yearly, with a monthly average of 227.8 h [43].

Lemnos is characterized by poor vegetation, nonetheless it exhibits an extraordinary flora that consists of about 681 plant taxa [41]. Particularly, the combination of halophytic and sand dune habitats is unique for the island area, hosting plants such as thyme, oleander and numerous species of wildflowers followed by the outstanding appearance of the sea lily on the sandy beaches, as well as remnants of *Quercus ithaburensis* subsp. *macrolepis* forest and extended phryganic vegetation all over the island (e.g., [41]). Due to the presence of extended wetlands and the island's location in the routes of migratory birds, Lemnos is featured by significant bird fauna, which includes at least 64 species. The island's fauna includes numerous reptiles and 12 rare or protected mammal species, including the Mediterranean seal *Monachus monachus*, the sea turtle *Caretta caretta* and the protected turtle species *Mauremys capsica* (e.g., [44]). The fish fauna is also rich, including at least 40 species of fish and shellfish and a large number of dolphin species. Due to its clean sandy beaches, the *Posidonia* meadows and the stunning reefal formations, Lemnos Island is considered one of the best preserved marine ecosystems of the Aegean Sea [43].

2.2. Geological Setting

The geology of the area (Figure 2) consists mainly of a clastic sedimentary sequence of the Oligocene and extensive volcanic rocks of the Early Miocene age [45,46]. The available geological maps [46,47] present a description of the dominant lithology. Previous stratigraphic and sedimentological studies (e.g., [48–50]) refer to a late Eocene–Oligocene stratigraphic range with a total thickness of less than 800 m. In particular, [46] have recognized three units within the sedimentary sequence: the Fissini-Sardes Unit that is the most extensive sequence and is characterized at its lower parts by sandstones, with greenish siltstones and shales, isolated flat blocks of cobblestone and gray nummulitic limestones and a layer of tuffs, while in its middle and upper part, thick layers of sandstone prevail over the silty clay; the Ifestia Unit consisting mainly of coarse-grained sandstones and conglomerates with its upper part mostly displaying silty clays, siltstones and marls, covered by an eroded surface associated with the volcanic activity; and the Therma Unit that consists of conglomerates restricted in the eastern and southeastern part of the island, presenting an abrupt change from the marine to continental environment, which corresponds to the beginning of the main volcanic cycle. [46] provided a rough age of middle Eocene to Early Miocene for the associated depositional intervals with the youngest age in accordance with [51] based on the plant fossils.

More recently, [52] have shown that the Lemnos volcano-sedimentary sequence exceeds 2200 m in thickness, extending from the late Eocene to the Oligocene/Miocene boundary, based on detailed calcareous nannofossil biostratigraphic analyses. The overall stratigraphic sequence of Lemnos, together with its equivalent sequence of Thrace, shows a deep marine environment of molassic type within a back-arc basin [53–55]. A blocky formation with olistolites of nummulitic neritic limestones is observed above the lowermost rhythmic alternations of turbiditic sandstones and pelites of the late Eocene, followed by several interlayering volcanic tuffs within a cyclical sedimentary sequence. Late Oligocene thick sandstones-conglomerates feature the middle part of the sequence, while impressive volcanic dikes and lava flows become frequent towards the upper horizons. The Early Miocene age has been documented at the sedimentary deposits of the northwestern part of the island [52]. The most extensive volcanic extrusions occur at the central–southern part of Lemnos and they are dated as Early Miocene (e.g., [56]). Plio-Pleistocene fluvial deposits are uncomfortably overlaying the pyroclastic succession, while aeolian sand deposits are recorded in the central–eastern part of the island [46]. Quaternary shallow marine limestones and calcarenites overlay the Eocene–Oligocene molasse-type sediments [47]. The Holocene coastal deposits at the archaeological settlement of Hephaistia and the Alyki Lagoon reveal a constant sea level rise during the last 7000 cal BP with fluctuations between temporary lagoon to shallow bay paleoenviroments; in particular, the determined shallow bay in the area of Hephaistia could have been used as a natural harbor before 4000 cal BP [57].

Innocenti et al. [46] recognized three units of volcanic rocks: the Romanou Unit pyroclastic sequence including a well-welded ignimbrite with gray and reddish pumice dated by radiometric K/Ar as 19.8 Ma [58] or 22.3 ± 0.7 Ma [59], and intercalations of continental sediments containing plant remains and silicified trunks; the Katalakkon Unit consisting mainly of lava domes that, according to the K/Ar radiometric dating by [58] and [45], formed after the pyroclastics of the Romano Unit (20–21 Ma); and the Myrina Unit, which represents the younger volcanics of the island aged between 19.3 and 18.2 Ma [45,58], mostly associated with lava domes.

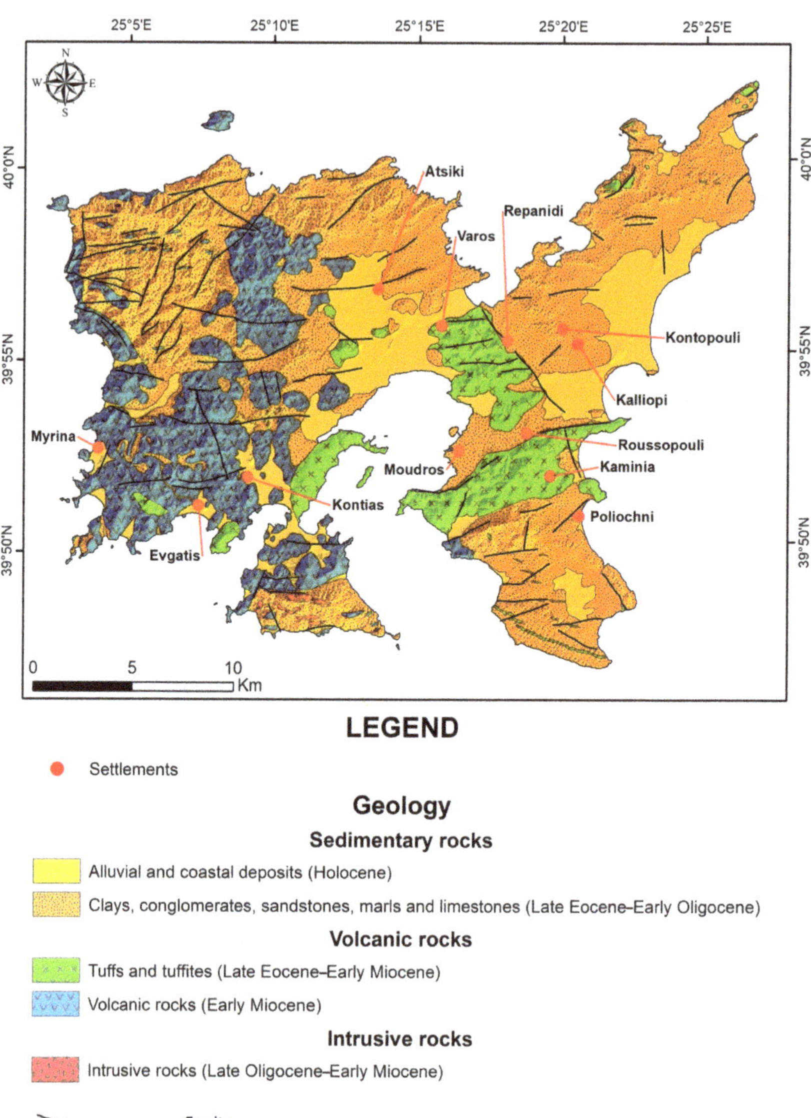

LEGEND

● Settlements

Geology

Sedimentary rocks

Alluvial and coastal deposits (Holocene)

Clays, conglomerates, sandstones, marls and limestones (Late Eocene–Early Oligocene)

Volcanic rocks

Tuffs and tuffites (Late Eocene–Early Miocene)

Volcanic rocks (Early Miocene)

Intrusive rocks

Intrusive rocks (Late Oligocene–Early Miocene)

Faults

Figure 2. Geological map of Lemnos Island and legend of the geological formations (modified from IGME map). The map shows the distribution of the main geological formations and faults.

Faulting is more intense in the SE of Moudros (Fanos-Agia Sofia fault) and in the NW part of the island with the Kondias/Kotsinas fault affecting the coastal zone close to the Hephaistia archeological site [60–62]. According to [63], all faults along the northern parts of the island are characterized by a dextral strike–slip component, while the faults on the southern parts of Lemnos, except for the Kaspakas and Moudros faults, are featured by an ENE–WSW strike. One of the most important faults on the island is that of Mourtzouflos, a strike-slip fault in the NE–SW direction defined as active by [64]. It belongs to an offshore fault zone and intersects with the mainland at Cape Mourtzouflos in the northwestern part of the island [63]. The Kaspakas fault is a normal NW–SE fault, which forms several fault scarps at the west side of the island, and the Kondias–Kotsinas fault zone represents a complex structure of multiple horizontal slip faults in a NE–SW direction, which crosses the island, affecting the coastline. Finally, the Moudros and Fanos-Agia Sofia faults are normal faults of the WSW–ENE direction that dip to the N–NW, located in the east and south of the island, respectively.

2.3. Socioeconomic Setting and Archaeological Values

Since 2011, the Municipality of Lemnos (capital city: Myrina) belongs to the regional unit of Lesbos, which in turn belongs to the region of the North Aegean. The Hellenic Statistical Authority keeps population data for the island of Lemnos beginning in 1920. The last census was made in 2011, while the next one was performed very recently (end of 2021). Despite the fact that the most recent census (Hellenic Statistical Authority, 2011) data available are relatively outdated, they are the only widely available and reliable population data so far.

The population evolution data (Table 1) show that there was a continual population increase until 1951, when the residents were about 24,000. Lemnos faced a significant population decline in the post-World War II period, when many residents migrated to the mainland Greece and abroad (Australia, Canada and USA). As a result, there was a gradual reduction until 2011, when the permanent residents of the island were about 17,000, while a decade later (2022), the current population has remained more or less stable (16,458). Lemnos is a relatively sparsely populated island with a population density of 35.7 residents per km^2.

Table 1. Population evolution of Lemnos for the last 100 yr and distribution of economic sectors, census 2011 (Source: Hellenic Statistical Authority).

Census Year	Population	Economic Sectors
1920	19,642	
1928	23,611	
1940	23,842	
1951	24,018	
1961	21,812	
1971	17,367	
1981	15,721	
1991	17,645	
2001	18,104	
2011	16,992	
2022	16,458	

According to the 2011 census, the island of Lemnos has a total of 5915 employed citizens, out of whom 11.6% were employed in the primary sector (agriculture, forestry and fishery), 12.8% were occupied in the secondary sector, while a total of 4268 citizens (72.16%) were engaged in the tertiary sector (transport and storage, information and communication, public sector, administration and services, hotels and restaurants etc.). The sector with the highest employment is that of public administration and defense (24.6%). In accordance, wholesale and retail trade considerably contribute to the total employment (11.4%). Hotels and restaurants uptake a 7.4% of the total employment, implying that there is significant potential for the development of this sector. In 2019, more than 97% of the 34,914 tourists who visited Lemnos have stayed at a hotel, while this was reduced to 93% in 2020. In that year, the total number of visitors was restricted to 13,645, reflecting the COVID19 negative effect on the touristic activity of the island.

Overall, the socioeconomic features of the Lemnos local economy reveal a well-established development of both the primary and tertiary sector on the island (Table 1). Interestingly, the 19th century administrative division of the island resulted to a rather equal proportion of farmland and grassland for all villages, therefore offering sufficient space for agricultural production for all the communities of Lemnos, enhancing the activities of the primary sector [65].

The local economy is still based on the primary sector, although tourism is increasingly becoming an important activity. Lemnos hosts many local varieties of grapevines, among them are ancient ones such as "Lemnio" including the red variety of "kalambaki", and imported ones such as "Muscat of Alexandria", which adapted very well to the island's microclimate [42], as they are rich in monoterpene content [66]. The roads of wine on Lemnos pass throughout the island from Myrina to Kaminia and Moudros, while in the area of Aghios Dimitrios Atsikis there exists over 45% of the Lemnos vineyards [67]. Cheese, meat and honey also compose the highlights of the Lemnos gastronomy. In particular, the famous "Katsikaki Limnou" (Lemnian goat) is delivered from local kids that graze mostly freely in the rich herbaceous vegetation of the island with its varieties of aromatic plants and scrub.

Lemnos exhibits a wealth of archaeological monuments. Among the most important ones, the Castle of Myrina dates back to the 4th millennium BC. Several sanctuaries distributed all over the island, such as that of Artemis in the area of Avlonas and the Sanctuary of Kaveirion dedicated to the Kaveirians, mystery male deities linked to Hephaestus, the ancient Greek god of fire, provide evidence of diachronous religiosity on the island. The Poliochni settlement is considered the oldest city in Europe that has existed since 5000 BC up to 1600 BC, when it was probably destroyed by an earthquake. The archaeological site of Hephaestia dates back to the Bronze era, while findings suggest that it was continuously inhabited until the Byzantine years. The exceptional findings coming from the excavations in Poliochni, Hephaestia and Kaveirion are exhibited in the Archaeological Museum of the island. Additionally, the Museum of Maritime Tradition and sponge-fishing and the Portianou Folklore Museum host pieces of evidence of the modern socioeconomic history of the island, such as traditional costumes and various objects from the daily life of the islanders.

3. Methodology

3.1. Evaluation Criteria of Geosites

A protocol to collect the most important information about the involved prominent geosites and archaeological sites was applied based on the calibration of a series of criteria for each location and the quantitative assessment of their scientific and touristic values. The criteria primarily covered the topics of geology in terms of its scientific and educational value, as well as ecology, culture and aesthetics; also, location, accessibility, services, territorial, morphological, socio-demographic characteristics and tourism infrastructure were considered. The outcome of the quantitative assessment was expected to enable decision making, thus providing a toolbox useful for sustainable Geo-Archaeo-tourism development at a local or regional level and form the basis for designing exciting *"Geo-Archaeo-Routes"*.

Despite the fact that there exists no single type of criteria for all geotopes [68], in the present study we adopted the evaluation method of [7,8], who proposed a series of 13 criteria (Table 2) covering 5 topics (geology, culture, aesthetics, tourism, ecology), incorporating the outcomes of [6,68–72]. Hence, the criteria used for the quantitative evaluation of Lemnos geosites in the topic of geology were defined as geological history, representativeness, geodiversity, rarity, (geo)conservation and education. In particular, the participation of the geosite in the geological history of the wider area, its representativeness, geodiversity, rarity and state of preservation were evaluated. Any educational value of the terrain belonged to the same category. By the term geodiversity we refer to the set of geological (rocks, minerals, fossils) and geomorphological (landscapes, natural processes) forms, while rarity quantifies the number of geosites on the island featured by analogous geological features.

Table 2. Evaluation criteria for geosite assessment according to [7]. Quantifications in geodiversity refer to the number of different geological characteristics of each site; quantifications in rarity refer to the number of geosites on Lemnos Island featured by analogous geological features and quantifications in visibility refer to the number of locations on the island from which the geosite is visible.

	1	2	3	4	5
Geological History	Small participation at local level	Moderate participation at local level	Great participation at local level	Moderate participation at regional level	Great participation at regional level
Representativeness	Not at all	Low	Medium	High	Unique
Geodiversity	1	<3	<5	<10	>10
Rarity	>20	>10	>5	>2	Unique
Conservation	Totally damaged	Low	Medium	High	Intact
Education	Not at all	-	Medium	-	High
History–Archaeology	Not at all	Existing—Low importance	Minor importance	Moderate importance	Great importance—Geohistoric site
Religion	Not at all	Existing—Low importance	Minor importance	Moderate importance	Great importance—Geohistoric site
Visibility	1	2	3	4	>4
Landscape Differentiation	Not at all	Low	Medium	High	Very high
Accessibility	Not accessible	Low	Medium	High	Very high
Tourist Infrastructure	Not at all	Low	Medium	-	High
Ecological Value	Not at all	-	Medium	-	High

The topic of culture included the sub-criteria of history–archaeology and religion. In this group, the human presence over the years was evaluated, particularly the association and connection of a geosite with regard to archaeological–historical findings, places of religious worship and other cultural monuments. The topic of aesthetics applied to the sub-criteria of visibility and relief differentiation, quantified as the number of locations on the island from which the geosite was visible; this is how the public recognizes a geosite according to its distinct visual characteristics. Finally, the topic of tourism was evaluated on the basis of the geosite accessibility and the tourist infrastructure of the wider area, whereas the topic of ecology was assessed by the sub-criterion of ecological value; namely, the contribution and integration of a geosite in the development of the surrounding ecosystems.

In order to use all criteria on a rational basis, a quantitative approach was required, therefore each sub-criterion was evaluated on a scale ranging 1 (low significance) to 5 (high significance).

The quantitative data produced when applying all the criteria set for the assessment of the geosites of Lemnos, led to an average value (Total Score = Sum of rating criteria/Number of used criteria), which determined their final classification as geosites at a global, national or regional/local level. More specifically, if Total Score > 3.5, the geosite was of global interest, Total Score values between 3.5 and 3.0 featured geosites at a national level, while values of Total Score < 3.0 marked geotopes of regional/local interest. A geodatabase was constructed in order to manage and analyze information with statistical and georeference tools.

3.2. Designing "Geo-Archaeo-Routes"

Specific geosites representative for the individual disciplines of geology (e.g., geomorphology, tectonics, stratigraphy, palaeontology, volcanology, etc.) blended with archaeological sites and other sites of cultural interest were selected for the creation of indicative *"Geo-Archaeo-Routes"*. The most important scientific information per site was compiled in the framework of each *"Geo-Archaeo-Route"* in a way such that anyone interested can follow them and enjoy natural beauty combined with cultural heritage. All data were imported in the G.I.S. ArcMap 10.4 software, and several maps were created, including the assessed geosites and the designed *"Geo-Archaeo-Routes"*. Additionally, satellite images from Google Earth Pro were auxiliary used for the determination of the *"Geo-Archaeo-Routes"*.

4. Results and Discussion

4.1. Distribution, Documentation and Assessment of the Geosites

Situated in an area of intense geological activity, the island of Lemnos hosts numerous geomorphosites (e.g., fluvial sites/rivers and waterfalls, gorges, small lakes, coastal plains and waterfalls, coastal landforms, karstic elements) and also volcanic and tectonic structures, fossiliferous sites and sites of mineral resources (e.g., [31]). In the present study, a total of 64 locations of high interest, including geosites (Table 3) and cultural sites (Table 3), were selected to construct specific *"Geo-Archaeo-Routes"* case studies. The involved geosites comprise volcanic, fossiliferous, fluvial, coastal karstic, hydrothermic and tectonic geological sites, as well as wetlands that were assessed based on the criteria proposed by [7] and, depending on the score they achieved, they were categorized into a local (54), national (6) and global (4) level of reference (Table 3, Figure 3).

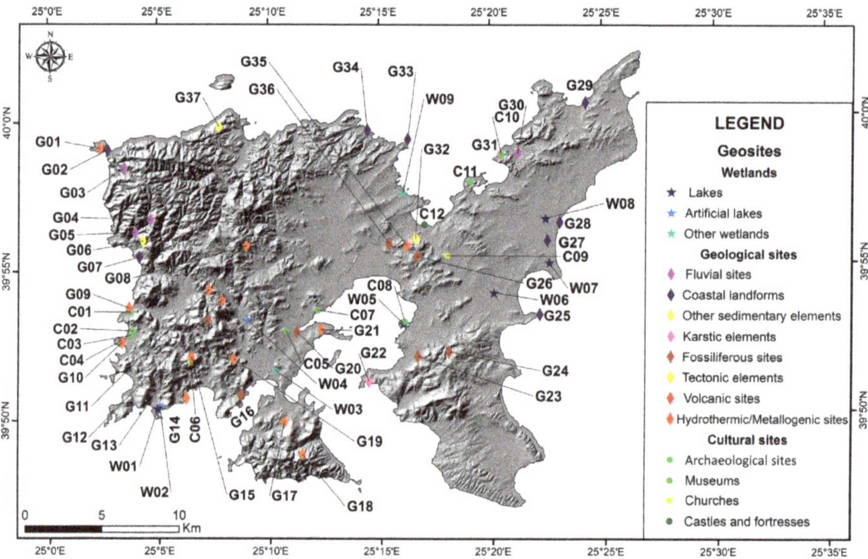

Figure 3. Location of selected geosites on Lemnos Island.

Table 3. Lemnos Island geosites assessment, following the evaluation criteria of Table 2.

Location Type	Fossil Sites						Natural Wetlands							Artificial Wetlands	
Name	Petrified Trunk of Profitis Ilias	Petrified Trunk of Paranisia	Petrified Forest of Lemnos (Tsimandria–Portiano)	Petrified Forest of Lemnos (Moudros–Roussopouli–Kaminia)	Petrified Forest of Lemnos (Romano–Varos)	Petrified Trunk of Varos	Petradi Lakes	Diapori Swamp	Koukonisi islet	Chortarolimni	Asprolimni	Alyki Lake	Sourladika	Thanos Reservoir	Kontias Artificial Lake
Code Name	G13	G16	G20	G23, G24	G26	G36	W01	W03	W05	W06	W07	W08	W09	W02	W04
Latitude	39°53′20″ N	39°50′47″ N	39°52′11″ N	39°52′09″ N	39°55′22″ N	39°55′48″ N	39°50′21″ N	39°51′36″ N	39°53′00″ N	39°54′05″ N	39°55′08″ N	39°56′43″ N	39°57′30″ N	39°51′35″ N	39°53′19″ N
Longitude	25°07′13″ E	25°08′40″ E	25°11′15″ E	25°18′05″ E	25°16′40″ E	25°15′24″ E	25°04′57″ E	25°10′15″ E	25°16′08″ E	25°20′04″ E	25°22′34″ E	25°22′21″ E	25°16′04″ E	25°07′00″ E	25°09′00″ E
Geological History	4	4	4	4	4	4	2	1	1	2	2	3	2	1	2
Representativeness	4	4	4	4	4	4	2	2	2	3	2	4	2	2	2
Geodiversity	1	1	5	5	5	1	2	2	2	2	2	3	3	1	2
Rarity	2	2	4	4	4	2	2	2	3	3	3	4	4	3	3
Conservation	5	5	4	4	4	5	3	4	4	5	4	4	4	3	4
Education	3	3	3	5	3	3	1	1	1	1	3	3	1	1	1
History–Archaeology	1	1	1	2	1	1	1	1	1	1	1	1	1	1	1
Religion	1	1	1	1	1	1	1	1	1	1	1	1	1	1	1
Visibility	2	2	4	4	4	5	5	5	5	5	5	5	5	5	5
Landscape Differentiation	1	1	3	3	3	1	4	4	5	5	4	5	3	5	5
Accessibility	5	5	5	5	5	5	5	5	5	5	5	5	5	5	5
Tourist Infrastructure	1	1	3	5	3	2	1	1	1	1	3	3	1	1	1
Ecological Value	5	5	5	5	5	5	5	5	5	5	5	5	5	5	5
Total Score	2.69	2.69	3.54	3.92	3.54	3.00	2.62	2.62	2.77	3.00	3.08	3.54	2.85	2.62	2.85

Location Type	Volcanic Structures														Tectonic Structures	
Name	Lava Dome of Mourtzouflo	Metal-Bearing Zone of Sardes	Lava Domes at Avlonas Beach	Lava Dome of Myrina Castle	Therma Springs	Lava Dome of Profitis Ilias	Lava Domes at Evgatis Beach	Lava Dome of Kakkavo	Lava Dome of Skopos	Metal-Bearing Zones of Fakos	Lava Dome of Kontias	Volcanic Vein of Portianou	"Terra Lemnia"	Volcanic Crater of Mosychlos	Kaspakas Waterfall	Active Fault of Kaspakas
Code Name	G01	G08	G09	G10	G11	G12	G14	G15	G17	G18	G19	G21	G32	G35	G05	G06
Latitude	39°59′07″ N	39°55′47″ N	39°53′45″ N	39°52′35″ N	39°54′19″ N	39°53′58″ N	39°50′42″ N	39°52′04″ N	39°49′32″ N	39°48′46″ N	39°52′00″ N	39°52′58″ N	39°55′58″ N	39°55′44″ N	39°56′16″ N	39°55′59″ N
Longitude	25°02′23″ E	25°08′58″ E	25°03′40″ E	25°03′20″ E	25°07′18″ E	25°07′51″ E	25°06′14″ E	25°06′28″ E	25°10′59″ E	25°11′28″ E	25°08′21″ E	25°12′20″ E	25°16′37″ E	25°16′13″ E	25°03′58″ E	25°04′19″ E
Geological History	4	3	4	4	4	4	4	4	4	3	4	4	4	4	3	4
Representativeness	3	2	3	4	3	3	3	5	3	2	3	3	5	4	2	3
Geodiversity	1	3	2	1	1	1	1	2	1	4	1	1	4	1	2	1
Rarity	1	3	1	1	2	1	1	5	1	3	1	3	5	4	1	3
Conservation	4	5	4	4	3	4	4	4	4	5	4	4	4	3	4	5
Education	1	1	1	5	3	1	1	5	1	1	1	3	5	3	1	1
History–Archaeology	1	1	1	5	1	1	1	5	1	1	1	4	5	4	1	1
Religion	1	1	1	2	1	1	1	5	1	1	1	2	1	1	1	1
Visibility	5	1	5	5	5	5	5	5	5	1	5	1	5	4	5	5
Landscape Differentiation	5	2	5	5	3	5	5	5	5	2	5	2	5	4	4	5
Accessibility	3	5	5	5	5	5	5	5	3	5	5	5	5	5	5	4
Tourist Infrastructure	1	1	2	5	5	1	1	3	1	1	2	1	1	1	2	1
Ecological Value	1	1	1	1	3	1	1	1	1	1	1	1	3	3	5	1
Total Score	2.38	2.23	2.69	3.62	3.00	2.54	2.54	4.15	2.38	2.31	2.62	2.62	4.00	3.15	2.77	2.69

Table 3. *Cont.*

Location Type	Geomorphological and Sedimentary Structures													
Name	Tombolo of Mourtzouflos	Neroviglia Gorge	Agios Ioannis Gorge	Coastal Geoforms in Agios Ioannis	Cave of the Seal	Tafoni Forms in Kavalaris	Coastal Sand Dunes of Alyki	Sandy Arm of Alyki	Honeycomb Weathering in Plaka	Hydrographic Network of Nettina	Cave of Philoctetes	Surficial Rectangular Forms in Trigies	Sedimentary Rocks of Faraklo	Inland Sand Dunes of Gomati
Code Name	G02	G03	G04	G07	G22	G25	G28	G29	G29	G30	G31	G33	G34	G37
Latitude	39°59′04″ N	39°58′27″ N	39°56′42″ N	39°55′27″ N	39°51′11″ N	39°53′22″ N	39°56′27″ N	39°55′51″ N	40°00′28″ N	39°58′48″ N	39°58′46″ N	39°59′19″ N	39°59′38″ N	39°59′28″ N
Longitude	25°02′45″ E	25°03′29″ E	25°04′40″ E	25°04′09″ E	25°14′27″ E	25°22′09″ E	25°23′05″ E	25°22′31″ E	25°24′16″ E	25°21′14″ E	25°20′26″ E	25°16′18″ E	25°14′28″ E	25°07′47″ E
Geological History	3	3	3	3	3	3	3	3	3	3	3	3	4	4
Representativeness	3	2	2	3	2	2	2	3	3	2	3	3	5	5
Geodiversity	2	2	2	4	3	2	2	2	3	2	2	2	3	4
Rarity	4	2	2	3	2	3	1	4	4	4	2	4	5	5
Conservation	5	5	4	4	5	5	4	4	5	5	5	5	4	4
Education	1	1	3	1	1	1	1	1	1	1	3	1	3	1
History–Archaeology	1	1	1	1	1	1	1	1	1	1	1	1	1	1
Religion	1	1	1	1	1	1	1	1	1	1	1	1	1	1
Visibility	5	5	5	5	5	4	4	3	5	5	5	4	5	5
Landscape Differentiation	5	4	4	5	5	3	3	4	4	3	5	3	4	4
Accessibility	3	3	5	5	4	4	4	5	4	4	4	4	5	5
Tourist Infrastructure	1	1	2	5	3	1	2	2	1	1	3	1	3	3
Ecological Value	3	3	3	3	3	3	5	5	3	3	3	3	3	5
Total Score	2.85	2.54	2.85	3.31	2.92	2.54	2.54	2.92	2.92	2.69	3.08	2.69	3.54	3.62

4.2. Selected "Geo-Archaeo-Routes" on Lemnos Island: A Spectacular "Nalture" Experience

The profound wealth of Lemnos is characterized by a variety of locations of geological, environmental and historical interest in its limited insular area (Figure 4), making it an ideal case study for developing and testing various types of *"Geo-Archaeo-Routes"* as a *"nalture"* experience in a geoethical perspective.

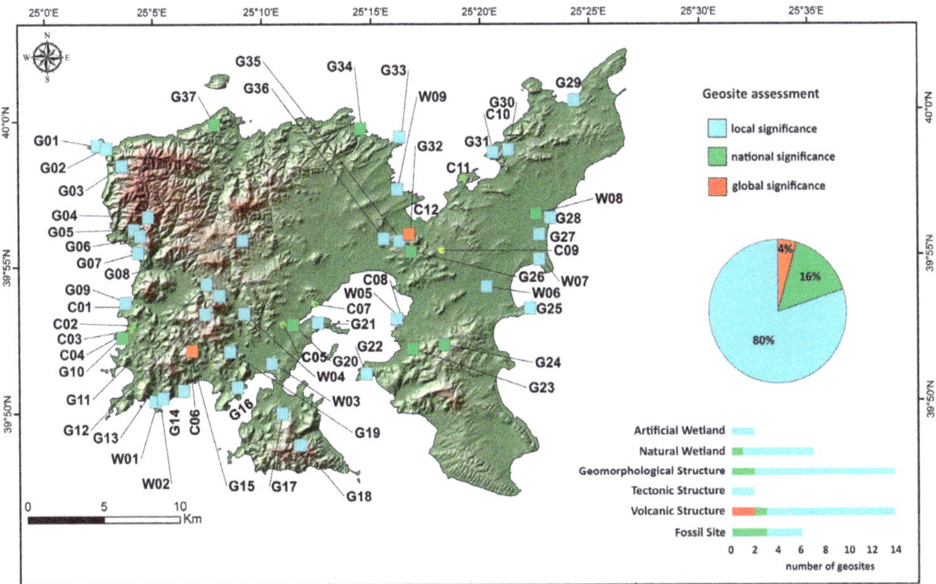

Figure 4. Categorization of the assessed geosites according to the evaluation criteria of [7], on Lemnos Island.

Through these *"Geo-Archaeo-Routes"*, visitors will be able to get to know the natural and cultural monuments of the area, the settlements, the traditional agricultural activities as well as the local products. The proposed routes are environmentally friendly, as either hiking trails or road trips using any means of transport (car, bike, local transport bus, etc.). A key element for the proper design of the route type is to serve the visitors in the best way covering a wide range of topics and to also be associated with good accessibility, safety and adequate available information. Thus, the hiking routes (Figure 5) were proposed to follow the already existing dirt roads network, the starting points being large villages and/or sites of ample interest. By establishing the *Geo-Archaeo-touristic* experience, we introduced a holistic geotouristic approach combining natural beauty, past civilizations and the history of the Earth in the unique geoenvironment of Lemnos Island.

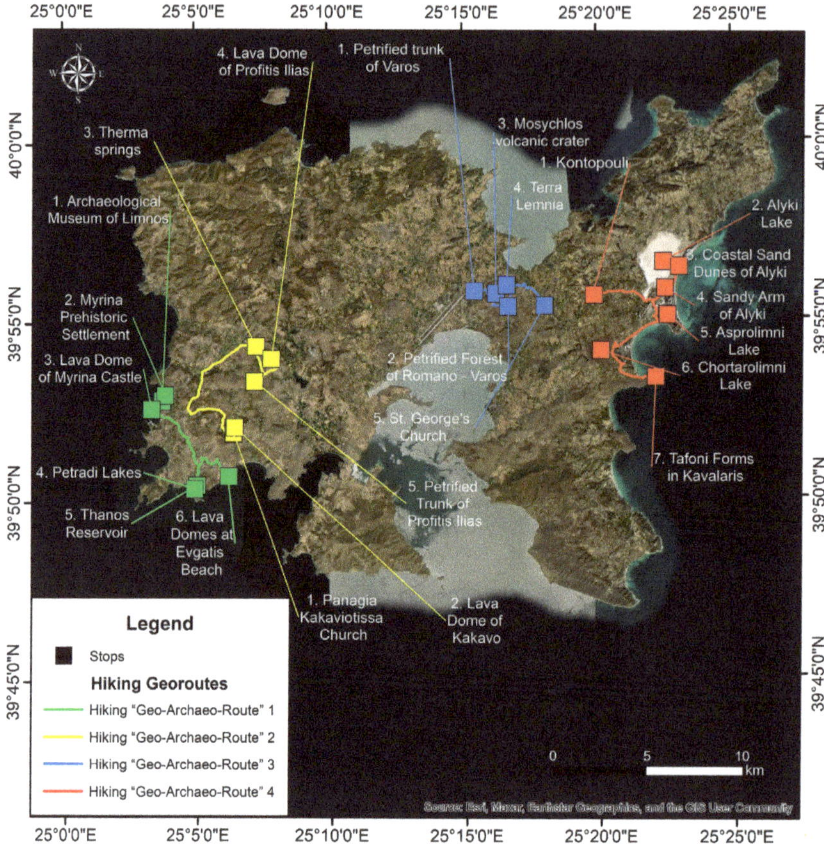

Figure 5. Designed hiking georoutes for Lemnos Island.

4.2.1. Hiking "Geo-Archaeo-Route" 1: Myrina–Thanos–Evgati Beach

This hiking route involves four cultural sites, one volcanic geological site and two geomorphosites (Figure 5). The starting point refers to the *Archaeological Museum of Lemnos* (C03; Table 4, Figures 3, 4 and 6a) in the center of Myrina, the capital of the island, where the visitors can admire stunning archaeological findings from all over the island as well as donations from private collections. The two-story neoclassical building that now houses the museum was built in the 19th century to accommodate the Turkish command post. In 1939, the building was sold by the community of Kastro to the Panlimniakos School Fund, on the condition that it would be donated to the public, in order to host the *Archaeological Museum of Lemnos*. During World War II, the archaeological treasures of Lemnos were transported to the Museum of Mytilene and to the National Archaeological Museum in Athens, and were returned back in 1961, after the restoration of the building. In the early 1990s a renovated exhibition was launched, mostly including findings from the excavations of the Italian Archaeological School and the Ephorate Prehistoric and Classical Antiquities in Hephaistia, Kaveirio, Poliochni and Myrina [73].

Table 4. Lemnos Island cultural sites.

Name	Code Name	Type
Sanctuary of Artemis	C01	Archaeological sites
Myrina Prehistoric Settlement	C02	Archaeological sites
Archaeological Museum of Limnos	C03	Museums
Myrina Castle	C04	Castles and fortresses
Portianou Folklore Museum	C05	Museums
Panagia Kakkaviotissa Church	C06	Churches
Museum of Maritime Tradition and Sponge-Fishing of Nea Koutali	C07	Museums
Koukonisi Prehistoric Settlement	C08	Archaeological sites
St. George's Church	C09	Churches
Kaveirion Archaeological Site	C10	Archaeological sites
Hephaestia Archaeological Site	C11	Archaeological sites
Kotsinas Fortress	C12	Castles and fortresses

Figure 6. Selected geosites on Lemnos Island, for coding see Figure 3. (**a**) *Archaeological Museum* (C03); (**b,c**) *Lava Dome of Myrina Castle* (G10); (**d**) *Lava Domes at Evgatis Beach* (G14).

Hiking "Geo-Archaeo-Route" 1 continues with a visit to the *Prehistoric Settlement of Myrina* (C02; Table 4, Figures 3 and 4), which is located only a few meters away from the Archaeological Museum and is considered an important residential center of the early Bronze Age. The *Prehistoric Settlement of Myrina* is located on the coastal site "Richa Nera", on a volcanic geological basement (Figure 2). It was developed slightly earlier than the Poliochni settlement in the eastern part of the island, with its earliest phases being assigned to the first half of the 4th millennium BC [30]. Together with the rest of the prehistoric settlements on the island, the site reflects the importance of Lemnos during the Bronze Age associated with its strategic location in the vicinity of Dardanelles [74]. The residential remains and the fortifications mostly built by the volcanic rocks of the Myrina Unit [45] denote the dense and regular habitation of the settlement, with urban planning proved by the existence of roads between the houses, sometimes paved, and sewage systems with stone pipes, which occupied an area of 80,000 m^2, and the area had a population of 3000–4000 inhabitants during its heyday [75]. Three building phases have been recorded with a total thickness of 5.10 m, reflecting a settlement with dynamic evolution in between the Late Neolithic and the Early Bronze age [76]. The *Prehistoric Settlement of Myrina* was destroyed many times potentially by fire or earthquakes, with the inhabitants rebuilding or repairing their houses according to their needs, as evidenced by the abandoned older buildings and the traces on the walls of all phases. The similarities in architecture and ceramic styles with the Poliochni settlement suggest the presence of a homogenous cultural group all over the island [76].

Afterwards, the route reaches the *Lava Dome of Myrina Castle* (G10) through the picturesque alleys of the city of Myrina. The impressive *Myrina Castle*, a geosite of national significance (Table 3, Figures 3, 4 and 6b,c), is located on the top of a lava dome of the volcanic Myrina Unit [45]. Lava domes are isolated hills of volcanic origin, formed mostly when the magma lost the dissolved gases while exiting the volcanic pore, resulting in an increase in its viscosity and thickness. Thus, the lava accumulated in the crater and cooled immediately, creating domes. The Myrina Unit consists of the youngest lava domes and occasional flows on Lemnos (21–18 Ma; [59]), suggesting a general evolution from shoshonitic to high-K calc-alkaline magmatism with time [45]. The *Myrina Castle* consists of both inner and outer surrounding walls, while it displays battlements, ramparts and 14 towers. The history of the castle dates back to the 13th century BC, when the Thessalians Minyans settled in Lemnos and fortified the hill with "Cyclopean" walls that are partially preserved outside the current fortress [77]. These were further enhanced by the Pelasgians, who conquered the island until the 6th century BC. The castle began to take the form in which it survives today during the 12th century AD, during the reign of the Byzantine emperor Andronikos I Komnenos. Over the years, the Byzantine castle experienced many conquerors, e.g., the great admiral of Romania, Philocalo Navigagioso, who built in 1207 a new castle on the middle plateau of the peninsula, and the Venetians who built the fortress to the north of the ancient castle and constructed the moat that surrounded the castle. In the 15th century the island was ceded by Ioannis VII Palaiologos to the Genoese Gateillusio family, who made repairs and some additions, giving the castle the form it has today. In 1479 and until 1912, the castle was under Ottoman occupation. Most of the buildings inside the castle date back to this period, which indicates the existence of a settlement [77]. Surprisingly, a restricted population of European fallow deer (*Dama dama dama* L.) lives in the fortress peninsula, as a result of a transferring of few individuals from the island of Rhodos in the early 1970s. The ground cover and forage availability is uneven throughout the peninsula, with the animals adapting their diet or searching for food outside the castle hill, therefore the principles of deer farming should be performed in order to maintain a healthy population [78].

The *"Geo-Archaeo-route"* 1 (Figure 5) then heads towards the village of Thanos, where the "twin" Petradi lakes are located. The water bodies on Lemnos include eighteen natural and three artificial wetlands [79], covering about 2.6% of the island's area. Out of them, *Petradi Lakes* (W01; Table 3, Figures 3 and 4), two neighboring seasonal wetland systems with rich succulent vegetation, are located south of the settlement of Thanos at Cape Asprokavos, being part of the protected insular small wetlands of Greece (Presidential Decree 2012, Government Gazette AAP 229/19.06.2012). Water is retained in them only during wet periods, while the natural continuation to the coastal zone is no longer existent [80]. Among the artificial water bodies, *Thanos Reservoir* (W02), 2.6 km from the homonymous settlement, was built in 1997 as an off-stream reservoir mostly fed by the Thanos torrent.

The route ends on the impressive *Lava Domes at Evgatis Beach* (G14; Table 3, Figures 3, 4 and 6d) that are composed by high-K rocks associated with the massive subduction of sedimentary material at the time of magmatic activity on Lemnos [81].

4.2.2. Hiking "Geo-Archaeo-Route" 2: Kakkavo–Therma–Profitis Ilias

Lava domes crop out in many areas on the island of Lemnos, as remnants of the intense volcanic activity that has shaped the island's morphology. Most of them are located in the southwestern part of the island with the characteristic *Lava Dome of Kakkavo* (G15; Table 3, Figures 3, 4 and 7a), which consists of medium-K rocks [81]. This dome, a geosite of global significance (Figure 4), is selected as the starting point of *"Geo-Archaeo-route"* 2 (Figure 5). The site, beyond the impressive dome, hosts the *Panagia Kakkaviotissa Church* (C06; Table 4, Figures 3, 4 and 7b,c), a unique construction, having the roof of a cave as a natural cover. The *Church of Panagia Kakkaviotissa* is a chapel near the now-deserted village of Zematas, at the top of Kakkavo hill, from which it took its name. It became the property of the monastery of Megisti Lavra in 1305, when monks settled in the cave to protect themselves from the raids of the Ottomans, to practice asceticism and to praise the Virgin Mary. Nowadays, stairs have been built to make the access to the temple easier and the chapel is open to the public. Traditionally, the church hosts a celebration every year on the first Tuesday after the Greek Orthodox Easter [82]. The cave is not of karstic origin, as it was formed by the intense weathering caused by the wind and the rainwater that infiltrates the discontinuities of the volcanic rocks. It can also be described as a very large tafoni form.

Figure 7. Selected geosites on Lemnos Island, for coding see Figure 3. (**a**) *Lava Dome of Kakkavo* (G15); (**b**,**c**) *Panagia Kakkaviotissa Church* (C06); (**d**) *Therma Springs* (G11); (**e**) *Profitis Ilias Lava Dome* (G12).

Afterwards, the route leads to the *Therma Springs* (G11; Figures 3, 4 and 7d), which represent a geological hydrothermic–metallogenic site. They are thermal springs located on the western part of the island, near the village of Kondias. The water of Therma gushes from two springs, coming from a depth of about 1200 m. It has a temperature of 42.3 °C, it is odorless, transparent, palatable and slightly alkaline with a pH of 8.55. It has a high content of sodium (83.9 mg/L), chlorine (75.1 mg/L) and bicarbonate (84.6 mg/L) and is characterized as "meteoric, low in minerals, hypotonic and slightly radioactive thermal water" [83,84]. Thermal baths in the area of Therma have been functioning since 1548.

Finally, the *hiking "Geo-Archaeo-Route" 2* ends at the volcanic site of the *Profitis Ilias Lava Dome* (G12; Table 3, Figures 3, 4 and 7e) and the outlying area where the *Petrified Truck of Profitis Ilias* is located (G13). Velitzelos et al. (2019) stated that Lemnos Island was covered during Late Eocene–Early Miocene by extended forests of conifers, of arboreal dicotyledons as Laurales, Fabales and Myrtales, and monocotyledons, especially Arecales. Few fossil plant but numerous fossil wood remains have been identified in Lemnos, with only two conifers described, *Glyptostrobus europaeus* (Brongniart) Unger and *Sequoia abietina* (Brongniart) Erw. Knobloch [85].

4.2.3. Hiking "Geo-Archaeo-Route" 3: Varos–Repanidi

The *Petrified trunk of Varos* (G36; Table 3, Figures 3, 4 and 8a) is the starting point of the *hiking "Geo-Archaeo-Route" 3* (Figure 5). In 2020, in the area of Varos, a huge, petrified tree trunk was discovered by members of the Natural History Museum of the Petrified Forest of Lesvos. It is a trunk of an angiosperm, probably a giant sequoia, more than eight meters high. As can be seen from the first phase of the excavations, the characteristics of the wood as well as the ends of its branches are preserved in excellent condition. Overall, the *Petrified Forest of Lemnos*, a geosite of national significance (Table 3, Figures 3 and 4), consists of three areas of about 45,000 acres. The main area of 24,400 acres is located between *Moudros, Roussopouli* (G23) and *Kaminia* (G24), while the other two are located in *Varos* (G26), with an area of 14,350 acres and in *Tsimandria–Portianou* (G20), with an area of 7000 acres. Moreover, besides the *Petrified trunk of Varos*, there are also isolated occurrences of petrified trunks in *Profitis Ilias* (G13) and *Paranisia* (G16), while it is considered that there are hundreds of trunks on the island that have not yet been discovered. There are some lignotaxa previously described from the *Petrified Forest of Lemnos*, such as *Laurinoxylon ehrendorferi, Cornoxylon pappi* and a problematic conifer, initially described as *Pinoxylon parenchymatosum*, later revised as a species of *Lesbosoxylon* [86]. More recently, some palm fossil taxa were described [87] and a series taxa of fossilized wood have been further identified [88], such as *Cupressinoxylon akdiki, Juniperoxylon acarcae, Tetraclinoxylon velitzelosii, Taxodioxylon gypsaceum, Taxodioxylon taxodii, Glyptostroboxylon rudolphii, Glyptostroboxylon tenerum, Pinuxylon pineoides* and *Pinuxylon halepensoides*.

Figure 8. Selected geosites on Lemnos Island, for coding see Figure 3. (**a**) *Petrified trunk of Varos* (G36); (**b**) *Volcanic Crater of Moschylos* (G35); (**c**) *Church of Agios Georgios* (C09).

The next highlight of the route is the *Volcanic Crater of Moschylos* (G35; Figures 3, 4 and 8b). To the east of the village of Varos, the main morphological feature of the area is an inactive volcanic crater, where several hills surround a small plateau with fertile volcanic soils covered by crops. According to mythology, the Greek god of fire Hephaestus, a characteristic example of the personification of fires and volcanoes, established his workshop in this area [89].

Later on, the route passes from a global significance geosite (Table 3; Figures 3 and 4), the hill where *"Terra Lemnia"* (G32) was mined. *"Terra Lemnia"* is a type of siliceous clayey mud derived from weathered volcanic tuffs [90] that outcrops only in Lemnos. From the Hellenistic Period until the beginning of the 20th century, *"Terra Lemnia"* was a widespread medicine throughout Europe, used as an "antidote" to poisons while also protecting against the plague. It was mined on Despotis hill between Repanidi, Kotsinas and Varos, which is located almost at the center of the island. *"Terra Lemnia"* comes in three different shades, which differ depending on their mineralogical composition. The red *"Terra Lemnia"* has a high content of illite (41%) and kaolinite (37.6%), while it also contains quartz (17.7%) and hematite (3.8%). The yellow–gray *"Terra Lemnia"* consists mainly of montmorillonite (66%), while it also contains illite (18.1%), albite (9%) and quartz (6.9%). Finally, the white *"Terra Lemnia"* consists mainly of dolomite (65.2%), while it also contains kaolinite (17.3%), illite (9.9%) and quartz (7.6%) [91].

The route ends at the village of Repanidi where the historical old *Church of Aghios Georgios* (C09; Table 4, Figure 8c) is located. This is a cemetery church declared as a historical monument, most probably founded in 1860. It is a typical example of the ecclesiastical architecture in Lemnos, a three-aisled basilica with neoclassical influences. The highly aesthetic wood-carved altarpiece of the church and the bright colors of the portable icons represent a unique art developed during the post-Byzantine era [92].

4.2.4. Hiking "Geo-Archaeo-Route" 4: Kontopouli–Kavalaris

This hiking route (Figure 5), starting from the picturesque village of Kontopouli with the traditional stone houses and churches, reaches the southern edge of the national significance geosite *Alyki Lake* (W08; Table 3, Figures 3, 4 and 9a,b), a natural salt pan, which is the largest seasonal water surface on an Aegean island, with annual salinity ranging between 3.5 and 25‰ and indigenous flora comprising of *Arthrocnemum glaucum, A. fruticosum, Chenopodium* sp.*, Cynodon dactylon, Phragmites australis, Plantago* sp.*, Salsola kali, Scirpus holoschoenus, Suaeda maritima, Trachomitum venetum, Allium bourgeaui* and *Phyla nodiflora* [79], and important avifauna, e.g., *Tadorna ferruginea* and *Burhinus oedicnemus* [93]. At its southeastern edge, a sand spit can be seen, inhabited by rich avifauna. The neighboring *Asprolimni Lake* (W07; Table 3, Figures 3 and 4) is a similar lake to Alyki but is much smaller in size and between them the area is covered by sand dunes. Then, the route follows the coastal road of the impressive beach of Keros, reaching the lake of *Chortarolimni* (W06; Table 3, Figure 9c), the largest seasonal swamp in the Aegean, with slightly brackish waters (annual salinity of around 0.5‰), surrounded by halophytic vegetation consisting of *Arthrocnemum glaucum, Salsola soda and Scirpus lacustris,* and extended sand dunes that protect the lake from the waves and the inflow of sea water [79].

Alyki (W08), *Asprolimni* (W07), the smallest out of the three lakes, and *Chortarolimni* (W06) represent a complex dynamic hydrogeological system of three lakes, developed on the eastern coast of the island. Due to the rare species it hosts (e.g., porpoise, bottlenose dolphin), as well as the habitats (Mediterranean grasslands with *Juncus*, saline steppes with shifting coastal dunes), the area is under the protection of the Natura 2000 network [94]. The wetland is featured by the development of sand dunes along the coast and the impressive presence of thousands of flamingos during the winter [79]. The paleoenvironmental evolution of the coastal plain has been studied by [57,95], who proved that a shallow marine environment with significant freshwater input was prevailing in between 5100 and 1040 BC, the paleo-shoreline being located 500 m in front of the present day shoreline. Later on, a mesohaline lagoon to shallow bay developed in the period between 1040 BC and 760 AD [95]. Overall, the paleoenvironmental evolution in this area has also been affected by the Black Sea outflow, starting at approx. 7500 cal. BP [96].

Figure 9. Selected geosites on Lemnos Island, for coding see Figure 3. (**a,b**) Alyki Lake (W08); (**c**) Chortarolimni (W06).

A special morphological feature of the area is the development of the *Coastal sand dunes of Alyki* (G28) along the Alyki coast. Their formation requires a large supply of sand and strong winds that blow for a long time, as well as the appropriate orientation of the coast in relation to the wind conditions [97]. Coastal sand dunes are the dominant landform on the eastern coasts, creating the barrier between the sea and Alyki lagoon [95]. The *Sandy Arm of Alyki* (G29; Table 3, Figures 3 and 4) represents a sand barrier that has been gradually created probably later than 820 AD [95].

Towards the southern tip of the gulf of Keros, the route ends at the spectacular *Tafoni forms in Kavalaris* (G25; Table 3, Figures 3 and 4), found in the volcanic rocks on the east of the island at cape Kavalaris. These weathering formations resembling small caves are often characterized as "aeolian erosion formations", however, their development is partly due to wind action, but is mainly due to chemical weathering by the salt contained in the sea water, which is transported by the wind [98,99]. They are developed especially in medium to coarse-grained silicate rocks with granular fabric, such as granites and granodiorites or even sandstones [99].

4.2.5. Road "Geo-Archaeo-Route" 1: Kotsinas–Faraklo–Gomati

The starting point of the road *"Geo-Archaeo-Route" 1* (Figure 10) is the *Kotsinas Fortress* (C12; Table 4, Figures 3, 4 and 11a,b), located in the village of Kotsinas, in the bay of Bournias. It was built in the 13th century probably by the Venetian Navigaiosi family who ruled the island. The fortress was built on an artificial hill and had a rectangular shape. Its three sides were surrounded by a fortification moat, while towards the sea there were four-sided defensive towers. Inside there was a chapel built in 1415 and dedicated to Virgin Mary [100]. Nowadays, only a few relics of the fortress remain, as the Venetians themselves destroyed it when they left Lemnos in 1657. The chapel has been replaced by the church of Zoodochos Pigi. The statue of Maroula of Lemnos in the church yard was made in honor of the heroine Maroula, who fought bravely in 1478 during the siege of the Kotsinas fortress by the Ottomans.

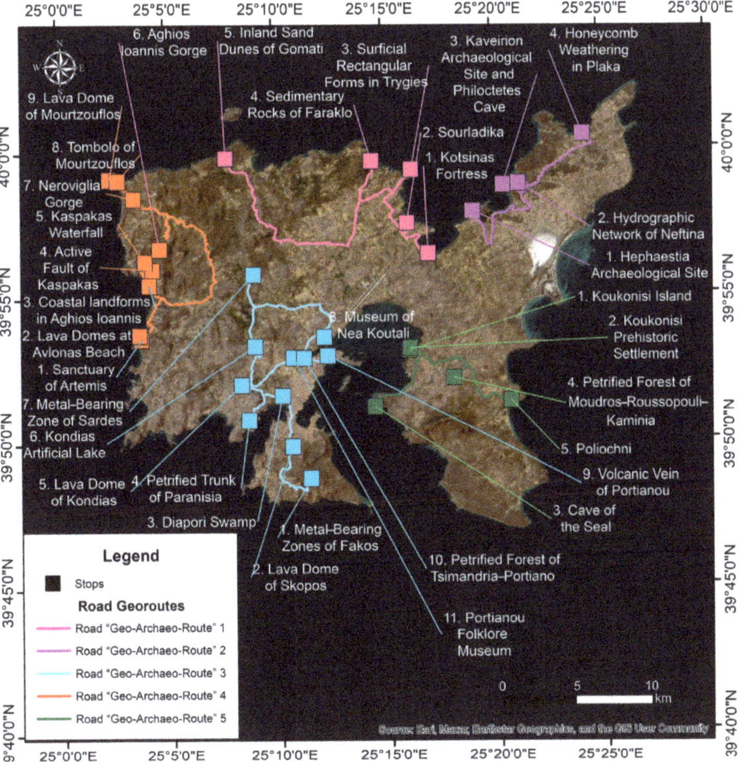

Figure 10. Designed road georoutes for Lemnos Island.

Northwest of the Kotsinas settlement, the route leads to *Sourladika* (W09; Table 3, Figures 3 and 4), which denotes a rare wetland system, consisting of several seasonal small brackish to fresh water lakes formed in cavities between the coastal sand dunes [93,101]. The area hosts rich vegetation, and among other faunal elements, a large number of wild rabbits.

Towards northeast the route reveals the *Surficial Rectangular Forms in Trygies* (G33), impressive erosional forms [102] similar to the *Sedimentary Rocks of Faraklo* (G34) that are unique outcrops with yellow spherical landforms (Figure 11c–e), a geosite assessed to be of national significance (Table 3; Figures 3 and 4). It has been erroneously believed that these landforms formed when volcanic lava came in contact with the sea water, but they most likely are the product of coastal erosion. In the area of Trygies and Faraklo, the geological background consists of clay–sandstone alternations, which constitute the majority of the outcropping sedimentary turbiditic rocks. Particularly in the area of Faraklo there is an increase in the thickness and frequency of sandstone benches, which indicates deposition in submarine fans. The sandstones are fine-grained to medium-grained, containing quartz, feldspars, lithic fragments of various rocks (schists, volcanics, limestones, etc.), titanite, chlorite, chromite grains and iron oxides–hydroxides [47,103]. In Cape Trygies, thick-stratified sandstones and cobbles are observed with flow structures and evidence of submarine sliding [47]. On the upper surface of the sandstone beds, spectacular forms are observed, such as sandstone spheres and surficial rectangular and honeycomb cavity features as the result of coastal erosion.

Finally, the route ends at the *Inland Sand Dunes of Gomati* (G37) or "Pachies Ammoudies", which are located at the northern part of the island [104]. With an area of about 70 acres, this ecosystem is characterized as a unique desert environment in the whole of Europe (Figure 11f–h) and a geosite of national significance (Table 3; Figures 3 and 4). These inland sand dunes, being important but also sensitive habitats, certainly comprise a phenomenon rare for an insular area, thus they are habitats protected by the Directive 92/43/EEC. They have been created by the strong NE winds that transport the sand from the nearby beaches as evidenced by the scarce presence of shallow marine environment benthic foraminifera (e.g., *Elphidium crispum* and fragments of echinoid spicules; Figure 11i), as well as by the aeolian erosion of the outcropping Oligocene sandstones. The local vegetation consists mostly of *Pancratium maritimum* (white lillies), *Ammophilala arenusus agnaria*, *Sarcopoterium spinosum*, *Thymus vulgaris*, *Nerium oleander* and *Olea europea* var. *oleaster*, etc.

Figure 11. Selected geosites on Lemnos Island, for coding see Figure 3. (**a**) The church of Zoodochos Pigi on the remnants of the *Kotsinas Fortress* (C12); (**b**) the statue of Maroula of Lemnos in the church yard (Kotsinas Fortress); (**c–e**) *Sedimentary Rocks of Faraklo* (G34); (**f–h**) *Inland Sand Dunes of Gomati* (G37); (**i**) detail of the sand from the dunes under a stereomicroscope: benthic foraminifera and a fragment of an echinoid spicule.

4.2.6. Road "Geo-Archaeo-Route" 2: Hephaistia–Kaveirion–Plaka

This route (Figure 10) begins at the *Archaeological Site of Hephaistia* (C11; Table 4, Figures 3 and 4), built north of the Pournias gulf, which was in historical times the second most important city of Lemnos after Myrina. Excavations carried out by the Italian School of Archeology and the Ephorate of Prehistoric and Classical Antiquities revealed antiquities that prove the habitation of the area from the Late Bronze Age to the Byzantine times. There are ancient ruins in almost 10 layers of a complex city displaying buildings, palaces, baths, Christian churches, a Hellenistic Roman theatre and an 8th to 6th century BC sanctuary, dedicated to the Great Goddess Lemnos, who corresponds to Goddess Artemis. The Late Bronze Age settlement is marked by Mycenean pottery and includes the so-called "walls of the isthmus" between the gulf of Pournias and the lagoon to the east, as also a small paved road running in a north–south direction [105]. The Mycenean village was abandoned abruptly, reflecting a time of new populations arriving in the area. The next chapter in the history of the settlement from the second half of the eighth to seventh century BC was featured by the Archaic Lemnian society with strong social stratification, the presence of an acropolis, a sanctuary and the imprint of the Mediterranean economy based on grain, wine and metal [105]. The sanctuary remains are preserved at two levels, while the central space had desks on both sides, on which votive offerings and parts of a cult statue of an ancient deity, the Great Goddess Lemnos, were found [106]. The ancient theater is the most important monument of the *Archaeological Site of Hephaistia* (C11; Figure 12a) and is associated with an Athenian settlement on the island. Its caveat was built on the semi-rocky slope of a hill, and on earlier sanctuaries of the 7th–6th century BC [107] stated that it is the most ancient Greek theatre, although [105] questions this. Other important findings within the archeological site are a cemetery dating from the mid-8th to the 5th century BC, two ceramic laboratory incinerators of the Hellenistic period (2nd–1st century BC) discovered near the sanctuary, while southeast of the city near the sea, bath facilities and remains of houses of the Hellenistic and Byzantine times have been discovered. Hephaistia was set on fire by the Persians in 511 BC but it was built again by the Athenians, being capital or co-capital of the island and the economic center until the Middle Ages, when its harbor was gradually embanked [108]. In the same location, the outdoor Ancient Quarry of Hephaistia (Figure 12b) at the northeastern part of the Paleopolis peninsula represents a site where porous limestone was mined during Classical to Hellenistic times and was widely used in the ancient city of Hephaistia [109]. Nowadays, a large part of the quarry is covered by the sea.

The road *Geo-Archaeo-Route 2* is continuous with the *Rejuvenated Drainage of Neftina* (G30) that represents the spectacular case of the final stage of the erosion cycle of a fully evolved hydrographic network (Figure 12c,d), indicating a local neotectonic uplift of a fault-bounded block in the area affected by the Kondias–Kotsinas fault zone [63]. Shallow riverbeds are furrowing the smooth morphological relief during the current stage of geomorphological relief rejuvenation.

Nearby at Cape Chloe, the *Sanctuary of Kaveirion* (C10; Figures 3, 4 and 12e) represents one of the oldest sanctuaries in the Aegean from the second half of the 7th century BC [110]; three telesteria have been discovered during the archeological excavations of the Italian Archaeological School, dating back to Archaic, Hellenistic and late-Roman times [111]. The religious ceremonies performed to honor the Kaveirians, mystery male deities sometimes called Hephaistoi, indicating their association with Hephaestus, the god of fire and blacksmiths and a secret cult, the so called Kaveirian Mysteries [112]. The oldest Archaic building consists of an irregular rectangular monument similar to the Acropolis of Hephaistia that was also abruptly abandoned at the end of the 6th century BC [105], featured by a circular protrusion used as an altar or a podium. The Hellenistic telesterion (200 BC) is a prostyle building with twelve Doric columns. It consists of a large central space, divided vertically into three parts by two rows of Ionic columns, with an opening to the north. Several items of the Classical and Hellenistic phase have been found in the buildings, including many offerings, small lamps, compasses, ceramics and pieces of sculptures, clay and copper statues, inscriptions, etc. During the years of the Roman Empire, between the 2nd and 3rd century AD, the telesterion was looted and burned and when the area was abandoned the site was used as a "quarry" for the construction of younger buildings. The late-Roman telesterion is built over the Archaic one clearly reminiscent of the destroyed Hellenistic temple, being a 17-m-long room, divided into three parts by two rows of five columns and a main hall separated from the sanctuaries by a corridor [111].

Figure 12. Selected geosites on Lemnos Island, for coding see Figure 3. (**a**) The *Archaeological Site of Hephaistia* (C11); (**b**) the *Ancient Quarry of Hephaistia*; (**c,d**) *Rejuvenated Drainage of Neftina* (G30); (**e**) *Sanctuary of Kaveirion* (C10); (**f**) *Cave of Philoctetes* (G31); (**g**) *Honeycomb Weathering formations in Plaka* (G29).

In the same area, the *Cave of Philoctetes* (G31; Table 3, Figures 3, 4 and 12f), a coastal cave located below the Kaveirion Archaeological Site, has been created by the effect of sea waves on the rocky coastal cliff consisting of sedimentary rocks, mostly sandstones, and has two entrances, the main one by the sea and a small one on the land side [95]. It took its name from the king of Thessaly Philoctetes, who according to mythology, resorted to it, when on the way to the Trojan War he was left on the island by the Achaeans because he had been attacked on the leg by a poisonous snake. He remained in the cave for about ten years and was healed with the help of *"Terra Lemnia"* [107].

The route ends towards the northwest coast with the *Honeycomb Weathering formations in Plaka* (G29; Table 3, Figures 3, 4 and 12g) that are developed on the exposed sandstones and tuffs. The phenomenon is intense, due to the wind action and the chemical weathering caused by the salty sea water that enters the rocks through the discontinuities.

4.2.7. Road "Geo-Archaeo-Route" 3: Diapori–Portianou–Sardes

The starting point at the *metal-bearing Zone of Fakos* (G18; Table 3, Figures 3 and 4) is a highlight of *Road Geo-Archaeo-Route 3* (Figure 10), developed in the periphery of a subvolcanic body of quartz monzonitic composition that intrudes the shoshonitic andesites, trachyandesites and trachytes volcanic rocks and the quartz-rich medium-grained sandstone sediments in the southern part of the Fakos peninsula, featured by intense hydrothermal alteration and quartz veining [113,114]. The area stands out due to the volcanic penetrations within the sedimentary rocks and the gold ores; i.e., the Fakos Cu–Mo–Au prospect, comprises the first occurrence of porphyry-related tourmaline in Greece [115]. The faults in the Fakos peninsula follow the general NE–SW and ENE–WSW directions featuring the whole Lemnos island, significantly contributing to the deposition of subvolcanic bodies and facilitating the flow of the hydrothermal magmatic liquids responsible for metallic mineralization [116]. Three metal-bearing zones are exposed on the peninsula. The first is located in the southern part, in the quartz veins within the sandstones and quartz monzonite. The second zone is located in the western part, while the third and largest is located in the eastern part, displaying a length of 1 km and a thickness of 10 m and is characterized by a high concentration of gold as an indication of a magmatic-hydrothermal contribution to the vein system [117].

To the north, the impressive *Lava Domes of Skopos* (G19; Table 3, Figures 3 and 4) and *Kondias* (G17; Table 3, Figures 3, 4 and 13a) are associated with the depression caused by the Kornos–Kondias fault that hosts the intensive magmatic activity of the Katalakkon and Myrina volcanic units, involving balloon-shaped domes of calc-alkaline shoshonitic affinity, emplaced between 22 and 18 Ma in the Early Miocene [45,81].

Figure 13. Selected geosites on Lemnos Island, for coding see Figure 3. (**a**) *Lava Dome of Kondias* (G17); (**b**) the *Pyramid of Lemnos* (G21) at the edge of the cape Punta.

In between the Lava Domes, the route meets the *Diapori Swamp* (W03) and the *Petrified Trunk of Paranisia* (G16). The *Diapori Swamp* (Table 3, Figures 3 and 4) is a Natura 2000 Network protected wetland system consisting of shallow freshwater swamps, scattered small seasonal lakes, salt marshes and extended sand dunes located in the area of Kondias bay [80]. The area hosts numerous amphibians and reptiles [44] and supports considerable amounts of threatened avifauna species (e.g., *Tadorna ferruginea, Burhinus oedicnemus*) on an annual basis [29]. The nearby *Kondias Artificial Lake* (W04; Figures 3 and 4) covering an area of 1,100,000 m² in the southeastern part of the island represents together with the Thanos reservoir the largest artificial water bodies on Lemnos [29], formed after the building of a soil dam in 1976.

Moving northwards, *Road Geo-Archaeo-Route 3* reaches the village of Sardes, where the *Metal-Bearing Zone of Sardes* (G08; Table 3, Figures 3 and 4), a system of quartz, has been developed in zones within the volcanic rocks and sandstones, in the same direction as that of the Fakos zones [118]. This interesting metal-bearing zone includes pyrite, marcasite and veinlets of silica [114].

The route continues to the southeast with a stop at the *Museum of Maritime Tradition and Sponge-Fishing of Nea Koutali* (C07; Table 4, Figure 3). This cultural site is blending nicely within *Road Geo-Archaeo-Route 3*, highlighting the history of the people of Koutali, by presenting the craft of sponge-fishing and also archeological findings brought up by the Koutalians during the sponge dives [119].

In the area of Portianou, the *Road Geo-Archaeo-Route 3* encounters several fossiliferous localities of in situ silicified wood hosted within pyroclastic rocks of the Romanou volcanic Unit [45], with the most important ones being the *Petrified Forest of Tsimandria–Portianou* (G20; Table 3, Figures 3 and 4). The Romanou Unit that outcrops in the area is the main host of the petrified woods on Lemnos island, consisting of up to 160 m thick pyroclastic flow deposits, classified as K-rich dacites to latites and are radiometrically dated as 19.8 Ma old [45]. The petrified trunks found could have been formed partly by devitrification processes by near neutral-pH fluids in swamp depressions, which were flooded by the upwelling hydrothermal fluids in the periphery of the magmatic-hydrothermal center of the Fakos peninsula [120].

After passing through the *Petrified Forest of Tsimandria–Portianou* (G20), the route heads towards the *Volcanic Vein of Portianou* (G21; Table 3, Figures 3 and 4), located at the cape of Punta. At the edge of the cape, the mysterious *Pyramid of Lemnos* (G21) represents according to a non-proven point of view, a 1920's Cossacks cenotaph, in honor of those who lost their lives in exile on the island during the Russian Civil War [121]. Another interpretation links this building with military constructions, potentially made by Australian mechanics during March of 1915 in the first world war [122].

Finally, a visit to the *Portianou Folklore Museum* (C05; Table 4, Figure 3) is a cultural must of *Road Geo-Archaeo-Route 3*. The *Folklore Museum* located in the Portianou village was founded in 1995 and is housed in a two-story traditional building, exhibiting the life-mode of the Lemnos inhabitants of the previous centuries, e.g., traditional costumes, various knitting tools, fireplace and cooking utensils, handmade wedding dresses and laces, agricultural tools, etc.

4.2.8. Road "Geo-Archaeo-Route" 4: Myrina–Avlona–Mourtzouflos

Road Geo-Archaeo-Route 4 (Figure 10) starts from the *Sanctuary of Artemis* (C01; Table 4, Figure 3) located in the area of Avlonas of Myrina at the Porto Myrina Hotel. The site was functioning as a temple of the goddess Artemis for more than eight centuries during the Archaic, Classical and Hellenistic Periods [123]. The temple complex included a large central paved yard and a sanctuary. Three wells were providing the sanctuary with water and a rectangular room at the northwest end of the enclosure was most probably acting as a gathering place for the initiation of the ceremonies. The worship of the Tauric Goddess was implanted at an early time in Lemnos, and also in Vravron (Attica) and Crimea. The legend of the Tauric Artemis was associated with human sacrifice, related to a bloodthirsty bull-devourer goddess, featuring the Attic-Lemnian worship [124]. The site excavations revealed many artefacts and relics, including pottery, figurines, perfume containers, jewelry, etc., all of which are on display at the Archaeological Museum of Myrina. Interestingly, several figurines of a bull, found together with the remains of a sacrificed bull, further indicate the relation to the worship of the Tauric Goddess Artemis in Vravron Attica.

Following the route to the north, the visitor encounters the impressive *Lava Dome at Avlonas Beach* (G09; Table 3, Figures 3 and 4), which belongs to the Myrina Unit consisting of the youngest lava domes and occasional flows with mainly dacitic composition and an age of 19.3 to 18.2 Ma [45].

Along the coastal road in the northwestern part of the island, the route comes across some unique *Coastal Landforms in Aghios Ioannis* (G07; Table 3, Figures 3, 4 and 14a,b) at the sandy beach of Aghios Ioannis near Kaspakas village. The outcropping landforms include spectacular sea stacks and tafoni forms of honeycomb weathering in volcanic rocks in the coastal zone of the Myrina volcanic Unit [46].

Figure 14. Selected geosites on Lemnos Island, for coding see Figure 3. (**a**,**b**) *Coastal Landforms in Agios Ioannis* (G07); (**c**) the *active Fault of Kaspakas* (G06) (photo courtesy of A. Chatzipetros); (**d**) the impressive *Tombolo of Mourtzouflos* (G02).

Nearby, the well-defined scarp of the important *Active Fault of Kaspakas* (G06; Table 3, Figures 3, 4 and 14c) marks the landscape, being one of the most famous faults on the island, as well as in the entire Aegean. It is a normal fault of about 11 km in length with a WNW–ESE strike and respective maximum expected earthquake magnitude of M6.4 [61]. It stands out for its well-defined fault scarp and polished fault mirrors, located at its westernmost tip, affecting mainly the volcanic basement during the Early Miocene [46,63]. In the area of Kaspakas, *Aghios Ioannis Gorge* (G04; Table 3, Figures 3 and 4), the most important gorge of the island bearing a length of about 3 km is associated with the tectonic activity of the Kaspakas fault combined with the continuous water erosion. The Katsaitis stream flows within the *Aghios Ioannis Gorge*, creating at its end the *Kaspakas Waterfall* (G05) near the beach of Aghios Ioannis; the fast-flowing waters fall from a height of about 15 m forming cavities and small lakes on the rocky substrate that provide shelter for the local fauna of frogs, turtles, eels and crabs [79].

Further to the north, the *Neroviglia Gorge* (G03; Table 3, Figures 3 and 4) with a length of about 1 km is located at the northwestern tip of the island, ending at a picturesque small sandy beach. Towards its end, the *Road Geo-Archaeo-Route 4* roams to reach its final destination at the cape of Mourtzouflos. The cape forms the *Lava Dome of Mourtzouflos* (G01; Table 3, Figures 3, 4 and 14d), which belongs to the Katalakkon volcanic Unit, intruded in the sediments of the Fissini-Sardes Unit [46]. The Katalakkon volcanics are exposed along a NW–SE trending zone that extends to the south extremity of Cape Fakos and has been radiometrically dated at 20–21 Ma [45]. The *Lava Dome of Mourtzouflos* is connected to the rest of the island with a sandy isthmus, the impressive *Tombolo of Mourtzouflos* (G02; Table 3, Figures 3, 4 and 14d).

4.2.9. Road "Geo-Archaeo-Route" 5: Koukonisi–Poliochni

The route starts (Figure 10) from the *Koukonisi islet wetland* (W05; Table 3, Figures 3, 4 and 15a,b) that is part of the Natura 2000 network and includes extended salt flats developing between the islet and the coastal zone at the northeastern part of the Moudros Gulf [80]. The *Prehistoric Settlement of Koukonisi* (C08; Table 4, Figure 3) located in the area, was first discovered in 1992 during exploratory excavations and together with the other prehistoric sites on the island pinpoints the importance of Lemnos during the Bronze Age associated with its strategic location in the Aegean at the vicinity of the Dardanelles entrance and opposite Troy. Human presence is testified in Lemnos from around 12,000 BC, with signs of communities of fruit pickers, fishermen and hunters located at Ouriakos, on the eastern outer coast of the gulf of Moudros [30]. Early Bronze Age (3200–2000 BC) evidence with signs of early urbanization appear in the northeastern part of the *Koukonisi* islet, in the settlement area on the Koukonos plateau [125–127]. The highest plateau on the islet had been artificially formed by continuous habitation for almost 2000 years, exhibiting Middle Helladic, Minoan and Mycenaean finds, with habitation lasting at least up to the Geometric and Archaic period. The Early Bronze Age figurines of human form made either of clay or of bone are thought to have been used in rituals and other ceremonies. The following Middle Bronze Age (2000–1650 BC) was marked by a hiatus detected in the habitation of the islet, being a period of prosperity for *Koukonisi* that was linked to the established trade activities with several Aegean regions [127]. During the Late Bronze Age (1650–1200 BC) early stages, the presence of Minoan civilization elements and culture coming from the southern Aegean strongly characterizes the southern part of the settlement. Later on, around the 14th century BC the Mycenaeans prevailed in *Koukonisi* in an attempt to maintain a balance of power in the Aegean area. Their installation was permanent, a fact that is attributed to the abundant appearance of Mycenaean pottery in the *Koukonisi* excavation site [127]; Mycenaean pottery has also been found in *Hephaistia* and other Lemnian sites, testifying the widespread Mycenaean presence on the island that proves commercial ties with Asia Minor.

Towards southeast, the *Road Geo-Archaeo-Route 5* heads to the *Cave of the Seal* (G22; Table 3, Figures 3, 4 and 15c), east of the bay of Moudros, at Mikro Fanaraki beach. The cave is a result of coastal erosion mainly of pyroclastic rocks such as tuffs of the Romanou volcanic Unit [46] and took its name from the *Monachus monachus* seals.

Turning back to the west, the route passes through the area of the natural monument of the *Petrified Forest of Moudros–Roussopouli–Kaminia* (G23, 24; Table 3, Figures 3 and 4) with silicified wood remains. The magmatic-hydrothermal systems that developed in the area about 20 million years ago provided the silica-enriched fluids that affected the tree trunks and plant remains, thus beginning the process of fossilization, the molecular replacement of organic plant matter by silica. The fossils that have been found in the area are: *Cedroxylon* sp., *Cornoxylon pappi, Daphnogene polymorpha*, Fagaceae, Lauraceae, *Glyptostrobus europaeus, Laurinoxylon ehrendorferi, Phragmites* sp., *Pronephrium stiriacum, Pinoxylon parenchymatosum, Sabal* sp., *Sequoia abietina*, as well as the roots, seeds and leaves of palm trees [85]. Most of the findings are exhibited in the old town hall of Moudros.

Figure 15. Selected geosites on Lemnos Island, for coding see Figure 3. (**a,b**) *Koukonisi islet wetland* (W05); (**c**) *The Cave of the Seal* (G22); (**d**) *Prehistoric Settlement of Poliochni* (C13).

The route ends at the famous *Prehistoric Settlement of Poliochni* (C13; Table 4, Figures 3 and 15d), which was discovered in the early 1930s by the Italian School of Archeology. The settlement is built on the east coast of Lemnos, in the area of Kaminia and was founded on a coastal terrace, in a coastal environment progressively flooded by the rising sea level [128]. Dated since the end of the Neolithic era and continuously developing mainly during the Early Bronze Age, *Poliochni* is considered to be one of the most ancient towns in Europe, preceding Troy I. Its development was due to the leading role it played in the transit trade with the islands of the northeastern Aegean, the coasts of Asia Minor, mainland Greece and the Cyclades islands. The evolution of the settlement is divided into seven periods, which are symbolized by colors [129,130]. The small settlement of circular huts of the "black period" (3700–3200 BC) has evolved into a larger and partially fortified settlement of rectangular elongated houses during the "cyan period" (3200–2700 BC). The "green period" (2700–2400 BC) is marked by population increases and a developed road network in contrast to the settlement shrink of the "red period" (2400–2200 BC) that nevertheless displayed the first monumental palaces. During the "yellow period" (2200–2100 BC), the settlement is associated with the Troy II [129] and is featured by important buildings. The irreversible destruction of the settlement took place in 2100 BC, when an earthquake struck Lemnos [131]. The following "brown" and "violet" (2000–1200 BC) periods are decline periods for the settlement associated with the rivalry with Troy.

5. Synthesis and Conclusions

Overall, Lemnos Island displays an authentic natural environment also rich in religious, archaeological, historical and cultural heritage, offering the opportunity for new, alternative forms of tourism that favor authentic experiences and a connection to local communities, thus setting the foundations for sustainable rural development in a geoethical perspective.

The final result of the quantitative assessment of the selected geosites combined with the distribution of the cultural sites on the island reveals the potential perspectives of the island for the development of *"Geo-Archaeo-Routes"* on Lemnos Island, towards a holistic geotouristic approach. In this way, the identification of even small-scale areas of interest is made possible, which can now be implemented in a broader network of touristic sites of the island to demonstrate in the most efficient way the linkages between local cultural context, archaeological monuments, biodiversity hotspots, landscapes and geological heritage.

In the present study, we propose *"Geo-Archaeo-Routes"* as authentic paths, a kind of time-capsules that integrate the geological features of the deep past of Lemnos Island together with the cultural archaeological and historical elements of the human societal imprints of the recent past, both being embodied in the modern natural geomorphological landscape. The involved sites are examples of mixed cultural–natural heritage providing cases where either geoheritage acts as an added value to the cultural heritage or vice versa [132].

Under this concept, the hiking *"Geo-Archaeo-Route"* 1 takes the geotourists from 19 million years ago when the volcanic rocks of the Myrina Unit were formed, to the Late Neolithic–Early Bronze age when these rocks were used as building stones for the *Prehistoric Settlement of Myrina* and later on, at the 13th century BC, for the "Cyclopean" walls of *Myrina Castle*, constructed by the Thessalians Minyans on the Lava Dome of Myrina. Out of the highlights of hiking *"Geo-Archaeo-Route"* 2, *Panagia Kakkaviotissa Church* of the 14th century AD, built in a large cavity shaped on the *Lava Dome of Kakkavo* volcanic rocks of 19 million years in age, further verifies the concept that natural rock formations and landforms often represent locations of religious or spiritual significance [15,132]. The *Petrified Forest of Lemnos*, with floristic evidence of about 20 million years in age, is scattered over a large part of Lemnos Island, comprising an important element of several *"Geo-Archaeo-Routes"*, enabling the geotourists to realize a totally different landscape for that time interval, with extended forests of coniferous trees petrified after extended volcanic eruptions. This experience has an apparent geoethical value as it provides the public with the opportunity to understand the consequences of potential natural hazards and also to raise awareness about the need for fossil preservation. The inactive *Volcanic Crater of Moschylos*, linked to the Greek god of fire and volcanoes God Hephaestus, in the hiking *"Geo-Archaeo-Route"* 3 and the *Cave of Philoctetes* in road *"Geo-Archaeo-Route"* 2 are typical cases of geomythology, a type of intangible heritage [132]. The hiking *"Geo-Archaeo-Route"* 3 offers one more travel in time at the geosite of *Terra Lemnia*, a weathered volcanic tuff clay of 20 million years in age that has been widely used as a medicine, particularly against the plague in medieval Europe. The modern landscape of *Alyki Lake* within hiking *"Geo-Archaeo-Route"* 4 keeps well hidden a rather complicated paleoenvironmental evolution of the coastal plain, providing evidence of the ancient shoreline (5100–1040 BC) to be located 500 m in front of the present-day shoreline, thus enabling geo-education about the hazard of sea level rise. In the same line, the *Inland Sand Dunes of Gomati*, a unique desert environment in the whole of Europe, along the road *"Geo-Archaeo-Route"* 1, present a perfect geo-educational example for the hazard of aridification and desertification in the current times of global warming. The *Ancient Quarry of Hephaistia* in road *"Geo-Archaeo-Route"* 2 represents an interface of geoheritage and cultural heritage [132], while the *Pyramid of Lemnos* in road *"Geo-Archaeo-Route"* 3 declares the early 20th century historical imprint on the natural environment. On the other hand, the temple of the Tauric Goddess and the road paved with volcanic stones from the 19-million-year-old yard of the *Sanctuary of Artemis* in road *"Geo-Archaeo-Route"* 4, imply the link between Lemnos and Crimea in the Black Sea, from the Archaic period. The prominent scarp of the *Active Fault of Kaspakas* not only marks the landscape of western Lemnos, but provides an excellent geoethical opportunity to contribute to the history of science and offer the public the scientific knowledge concerning the geology of the whole North Aegean to also raise awareness about the seismic hazards (e.g., [132]). The *Prehistoric Settlement of Koukonisi* at the vicinity of the homonymous wetland within *"Geo-Archaeo-Route"* 5 further verifies the development of Lemnos during the Bronze Age, while increasing the public awareness for sustainable environmental management. Finally, the Poliochni settlement, considered as one of the oldest towns in Europe, takes the visitors back not only to the late Neolithic age, but even as old as more than 20 million years ago, when the area was experiencing intense volcanic activity, as the building stones of the site mostly come from the surrounding Romanou volcanic Unit.

This study resulted in the extensive documentation and assessment based on the scientific literature and evaluation criteria of 46 geosites on Lemnos Island. All of them have been integrated with 12 cultural sites in 9 hiking and road *"Geo-Archaeo-Routes"* designed to serve the visitors in the best way, covering a wide range of topics also associated with good accessibility, safety and adequately available information. The proposed *"Geo-Archaeo-Routes"* have been followed and mapped in the field. Afterwards, all data were imported in the G.I.S. ArcMap 10.4 software to produce the relevant maps. Both hiking and road routes are proposed to follow the already existing dirt and paved roads network, the starting points being large villages and/or sites of ample interest so that besides hiking where feasible, they can be realized by any means of transport (car, bike, local transport, bus, etc.) in a holistic geotouristic approach.

As a next step, there will be an effort to make the touristic operators, local authorities and society of the island interested in the promotion of the proposed *"Geo-Archaeo-Routes"* to the touristic audience. In parallel, technology and innovation and a bridging approach between the cultural and creative sectors will be applied for the construction of story maps and 3D models that will be used as inputs for Augmented Reality/Virtual Reality (AR/VR) applications or even for the production of 3D prints at various scales. By transforming the natural and cultural touristic experience into an AR/VR product, we may transfer natural beauty, past civilizations and the history of Lemnos to people from all over the world, providing an opportunity to take part in socially, historically and culturally immersive virtual tours.

In all cases, there will be a particular effort concerning the geo-educational part; namely, coaching the touristic masses to behave according to the geoethical values and to be conscious about the preservation of the unique Lemnos geoenvironment.

Author Contributions: Conceptualization, M.V.T., N.F. and A.S.; methodology, M.V.T., N.F. and A.S.; software, E.V., T.T. and E.S.; validation, M.V.T., N.F., A.O. and A.S.; formal analysis, M.V.T., N.F. and A.O.; investigation, M.V.T., N.F., O.K. and A.O.; resources, M.V.T., N.F. and A.O.; data curation, N.F.; writing—original draft preparation, M.V.T. and N.F.; writing—review and editing, M.V.T., T.T. and E.S.; visualization, T.T., E.V., E.S. and O.K.; supervision, M.V.T.; project administration, M.V.T. and N.F. All authors have read and agreed to the published version of the manuscript.

Funding: This research received no external funding.

Data Availability Statement: The data presented in this study are available on request from the corresponding author.

Conflicts of Interest: The authors declare no conflict of interest.

References

1. Dowling, R.K. Global Geotourism—An Emerging Form of Sustainable Tourism. *Czech J. Tour.* **2014**, *2*, 59–79. [CrossRef]
2. Theodosiou, E.; Fermeli, G.; Koutsouveli, A. *Our Geological Heritage*; Kalidoskopio Publications: Athens, Greece, 2006.
3. Stürm, B. Geotop. Grundzüge einer Begriffsentwicklung und Definition. *Materialien* **1993**, *1*, 13–14.
4. Ielenicz, M. Geotope, Geosite, Geomorphosite. *Ann. Valahia Univ. Târgovişt. Geogr. Ser.* **2009**, *9*, 7–22.
5. Reynard, E.; Panizza, M. Geomorphosites: Definition, assessment and mapping. *Geomorphol. Reli. Process. Environ.* **2005**, *11*, 177–180. [CrossRef]
6. Triantaphyllou, M. Greek caves and touristic development. *Bull. Speleol. Soc. Greece* **1992**, *20*, 28–76.
7. Skentos, A. *Geosites of Greece: Record, Schematic, Geological Regime and Geotouristic Assessment*; National and Kapodistrian University of Athens: Athens, Greece, 2012.
8. Skentos, A.; Triantaphyllou, M. Registration and tourism assessment of Greek geosites using G.I.S. techniques. In Proceedings of the 32nd International Geographical Congress, Cologne, Germany, 26–30 August 2012. IGU 2012.
9. Pica, A.; Fredi, P.; Del Monte, M. The ernici mountains geoheritage (Central Apennines, Italy): Assessment of the geosites for geotourism development. *Geoj. Tour. Geosites* **2014**, *14*, 193–206.
10. Ruban, D.A. Geotourism—A geographical review of the literature. *Tour. Manag. Perspect.* **2015**, *15*, 1–15. [CrossRef]
11. Ólafsdóttir, R.; Tverijonaite, E. Geotourism: A systematic literature review. *Geosciences* **2018**, *8*, 234. [CrossRef]
12. Newsome, D.; Dowling, R.; Leung, Y.F. The nature and management of geotourism: A case study of two established iconic geotourism destinations. *Tour. Manag. Perspect.* **2012**, *2–3*, 19–27. [CrossRef]
13. Fassoulas, C.; Zouros, N. Evaluating the influence of Greek Geoparks to the local communities. *Bull. Geol. Soc. Greece* **2017**, *43*, 896. [CrossRef]
14. Dowling, R.; Newsome, D. *Handbook of Geotourism*; Edward Elgar Publishing Ltd.: Cheltenham, UK, 2018.
15. Gordon, J.E. Geoheritage, geotourism and the cultural landscape: Enhancing the visitor experience and promoting geoconservation. *Geosciences* **2018**, *8*, 136. [CrossRef]
16. Bithas, K. A bioeconomic approach to sustainable development: Incorporating ecological thresholds within intergenerational efficiency. *Sustain. Dev.* **2020**, *28*, 772–780. [CrossRef]

17. Antić, A.; Peppoloni, S.; Di Capua, G. Applying the Values of Geoethics for Sustainable Speleotourism Development. *Geoheritage* **2020**, *12*, 73. [CrossRef]
18. Peppoloni, S.; Di Capua, G. The Meaning of Geoethics. In *Geoethics: Ethical Challenges and Case Studies in Earth Sciences*; Wyss, M., Peppoloni, S., Eds.; Elsevier: Amsterdam, The Netherlands, 2015; pp. 3–14.
19. Silvia Peppoloni; Giuseppe Di Capua Geoethics and geological culture: Awareness, responsibility and challenges. *Ann. Geophys.* **2012**, *55*. [CrossRef]
20. Peppoloni, S.; Di Capua, G. The significance of geotourism through the lens of geoethics. In Proceedings of the EGU General Assembly 2022, Vienna, Austria, 23–28 May 2022; p. 1756.
21. May, V. Integrating the geomorphological environment, cultural heritage, tourism and coastal hazards in practice. *Suppl. Geogr. Fis. Din. Quat.* **2008**, *31*, 187–194.
22. Grunwald, A. The imperative of sustainable development: Elements of an ethics of using geo-resources responsibly. In *Geoethics: Ethical Challenges and Case Studies in Earth Sciences*; Wyss, M., Peppoloni, S., Eds.; Elsevier: Amsterdam, The Netherlands, 2015; pp. 26–35.
23. Velitzelos, E.; Mountrakis, D.; Zouros, N.; Soulakelis, N. *Atlas of the Geological Monuments of the Aegean Sea*; Publications of the Ministry of the Aegean: Aegean, Greece, 2002.
24. Zafeiropoulos, G.; Drinia, H.; Antonarakou, A.; Zouros, N. From geoheritage to geoeducation, geoethics and geotourism: A critical evaluation of the Greek region. *Geosciences* **2021**, *11*, 381. [CrossRef]
25. Spyrou, E.; Triantaphyllou, M.V.; Tsourou, T.; Vassilakis, E.; Asimakopoulos, C.; Konsolaki, A.; Markakis, D.; Marketou-Galari, D.; Skentos, A. Assessment of Geological Heritage Sites and Their Significance for Geotouristic Exploitation: The Case of Lefkas, Meganisi, Kefalonia and Ithaki Islands, Ionian Sea, Greece. *Geosciences* **2022**, *12*, 55. [CrossRef]
26. Drinia, H.; Tripolitsiotou, F.; Cheila, T.; Zafeiropoulos, G. The Geosites of the Sacred Rock of Acropolis (UNESCO World Heritage, Athens, Greece): Cultural and Geological Heritage Integrated. *Geosciences* **2022**, *12*, 330. [CrossRef]
27. Official Government Gazette of the Hellenic Republic. Issue 2 no 1278 2000. Free access: The National Printing House of Greece. Available online: http://www.et.gr (accessed on 7 May 2023).
28. Official Government Gazette of the Hellenic Republic. Issue 4 no208 2002. Free access: The National Printing House of Greece. Available online: http://www.et.gr (accessed on 7 May 2023).
29. Kakalis, E. *Report on the ornithological evaluation of area G132 Lakes Chortarolimni and Alyki, Moudros Bay, Diapori Swamp, Fakos Peninsula*; Athens, Greece, 2009.
30. Efstratiou, N. The final Paleolithic hunting camp of Ouriakos on the island of Lemnos. *Eurasian Prehistory* **2014**, *11*, 75–96.
31. Zouros, N.; Soulakellis, N.; Valiakos, I.; Grimpylakos, G.; Lamprakopoulos, A.; Panidou, P.; Papadopoulou, E.; Tataris, G. Geological and Geomorphological Heritage Sites in Lemnos Island, Greece and their Geotourism Potential. *Bull. Geol. Soc. Greece* **2021**, *Sp. Pub. 8*, 64–68.
32. Natural History Museum of Lesvos Petrified Forest Geomonuments and Geosites of Lemnos. Available online: https://www.lemnosgeomonuments.gr/en/ (accessed on 10 February 2023).
33. Panizza, M.; Piacente, S. Geomorphosites: A bridge betwenn scientific research, cultural integration and artistic suggestion. *Alp. Mediterr. Quat.* **2005**, *18*, 3–10.
34. Coratza, P.; Gauci, R.; Schembri, J.; Soldati, M.; Tonelli, C. Bridging Natural and Cultural Values of Sites with Outstanding Scenery: Evidence from Gozo, Maltese Islands. *Geoheritage* **2016**, *8*, 91–103. [CrossRef]
35. Possamai, F.V. Nature and Culture: Genesis of an Obsolete Dichotomy. *Philos. Study* **2013**, *3*, 836–842.
36. Roston, H.I. Nature and culture in environmental ethics. In *the Proceedings of the Twentieth World Congress of Philosophy*; Brinkmann, K., Ed.; Philosophy Documentation Center: Bowling Green, OH, USA, 1999; pp. 151–158.
37. Sauer, C.O. The morphology of landscape. *Publ. Geogr.* **1925**, *2*, 19–54.
38. Reynard, E.; Giusti, C. The Landscape and the Cultural Value of Geoheritage. In *Geoheritage*; Elsevier: Amsterdam, The Netherlands, 2018; pp. 147–166.
39. Dimopoulos, T.; Dimitropoulos, G.; Georgiadis, N. The Land Use Systems of Lemnos Island. Terra Lemn. Rep. 2018. terra lemnia project. Available online: https://med-ina.org/wp-content/uploads/2020/12/Terra-Lemnia-1.1.1-Land-Use-Systems-of-Lemnos-Dec-2018.pdf (accessed on 10 February 2023).
40. Sidiropoulou, M. *Geomorphological and Palaeogeographical Evolution of Lemnos Island (NE Aegean), the Last 6000 Years*; Harokopeion University: Athens, Greece, 2016.
41. Panitsa, M.; Snogerup, B.; Snogerup, S.; Tzanoudakis, D. Floristic investigation of Lemnos island (NE Aegean area, Greece). *Willdenowia* **2003**, *33*, 79–105. [CrossRef]
42. Thomas, K.; Thanopoulos, R.; Knüpffer, H.; Bebeli, P.J. Plant genetic resources of Lemnos (Greece), an isolated island in the Northern Aegean Sea, with emphasis on landraces. *Genet. Resour. Crop Evol.* **2012**, *59*, 1417–1440. [CrossRef]
43. Municipality of Limnos. Available online: www.limnos.gov.gr/files/007/epitropidiabouleusis/epixsxed2015-19.docx (accessed on 10 February 2023).
44. Broggi, M.F. The "wetland island" of Lemnos (Greece): Herpetofauna and nature conservation. *Herpetozoa* **2018**, *30*, 169–178.
45. Innocenti, F.; Manetti, P.; Mazzuoli, R.; Pertusati, P.; Fytikas, M.; Kolios, N. The geology and geodynamic significance of the Island of Limnos, North Aegean Sea, Greece. *Neues Jahrb. Geol. Paläontologie* **1994**, *1994*, 661–691. [CrossRef]

46. Innocenti, F.; Manetti, P.; Mazzuoli, R.; Pertusati, P.; Fytikas, M.; Kolios, N.; Vougioukalakis, G.E.; Androulakakis, N.; Critelli, S.; Caracciolo, L. Geological map (scale 1:50,000) of Lemnos island (Greece): Explanatory notes. *Acta Vulcanol.* **2009**, *21*, 123–134. [CrossRef]

47. Roussos, N.; Katsaounis, A.; Tsaila-Monopoli, S.; Ioakeim, X.; Karadasi, S.; Davi, E. *Geological Map of Lemnos Island*; Institute of Geology and Mineral Exploration: Athens, Greece, 1993.

48. Caracciolo, L.; Critelli, S.; Innocenti, F.; Kolios, N.; Manetti, P. Unravelling provenance from Eocene-Oligocene sandstones of the Thrace Basin, North-east Greece. *Sedimentology* **2011**, *58*, 1988–2011. [CrossRef]

49. Maravelis, A.; Zelilidis, A. Paleoclimatology and Paleoecology across the Eocene/Oligocene boundary, Thrace Basin, Northeast Aegean Sea, Greece. *Palaeogeogr. Palaeoclimatol. Palaeoecol.* **2012**, *365–366*, 81–98. [CrossRef]

50. Kostopoulou, S.; Maravelis, A.G.; Zelilidis, A. Biostratigraphic analysis across the Eocene-Oligocene boundary in the southern Hellenic Thrace basin (Lemnos Island, north Aegean Sea). *Turk. J. Earth Sci.* **2018**, *27*, 232–248. [CrossRef]

51. Papp, A. Erla uterungen zur Geologie der Insel Lemnos. *Ann. Géol. Pays Hell.* **1953**, *5*, 1–25.

52. Papanikolaou, D.; Triantaphyllou, M. New stratigraphic data of the Lemnos volcano-sedimentary sequence and correlations with the Thrace Basin. 15th International Congress of the Geological Society of Greece. *Bull. Geol. Soc. Greece* **2019**, *Sp. Pub. 7*, GSG2019-240.

53. Papanikolaou, D. Geotectonic evolution of the Aegean. *Bull. Geol. Soc. Greece* **1993**, *28*, 33–48.

54. Papanikolaou, D. Tectonostratigraphic models of the Alpine terranes and subduction history of the Hellenides. *Tectonophysics* **2013**, *595–596*, 1–24. [CrossRef]

55. Papanikolaou, D.I. *The Geology of Greece*; Regional Geology Reviews; Springer International Publishing: Cham, Switzerland, 2021; ISBN 978-3-030-60730-2.

56. Pe-Piper, G.; Piper, D.J.W. *The Igneous Rocks of Greece. The Anatomy of an Orogen*; Gebruder Borntraeger: Berlin, Germany, 2002.

57. Pavlopoulos, K.; Fouache, E.; Sidiropoulou, M.; Triantaphyllou, M.; Vouvalidis, K.; Syrides, G.; Gonnet, A.; Greco, E. Palaeoenvironmental evolution and sea-level changes in the coastal area of NE Lemnos Island (Greece) during the Holocene. *Quat. Int.* **2013**, *308–309*, 80–88. [CrossRef]

58. Fytikas, M.; Giuliani, O.; Innocenti, F.; Manetti, P.; Mazzuoli, R.; Peccerillo, A.; Rita, F.; Villari, L. Neogene Volcanism of the Northern and Central Aegean region. *Ann. Géol. Pays Hell.* **1979**, *30*, 106–129.

59. Pe-Piper, G.; Piper, D.J.W.; Koukouvelas, I.; Dolansky, L.M.; Kokkalas, S. Postorogenic shoshonitic rocks and their origin by melting underplated basalts: The Miocene of Limnos, Greece. *Geol. Soc. Am. Bull.* **2009**, *121*, 39–54. [CrossRef]

60. Pavlides, S.; Mountrakis, D.; Kilias, A.; Tranos, M. The role of strike-slip movements in the extensional area of Northern Aegean (Greece). *Ann. Tecton.* **1990**, *4*, 196–211.

61. Pavlides, S.; Tsapanos, T.; Zouros, N.; Sboras, S.; Koravos, G.; Chatzipetros, A. Using active fault data for assessing seismic hazard: A case study from NE Aegean Sea, Greece. In Proceedings of the Earthquake Geotechnical Engineering Satellite Conference XVIIth International Conference on Soil Mechanics & Geotechnical Engineering, Alexandria, Egypt, 2–3 October 2009; pp. 1–14.

62. Tranos, M.D. Faulting of Lemnos Island; a mirror of faulting of the North Aegean Trough (Northern Greece). *Tectonophysics* **2009**, *467*, 72–88. [CrossRef]

63. Chatzipetros, A.; Kiratzi, A.; Sboras, S.; Zouros, N.; Pavlides, S. Active faulting in the north-eastern Aegean Sea Islands. *Tectonophysics* **2013**, *597–598*, 106–122. [CrossRef]

64. Pavlides, S.B.; Tranos, M.D. Structural characteristics of two strong earthquakes in the North Aegean: Ierissos (1932) and Agios Efstratios (1968). *J. Struct. Geol.* **1991**, *13*, 205–214. [CrossRef]

65. Bakalis, C. *Lemnos: Urban Structure, Social Restructuring, Migratory Networks and Urban Reflections in the 19th and 20th Century*; University of Aegean, School of Social Sciences: Aegean, Greece, 2007.

66. Lanaridis, P.; Salaha, M.-J.; Tzourou, I.; Tsoutsouras, E.; Karagiannis, S. Volatile compounds in grapes and wines from two Muscat varieties cultivated in Greek islands. *OENO One* **2002**, *36*, 39. [CrossRef]

67. Makris, G. Wine Surveyor. Available online: https://winesurveyor.weebly.com/tour1801.html (accessed on 15 February 2023).

68. Fassoulas, C.; Mouriki, D.; Dimitriou-Nikolakis, P.; Iliopoulos, G. Quantitative Assessment of Geotopes as an Effective Tool for Geoheritage Management. *Geoheritage* **2012**, *4*, 177–193. [CrossRef]

69. Grandgirard, V. Methode pour la realisation d'un inventaire de geotopes geomorphologiques. *UKPIK Cah. l'Institute Geogr. l'Universite Fribg.* **1995**, *10*, 121–137.

70. Rivas, V.; Rix, K.; Francés, E.; Cendrero, A.; Brunsden, D. Geomorphological indicators for environmental impact assessment: Consumable and non-consumable geomorphological resources. *Geomorphology* **1997**, *18*, 169–182. [CrossRef]

71. Zouros, N.C. Geomorphosite assessment and management in protected areas of Greece Case study of the Lesvos island—Coastal geomorphosites. *Geogr. Helv.* **2007**, *62*, 169–180. [CrossRef]

72. Reynard, E.; Fontana, G.; Kozlik, L.; Pozza, C.S. A method for assessing «scientific» and «additional values» of geomorphosites. *Geogr. Helv.* **2007**, *62*, 148–158. [CrossRef]

73. Archontidou, A. Archaeological Museum of Lemnos, Myrina. Available online: http://odysseus.culture.gr/h/1/gh152.jsp?obj_id=3413 (accessed on 10 February 2023).

74. Privitera, S. Hephaestia on Lemnos and the Mycenean presence in the islands of the northeastern Aegean. In *Emporia: Aegeans in the Central and Eastern Mediterranean, Proceedings of the 10th International Agean Conference, Athens Italian School of Archaeology, Athens, Italy, 14–18 April 2004*; Université de Liège, Histoire de l'art et Archéologie de la Grèce Antique: Liège, Belgium, 2005; pp. 227–236.

75. Tsolaki, D. Prehistoric Settlement of Myrina. Available online: http://odysseus.culture.gr/h/3/gh351.jsp?obj_id=7461 (accessed on 10 February 2023).

76. Avgerinou, P. The settlement at Myrina: A preliminary assessment. In *Poliochni and the Early Bronze Age in the Aegean*; Scuola Italiana di Atene: Athens, Greek, 1997; pp. 273–281.

77. Ephorate of Antiquities of Lesvos The Medieval Castle of Myrina, Lemnos Island. Available online: https://www.efales.gr/sight/mesaioniko-kastro-myrinas-n-limnoy (accessed on 10 February 2023).

78. Mattila, M.; Hadjigeorgiou, I. Conservation and management of fallow deer (*Dama dama dama* L.)on Lemnos Island, Greece. *Turk. J. Vet. Anim. Sci.* **2015**, *39*, 560–567. [CrossRef]

79. Kousouris, T. *A Recent Survey for the Small Natural Lakes in Greece with Their Features*; Independently Published: Athens, Greece, 2016.

80. WWF Greece Ygrotopio Islands—Database of the Greek Island Wetlands. Available online: https://www.ygrotopio.gr/general/search.php?lang=en&action=map (accessed on 15 February 2023).

81. Gläser, L.; Grosche, A.; Voudouris, P.C.; Haase, K.M. The high-K calc-alkaline to shoshonitic volcanism of Limnos, Greece: Implications for the geodynamic evolution of the northern Aegean. *Contrib. Mineral. Petrol.* **2022**, *177*, 73. [CrossRef]

82. Liapi, E. Panagia Kakaviotissa. Available online: https://stirouga.blogspot.com/2011/05/blog-post_23.html?m=1 (accessed on 15 February 2023).

83. Hellenic Republic, Ministry of Tourism. Available online: https://mintour.gov.gr/meletes-pinakes/stoicheia-iamatikon-pigon/ (accessed on 20 February 2023).

84. Aggelidis, Z.; Mitrakas, M. *Results of the Chemical Analysis of a Sample of Therma Water*; Faculty of Engineering, Aristotle University of Thessaloniki: Thessaloniki, Greece, 2005; Available online: http://users.sch.gr/dipap/analisi_neroy.htm (accessed on 20 February 2023).

85. Velitzelos, D.; Bouchal, J.M.; Denk, T. Review of the Cenozoic floras and vegetation of Greece. *Rev. Palaeobot. Palynol.* **2014**, *204*, 56–117. [CrossRef]

86. Süss, H.; Velitzelos, E. Lesbosoxylon gen. nov., eine neue Morphogattung mit dem Typus Lesbosoxylon ventricosuradiatum sp. nova aus dem Tertiär der Insel Lesbos, Griechenland. *Feddes Repert.* **2010**, *121*, 18–26. [CrossRef]

87. Velitzelos, D.; Iamandei, S.; Iamandei, E.; Velitzelos, E. Palaeoxylotomical studies in the Cenozoic petrified forests of Greece. Part one—Palms. *Acta Palaeobot.* **2019**, *59*, 289–350. [CrossRef]

88. Iamandei, S.; Iamandei, E.; Velitzelos, D.; Velitzelos, E. Palaeoxylotomical Studies In The Cenozoic Petrified Forests Of Greece. Part Two—Conifers. *Acta Palaeontol. Rom.* **2021**, 65–111. [CrossRef]

89. Luce, S.B. Archaeological News and Discussions. *Am. J. Archaeol.* **1941**, *45*, 283–301. [CrossRef]

90. Photos-Jones, E.; Edwards, C.; Häner, F.; Lawton, L.; Keane, C.; Leanord, A.; Perdikatsis, V. Archaeological medicinal earths as antibacterial agents: The case of the Basel Lemnian sphragides. *Geol. Soc. London Spec. Publ.* **2017**, *452*, 141–153. [CrossRef]

91. Venieri, D.; Gounaki, I.; Christidis, G.E.; Knapp, C.W.; Bouras-Vallianatos, P.; Photos-Jones, E. Bridging the Gaps: Bole and Terra Sigillata as Artefacts, as Simples and as Antibacterial Clays. *Minerals* **2020**, *10*, 348. [CrossRef]

92. Loupou, A. Church of Agios Georgios in Repanidi. Available online: http://odysseus.culture.gr/h/2/gh251.jsp?obj_id=1670 (accessed on 10 February 2023).

93. Georgiadis, N.M.; Bergmeier, E.; Dimitropoulos, G.; Kairis, O.; Kakalis, E.; Kosmas, K.; Meyer, S.; Bembeli, P.; Panitsa, M.; Perdikis, D.; et al. The Natural Environment of Lemnos with Emphasis to Biodiversity and Soil. 2020. Available online: https://med-ina.org/wp-content/uploads/2021/11/Terra-Lemnia-Biodiversity-and-Soil-Synthesis-Report-GR.pdf (accessed on 10 February 2023).

94. Katsadorakis, G.; Paragamian, K.; Georgiadis, N. The status of the wetlands in the island of the Aegean Sea. In Proceedings of the Panhellenic Conference «Biosciences in the 21st Century», Athens, Greece, 2006.

95. Sidiropoulou, M.; Fouache, E.; Pavlopoulos, K.; Triantaphyllou, M.; Vouvalidis, K.; Syrides, G.; Greco, E. Geomorphological evolution and paleoenveronment reconstruction in the northeastern part of Lemnos Island (North Aegean Sea). *Aegaeum* **2014**, *37*, 49–55.

96. Triantaphyllou, M.V.; Gogou, A.; Dimiza, M.D.; Kostopoulou, S.; Parinos, C.; Roussakis, G.; Geraga, M.; Bouloubassi, I.; Fleitmann, D.; Zervakis, V.; et al. Holocene Climatic Optimum centennial-scale paleoceanography in the NE Aegean (Mediterranean Sea). *Geo. Mar. Lett.* **2016**, *36*, 51–66. [CrossRef]

97. Davis, R.A.; Fitzegerald, D.M. *Beaches & Coasts*; Blackwell Science Ltd.: Oxford, UK, 2004.

98. Soukis, K.; Koufosotiri, E.; Stournaras, G. Special landforms on Tinos Island: Spheroidal weathering "Tafoni" forms. In Proceedings of the 3rd International Scientific Symposium of Protected areas and Natural Monuments, Molyvos, Greece, 13–15 July 1998.

99. Evelpidou, N.; Leonidopoulou, D.; Vassilopoulos, A. Tafoni and Alveole Formation. An Example from Naxos and Tinos Islands. In *Natural Heritage from East to West*; Evelpidou, N., Figueiredo, T., Mauro, F., Tecim, V., Vassilopoulos, A., Eds.; Springer: Berlin Heidelberg, Germany, 2010; pp. 35–42, ISBN 978-3-642-01576-2.

100. Gkioles, N.; Pallis, G. *Atlas of the Christian Monuments of the Aegean. From the Early Christian Years to the Fall of Constantinople*; General Secreteriat of the Aegean & Island Policy: Athens, Greece, 2014.
101. WWF Greece WWF Greece, LEM037-Sourladika. Available online: http://www.oikoskopio.gr/ygrotopio/gallery/view-gallery.php?action=view_images&alb_category_id=57&alb_album_title=LEM037-Σουρλάδικα/Sourladika&alb_album_id=461&lang=en_US (accessed on 10 March 2023).
102. Erosion Structures—Trigies Cape. Available online: https://www.lemnosgeomonuments.gr/en/simeio/erosion-structures-trigies-cape/ (accessed on 10 March 2023).
103. Georgopoulos, P. *Environmental Impact Study of the Shooting Range in the Wider Falakro Area of the Municipality of Atsiki on the Island of Lemnos*; University of the Aegean: Lesvos, Greece, 2010.
104. Chatzidimitriou, A.; Pantekas, T.; Papanastasiou, A.; Davas, E.; Kaltsidis, A. Spatial and Residential Organization Plan of An Open City D.E. Atsiki Municipality of Lemnos—Strategic Environmental Impact Study; 2016, Municipality of Lemnos. Available online: https://limnos.gov.gr/files/keimenob1fasi.pdf (accessed on 10 March 2023).
105. Greco, E. For an archaeological phenomenology of the society of Hephestia (Lemnos) from the Late Bronze Age to the end of Archaism. *Ann. Archeol. Estoria Antica.* **2018**, *25*, 11–31.
106. Sanctuary of the Great Goddess at the Archaelogical Site of Hephaistia. Available online: http://odysseus.culture.gr/h/2/gh251.jsp?obj_id=19985 (accessed on 10 March 2023).
107. Archontidou, A.; Limnaiou, E.; Panatsi, A.; Pylarinou, N. *Myrina of the Early Bronze Age*; Ministry of Culture and Sports, Ephorate of Antiquities: Lemnos, Greece, 2004.
108. Hephaistia. Lemnos Island Travel Guide. Available online: https://www.mylemnos.gr/greece/ifaistia.html (accessed on 10 March 2023).
109. Kokkorou-Alevra, G.; Poupaki, E.A.E.; Chatzikonstantinou, A. *Corpus of Ancient Quarries*; Section of Archaeology and History of Art, National and Kapodistrian University of Athens: Athens, Greece , 2014.
110. Beschi, L. Culto e riserva delle acque nel santuario arcaico di Efestia. *Annu. Sc. Archeol. Atene Mission. Ital. Oriente* **2006**, *83*, 95–234.
111. Ministry of Education and Religious Affairs Culture and Sports Odysseus. Available online: http://odysseus.culture.gr/index_en.html (accessed on 10 March 2023).
112. Campbell, J. *The Mysteries*; Princeton University Press: Princeton, NJ, USA, 1978.
113. Arikas, K.; Voudouris, P. Hydrothermal alteration and mineralizations of magmatic rocks in the southern Rhodope Massif. *Acta Vulcanol.* **1998**, *10*, 353–365.
114. Voudouris, P.; Skarpelis, N. Epithermal gold-silver mineralizations in Perama (Thrace) and Lemnos. *Bull Geol Soc Greece* **1998**, *32*, 125–135.
115. Voudouris, P.; Alfieris, D. New porphyry—Cu±Mo occurrences in the north-eastern Aegean, Greece: Ore mineralogy and epithermal relationships. In Mineral Deposit Research: Meeting the Global Challenge, Proceedings of the Eighth Biennial SGA Meeting, Beijing, China, 18—21 August 2005; Springer: Berlin/Heidelberg, Germany, 2005; pp. 473–476.
116. Papoulis, D.; Tsolis-Katagas, P. Formation of alteration zones and kaolin genesis, Limnos Island, northeast Aegean Sea, Greece. *Clay Min* **2008**, *43*, 631–646. [CrossRef]
117. Fornadel, A.P. *Mineralogical, Petrological, Stable Isotope and Fluid Inclusion Studies of the Palea Kaval Reduced Intrusion-Related and the Transitional Fakos Porphyry Cu-Mo to Epithermal Au-Te Ore Systems*; Iowa State University: Ames, IA, USA, 2010.
118. Anyfadi, A.; Parcharidis, I.; Sykioti, O. Hidrothermal alteration zones detection in Limnos island, through the application of remote sensing. *Bull. Geol. Soc. Greece* **2016**, *50*, 1595–1604.
119. Museum of Maritime Trandition and Sponge Fishing MUSEUM of Maritime Trandition and Sponge Fishing. Available online: https://www.facebook.com/profile.php?id=100069949720331 (accessed on 20 March 2023).
120. Voudouris, P.; Velitzelos, D.; Velitzelos, E.; Thewald, U. Petrified wood occurrences in western Thrace and Limnos island: Mineralogy, geochemistry and depositional environment. *Bull. Geol. Soc. Greece* **2007**, *40*, 238–250. [CrossRef]
121. Ioannou, A. The Oddities of Lemnos. Available online: https://www.geografikoi.gr/limnos-perierga-1/ (accessed on 10 March 2023).
122. Claven, J. The Mysterious Lighthouse of Lemnos and Its Connection with Australia. Available online: https://neoskosmos.com/el/2020/06/11/features/o-mystirios-faros-tis-limnou-kai-i-syndesi-tou-me-tin-afstralia/ (accessed on 10 March 2023).
123. Pappa, M. Sanctuary of Artemis at Avlonas in Myrina. Available online: http://odysseus.culture.gr/h/2/gh251.jsp?obj_id=2489 (accessed on 10 March 2023).
124. Farnell, L.R. *The Cults of the Greek States, vol. II.*; Clarendon Press: Oxford, UK, 1896.
125. Boulotis, C. Koukonisi on Lemnos: Reflections on the Minoan and Minoanising evidence. In *The Minoans in the Central, Eastern and Northern Aegean—New Evidence*; Macdonald, C.F., Hallager, E., Niemeier, W.-D., Eds.; Monographs of the Danish Institute at Athens: Athens, Greece, 2009; Volume 8, pp. 175–218.
126. Boulotis, C. Koukonisi (Lemnos), un site portuaire florissant du Bronze Moyen et du début du Bronze Récent dans le Nord de l'Égée. In *Mesohelladika: La Gréce Continentale au Bronze Moyen, Colloque International Organisé par l'École Française d'Ahènes en Collaboration avec l'American School of Classical Studies at Athens et le Netherlands Institute in Athens, Athènes, EFA/Cotsen Hall*; Philippa-Touchais, A., Touchais, G., Voutsaki, S., Wright, J., Eds.; École française d'Athènes: Athens, Greece, 2010; pp. 891–907.

127. Boulotis, C. The Prehistoric settlement of Koukonisi, Lemnos. Available online: https://www.archaiologia.gr/blog/2012/03/12/o-προϊστορικός-οικισμός-στο-κουκονήσι/ (accessed on 10 March 2023).
128. Vouvalidis, K.; Kapsimalis, V.; Tsourlos, P.; Koukousioura, O.; Almpanakis, K.; Czebreszuk, J.; Aidona, E.; Doani, S.; Vandarakis, D.; Dimou, G.-V. Ancient Poliochni Geoarchaeological Research: Its proximity to the sea and the possible location of an ancient harbor (North Aegean Sea, Greece). In Proceedings of the 10th International Conference on Geomorphology, Coimbra, Portugal, 12–16 September 2022; p. ICG2022-442.
129. Brea, L.B. A Bronze Age house in Poliochni. *Proc. Prehist. Soc.* **1955**, *21*, 145.
130. Kyriakopoulou, T. Archaeological site of Poliochni at Kaminia Area. Available online: http://odysseus.culture.gr/h/3/eh351.jsp?obj_id=2534 (accessed on 10 March 2023).
131. Cultrato, M. Domestic architecture and public space in Early Bronze Age Poliochni (Lemnos). *Br. Sch. Athens Stud. Build. Communities House, Settl. Soc. Aegean Beyond* **2007**, *15*, 55–64.
132. Pijet-Migoń, E.; Migoń, P. Geoheritage and Cultural Heritage—A Review of Recurrent and Interlinked Themes. *Geosciences* **2022**, *12*, 98. [CrossRef]

Article

Ghost Mines for Geoheritage Enhancement in the Umbria Region (Central Italy)

Laura Melelli *,†, Massimo Palombo † and Sabrina Nazzareni *,†

Department of Physics and Geology, University of Perugia, Via A. Pascoli, s.n.c., 06123 Perugia, Italy
* Correspondence: laura.melelli@unipg.it (L.M.); sabrina.nazzareni@unipg.it (S.N.);
  Tel.: +39-(0)75-584-9579 (L.M.); +39-(0)75-585-2616 (S.N.)
† These authors contributed equally to this work.

Abstract: The paper proposes a method to valorize abandoned mines whose traces were lost in the territory and in the collective memory. We selected two case studies in the Umbria region (central Italy) that were used as examples. The evidence of the presence of lignite mines on the Upper Tiber River Valley (northern Umbria) has been completely erased, and since they were located in rural areas, they represent an interesting challenge regarding recovering the memory of the places and proposing a no-longer-existent site as a geosite. The recovery and valorization of historical documents of the two lignite mines (Caiperino–Terranera and Carsuga) and their conversion into a digital format was carried out before constructing a geolocalized database in a GIS environment. This framework is the starting point for a promising dissemination process via a digital media app, using multimedia contents as video, 3D models and the principles of augmented reality (AR) to enhance the touristic or didactic experience and promote the cultural heritage of the territory by keeping the memory of 'ghost places'.

Keywords: mines; mining; geosite; geoheritage; landscape; coal

Citation: Melelli, L.; Palombo, M.; Nazzareni, S. Ghost Mines for Geoheritage Enhancement in the Umbria Region (Central Italy). *Geosciences* **2023**, *13*, 208. https://doi.org/10.3390/geosciences13070208

Academic Editors: Jesus Martinez-Frias, Deodato Tapete, Hara Drinia, Panagiotis Voudouris and Assimina Antonarakou

Received: 23 May 2023
Revised: 13 June 2023
Accepted: 3 July 2023
Published: 11 July 2023

1. Introduction

Geoheritage is a part of the identified geodiversity of the Earth, where a value is recognized and conservation is needed [1–4]. In order to assess geoheritage, the evaluation of geodiversity [5–8] and the identification of geosites [9–11] are crucial steps. A geosite is always the expression of the landscape where it is located; this is particularly true when the geosite corresponds to a landform (geomorphosite, [7]).

As a matter of fact, geosites can be damaged or destroyed by vegetation encroachment and natural erosion but also by degradation by infrastructure and housing buildings. In addition, some geosites may be destroyed simply because local communities do not know or understand their value. A valid example of risk analysis of degradation is applied in La Rioja, Spain. In this case study, the risk of the degradation of the geological characteristics was attributed to intrinsic factors (fragility) and to extrinsic factors (vulnerability), both natural and anthropogenic, and was affected by public use. According to this preliminary analysis, it is possible to plan the conservation actions, such as recording and rescuing the geosites where damage is inevitable and protecting and promoting the areas where there are preliminary conditions, allowing for the preservation of the geosites [12].

As the landscape is the result of several factors (endogenous, exogenous, natural and anthropic) developing at different time steps, the evolution of geomorphosites also changes over time as well [13]. In those areas where the human presence is widespread and where man becomes the most relevant morphogenetic factor, the landscape may evolve in a short time, if compared with natural evolution. In such contexts, the probability of losing track of the geoheritage increases. Nevertheless, also in these cases, if the natural landscape evolution together with the anthropogenic impact may partially or totally erase the evidence of geosites [14], their contribution in the current landscape layout is undeniable and should be addressed.

Existing geosites can be subjected to physical and biological hazards such as geomorphological erosion and deposition activity or vegetation growth with a possible increase in the efficiency of these natural processes due to climate change. Moreover, anthropic hazards due to human activity, such as infilling or earthwork, river management or coastal protection, may permanently delete geosites. Conservation is a fundamental key to the enhancement and dissemination of the geoheritage [12]. If the preservation of an existing geosite might be complex, but feasible, the protection of a geosite that has been jeopardized by the aforementioned factors is a much more complex issue. Key considerations include: why does it make sense to keep the memory of something that no longer exists? How is it possible to evaluate the role of a site in the current landscape configuration? An important issue is with regard to retrieving the necessary documentation for defining the right value of the extinct site, especially in the case where the geosite only has a local value. Key clues might be derived from the inherited landforms that are witnesses of the Earth's evolution; for this reason, they should be recorded and taken into account in the geoheritage evaluation [13].

In the scientific literature, geosites are classified and described using different principles and values [15], including selected closed mines that have been considered as geosites. Nevertheless, the roles of mining activity and, more in general, of the extractive industries (ore mineral, oil and gas exploitation) have been identified as a negative factor on the environment and on the World Heritage. The UNESCO World Heritage Committee established the incompatibility between the mineral exploitation and the World Heritage status—although some old mines were declared World Heritage—since the global exploration and exploitation of mining, oil and gas has been recognized as the seventh most relevant reason affecting the areas declared as World Heritage. On the other hand, numerous actions have been promoted in many countries for the recovery of dismissed mining areas [16–18] . Mining closure is not necessarily the final step of the mining activity but, on the contrary, should be the first phase of the inclusion of mines in the geoheritage list of a country [6,19]. The mining activities collect different types of values, where the first and most direct are the minerals and lithology that remain along the excavation tunnels and in the surrounding areas. These materials can be observed in situ in open pits, along unfilled tunnels or in exhibitions and museums close to the tailing piles. Moreover, the information provided by the mined material is not only limited to the description of the single mineral or lithotype but is the trigger for illustrating the geological setting of a wider area around the mining site.

In addition to the geological value, several values can be associated to the mining sites. A first group includes values related to the activity: aesthetic and architectural, antiquity, duration of mining activity, economic, technological. As an example, the Zollverein Coal Mine Industrial Complex (Essen Germany) has been included in 2011 in the World Heritage List for the excellent architectural quality of its buildings [20]. In addition, the duration of the mining activity and the starting period determine the historical value together with the scale of the operation, the social impact and also the economic impact of the mine in the area, in the country or worldwide. The technological value is associated with innovations due to the industrial and technical improvements derived from the mining activity, such as the need to reach deeper layers, or to have faster extraction and processing times. The memory preservation of the past mining sites also has a social value since abandoned mines have a social context, especially in rural communities [21]. Last but not least, the mining sites are a very important factor as landscape modifiers. The elements characterizing the mining activity as extraction sites, areas of accumulation of waste, infrastructure and mining villages change the natural and original landscape morphology quickly and strongly.

The mining landscape can therefore be seen as a cultural landscape, and a peculiar mix of natural and man-made landforms. This landscape evolves in a very short time if we consider, for example, the open cast mine. On the contrary, in subterranean, long-term activities, the changes are not fast; however, the infrastructure necessary for the surface extraction and transport of materials rapidly alters the original topography and must be taken into account in the analysis of the evolution of the anthropic landscape. Due to all these values, the mine closure is only the beginning for the post-mining stage and thus the beginning for a new cycle of initiatives aimed at considering the mining sites as a resource for the territory.

Sustainability is a value that should be ensured during mining exploitation. After closing, these areas have potential interest together with environmental remediation and social reintegration [22]. Following this process, the initial assumption that a mine is a destabilizing factor for the natural heritage was partially reversed to the point that UNESCO declared mining sites as also belonging to the World Heritage List [23]. In addition, geomining heritage is present in the UNESCO Global Geoparks list, proving the possibility of including mining areas in the geological and cultural heritage of a territory.

In this paper, we proposed two abandoned and no-longer-existent mines in northern Umbria (central Italy) as a case study for their promotion as potential geosites. The study cases are a clear example of mining sites whose traces have been lost almost completely in the current landscape. For this reason, the selection of the data input, the proposed methodology for the analysis step and the solutions adopted for the output dissemination could be a valid guideline for these contexts.

In the Umbria region, the proposed study cases can be a promising approach for geotourism since some other examples of abandoned mining sites converted into didactic and touristic destinations are already present (see http://www.parcogeologicovalnerina.it/ (accessed on 17 December 2020)). In particular, since these mining areas no longer exist, it is necessary to insert the sites as points of interest along touristic routes with other values (cultural, artistic, historical). In the study area, other paths are present, such as the Tiber route (https://www.umbriatourism.it/it/-/strada-14-borghi-e-castelli-nell-alta-valle-del-tevere-it (accessed on 17 December 2020)), connecting museums, towns with a high cultural heritage and naturalistic points of interest. The proposed mining areas could take advantage of the well-consolidated tourist offer and enter as additional information for visitors.

Lignite Deposits in the Italian Mining Heritage

According to the Italian legislation (R.D. 1443/1927), a mine concerns the exploitation of materials of high value and strategic importance, such as ore minerals and fuel. Since 2006, ISPRA (Italian Institute for Environmental Protection and Research), which is responsible for the conservation and enhancement of the technical–scientific, historical–cultural and environmental heritage of mines, carried out (and constantly update) the "National Census of 2990 abandoned mining sites (from 1870 to 2006)" (Figure 1).

In the census, the number of sites, the extracted minerals, the activity period, the type of mining (underground, open-pit or mixed), the last dealer, the location of sites at municipal level, the current situation (abandoned, inactive or active) and the claim term, plus additional sources and references, are considered. Mining activity is widespread throughout Italy (Figure 1), and, from a geographical point of view, 74,78% of mines are present in 5 regions: Sicily (724 sites), Sardinia (427 sites), Tuscany (416 sites), Piedmont (375 sites) and Lombardy (294 sites).

At present, promotion and musealization projects interested Italian abandoned mines, producing mining museums, science and technology museums, ecomuseums and mining parks. Furthermore, thematic itineraries and mining trekking paths have been defined. This interlaces with other international initiatives and the promotion of the connected territories aimed at triggering further economic, social and touristic development. The common idea of several projects is connecting these sites along mineral routes, such as the RUMYS project [24]. The RUMYS project defines fifteen routes along a network in ten Latin American countries. The acronym derives from the Quechua word 'Rumi', which means 'stone'. As a matter of fact, in this project, the stones and the minerals that had a key role in the local social development are the point of interests and the nodes of these routes.

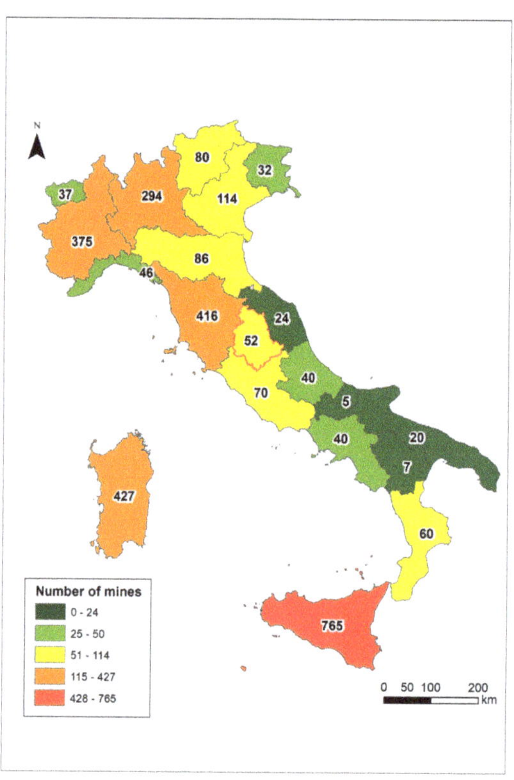

Figure 1. Number of mines for region in Italy recorded for the period from 1870 to 2006 (data from https://www.isprambiente.gov.it/files/miniere/i--siti--minerari--italiani--1870--2006.pdf).

In Italy, the ReMi project (Rete Nazionale dei Parchi e Musei Minerari Italiani - National Network of Italian Mining Parks and Museums) is the reference for a national database. Thanks to this activity, geo-mining parks have been introduced as destinations of sustainable tourism in other projects, such as the Atlas of the Soft Mobility created by the Italian Railway Network (https://experience.arcgis.com/experience/805005081da841bfb4 0120cd96290bcd (accessed on 17 December 2021)).

An important product of the Italian mining industry is lignite, of which one hundred and forty mines were accounted for in the national territory, mainly in Tuscany (seventy-one mines, 51%). In Italy, this fossil fuel had its maximum exploitation around 1920; then, due to the excessive cost of extraction activity and the low calorific value compared to other fuels, most of the lignite mines were closed.

Lignite is the lowest grade term of coal composed by organic material (vegetable origin), and the carbonization stage is intermediate between peat and bituminous coal. Lignite has a carbon content of approximately 60–70% on a dry-ash-free basis that is reduced to 25–35% on an ash-received basis that encompasses both inherent moisture and mineral matter, and has a calorific value of 18.8–25.1 MJ/kg (4500–6000 kcal/kg). The humic substances (organic compounds contained in the soil) are more abundant when compared to peat and less abundant when compared to bituminous coal (Figure 2) due to different temperatures and pressures to which they have been subjected to and/or the shortest time elapsed since the beginning of organic matter decomposition [25,26].

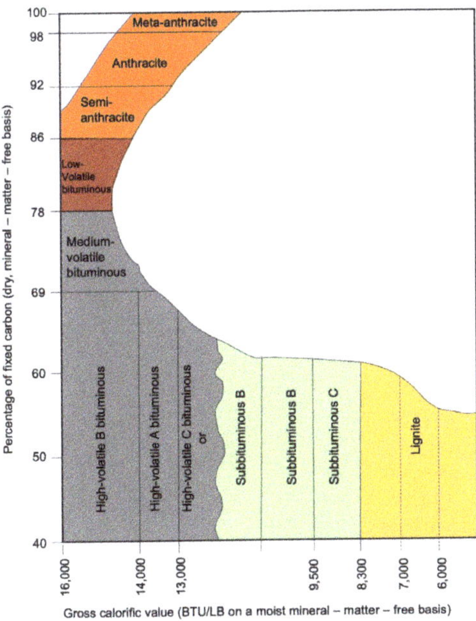

Figure 2. Scheme showing classification of coals by rank in the United States as a measure of the progressive alteration from lignite to anthracite. Modified from [27].

Different types of lignite have been identified, with xyloid lignite as the most common type. It consists mainly of tree trunks with woody texture and a light brown color. Picea lignite is black, amorphous, compact and shiny, with conchoidal fracture; there is no plants evidence in the texture. It is the most valuable and can be considered as the intermediate term between bituminous coal and lignite. Peat lignite, also called brown lignite, is rich in minute plant elements (residues of herbaceous plants) cemented in a dark and compact matrix. Italian lignite of xyloid and picea types is almost completely included in Tertiary deposits, whereas the peat type is confined to Quaternary deposits [28]. The most important deposits in Italy are in Liguria, Veneto, Tuscany, Umbria and Basilicata regions.

Even if the lignite beds are close to the topographic surface and, therefore, readily exploitable, they crumble very easily and are difficult to be transported. In addition, the carbonification process is not complete in lignite and its calorific value is lower than bituminous coal. For these reasons, currently, the extraction costs do not justify its cultivation. On the contrary, in the past, lignite was extensively mined and used for steam–electric power generation in production plants close to the mining sites.

2. Lignite Deposits in Umbria Region (Central Italy)

The studied cases of this work were located in the Umbria region, central Italy. Despite its limited areal extension (8.456 km^2), the Umbria region has a high geodiversity and relevant geoheritage [8] due to a complex geological history that is still well evident in the morphological and geographical setting. The region is characterized by a prevalent hilly and mountainous morphological setting with the flat areas limited to some intermontane basins [29]. The fold mountain system of the Apennine, with anticlines and synclines elongated in the N-S or NNW-SSE direction, is derived from a compressional tectonic phase (Late Miocene [30,31]). Since the Late Pliocene, a new extensional tectonic phase started, and is still active, producing Quaternary basins delimited by normal fault systems [32–34] that interrupt the hilly and mountainous morphological features.

The Tiberino Basin (TB) is the largest of the intermountain basins (Figure 3), with an area of approximately 1800 km^2 and a shape similar to an overturned Y.

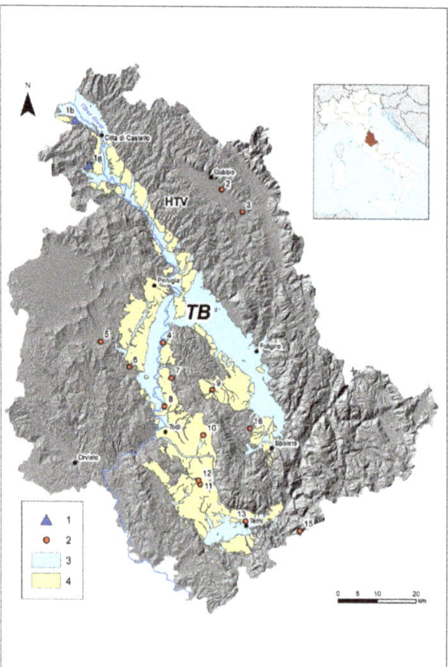

Figure 3. Abandoned mines in Umbria region and the Tiberino Basin. (1) Investigated sites: 1a Caiperino–Terranera; 1b Carsuga. (2) Known mines: 2 Padule; 3 Branca Galvana; 4 Deruta; 5 Pietrafitta; 6 Morcella; 7 Collazzone; 8 Ilci; 9 Bastardo; 10 Massa Martana; 11 Monte Castrilli; 12 Collesecco; 13 Colle dell'Oro; 14 Bonacquisto; 15 Morgnano; 16 S. Angelo in Mercole. (3) Alluvial deposits in the Tiberino Basin (Holocene). (4) Fluvial and lacustrine complex (Pliocene and Pleistocene).

TB crosses the entire region in the longitudinal direction, with the northern section crossed by the Tiber River flowing from north to south. The TB has been active since the Late Pliocene and is limited by a segmented system of both ENE-dipping and WSW-dipping normal faults [32]. The basin is infilled by a Late Pliocene–Pleistocene continental sequence accumulated in a depositional environment where braided rivers and shallow lakes alternated and changed in time [35–37]. Above the sequence, the alluvial deposits of the current hydrographic network are present. The thicknesses of the Late Pliocene–Pleistocene continental sequence and of the overlaid alluvial deposits depend on the relationship between uplift and tectonic activity along the fault planes [34]. In the areas where the lacustrine and palustrine sedimentation prevails, the environment was favorable for the formation of lignite levels within clays. In most cases, lignite originated from accumulations of marsh vegetation developed in shallow water (peat bogs). Sporadically, there are remains of tree trunks, testimony of tall plants in the marshy vegetation when the level of flooding increased very slowly [25].

In particular, given the need for self-sufficiency during wartime, lignite mining was very common in Umbria from the First World War until the middle of the twentieth century. Lignite deposits are widespread and were the starting point for most of the Umbrian mining in the TB, where the deposits were recognized along the west and east basin side in the northern and southwestern sector; on the contrary, in the southeastern sector, the deposits were present only along the western margin [25].

In the past, the mining exploration was concentrated along the margins of the TB mainly because, previously, the basin was interpreted as a single large lake [38] and not, as more correctly interpreted today, as a set of interconnected river and lake environments. The mining exploration based on this wrong interpretation was oriented at the paleo-shores of the hypothetical Tiberino Lake mainly for two reasons: (1) the lignite outcrops should be more evident along the limit of the lacustrine environment, where the deposit layers should have a lower thickness, and (2) because the accumulation of plants, transported to the lake from the surrounding mountains area by the drainage network, could have accumulated in the inlets of the coast where the lake currents were less strong.

Mining activities have always been limited by the low reserve of the deposits, but it was still enough to guarantee the life of the mine for a number of years, without planning exploration activities to assess the true reserve of lignite. However, considering the census of the abandoned mines in Umbria, the lignite and, more in general, the fossil fuel were the focus of the regional mining. In a total number of 52 mining sites, 30 were devoted to fossil fuel (lignite and/or lignite xyloid and/or peat) and, among them, 20 were of lignite (Table 1, Figure 3). The remaining sites were dedicated to marl for cement.

Table 1. Number of extraction sites in Umbria divided by type of mineral.

Extracted Minerals	Number of Sites
Common lignite	20
Marl for cement	17
Xyloid lignite	10
Clay	2
Leucite	1
Iron	1
Peat	1

In the northern segment of the TB, also called the High Tiber Valley (HTV), lignite has been identified in the Fighille Unit (Lower Pleistocene), which is at least 120 m thick [34,39], where clays and bluish grey marly clayey silts and light brown, grey and orange brown sandy and clayey silts prevail in one facies. The clayey silts enclose abundant plant remnants, peats and lignite levels, testifying a palustrine environment [34]. The depositional environment was a periodically flooded floodplain, with ponds, wetlands, braided rivers and alluvial fans with ephemeral channels. The climate was characterized by the alternation of dry and moist, with drained soils (calcareous nodules), or was developed under water saturation conditions (dark color) and ephemeral or highly erosive streams during heavy rains.

Some deposits have been found in the Citerna Unit (lower Early Pleistocene), superimposed to the Fighille one and with a thickness of around 200–400 m. The Citerna Unit [34,39] is characterized by two facies: (1) a coarse-grained texture with clast-supported gravels in a sandy matrix and with sandy lenses or layers; (2) a fine-grained facies associated with prevailing sandy and clay silts. The Citerna Unit refers to a braided river channel environment. We focused on the abandoned mines of the HTV, where the benches of lignite vary between 1.5 and 7 m. In detail, two little-known and only partially investigated sites were taken into account: the Caiperino and Terranera mine (Lat–Long 43.39107, 12.19578) and the Carsuga mine (Lat–Long 43.49422, 12.15480, both in Figure 3.).

3. 3. Data and Methods

3.1. Documentary Search

Documents related to the exploitation of lignite in HTV are scarce and difficult to find. The scientific literature is not updated and has many gaps [40–42]; in addition, some sites have never even been listed in the official census and only historical documents prove their existence. We performed an extensive search on the scientific literature database and only a minimal documentation for Caiperino and Terranera mines was found. For this reason, the research method has some mandatory steps. The first one is the collection of multitemporal and multiscale maps and documents preserved in local archives. This activity is often time-consuming since the minor archives do not have available digital databases but only documentation in analog format. Three local historical archives (Archivio Storico Comunale e Archivio Notarile di Città di Castello, Archivio Storico Diocesano di Città di Castello, Archivio Storico del Comune di Monte Santa Maria Tiberina), a local library (Biblioteca Comunale di Città di Castello "Giosuè Carducci") and a photo library ("Fototeca Tifernate", photographic archive by Giuseppe Tacchini) were consulted together with two national libraries (Library of Mathematical, Physical and Geological Sciences of the University of Perugia, and "Mario Marte" Agricultural Library of the University of Perugia) and a regional and Italian digital library (MLOL—Biblioteca digitale italiana - Italian digital library). Ancient maps and documentation related to the operation of studied mine were collected.

3.2. Field Work

Once the first multitemporal and multiscale analysis on the cartographic documents and the analysis of the geological information were concluded, a field survey on the potential areas of interest was carried out. We aimed to verify and compare the actual landscape with the archival documents. Since field evidence of the mines' existence is no longer visible, photographs, soil samples and observations were complemented by interviews with the inhabitants living in the surroundings of the extraction areas in order to build the most complete information dataset. In order to have an azimuthal vision of the landscape, in addition to the observation of multiscale and multitemporal aerial photos and considering the narrow range of the areas investigated, a survey with a drone was planned. An unmanned aerial vehicle, commonly known as drone, is a new technology for acquiring remote-sensed data used for several purposes and has been tested for geological aims, particularly for promoting areas difficult to reach or landscape contents that are hardly recognized from the ground [43–45]. Even if human and natural processes modified the landscape, some features are still present and may be used as 'control point' to link the historical documents to the current landscape in order to find the location of the mining sites with the greatest possible accuracy. The observations on the field together with the drone footage allowed us to identify the reference points and to obtain a precise temporal reconstruction.

3.3. Digitalization

The obtained data were used as an input in a geographical information system (G.I.S.) in order to digitize and georeference the spatial information. The geolocalization of several multitemporal maps has some limits due to the uncertainty of localization and the approximation in the realization of maps, especially the ancient ones. Moreover, the different map scales may limit the effectiveness of the comparison [46]. Historical documents were scanned and integrated as hyperlinks to geolocalized point data. The photographic documents derived from old photographic plates were converted to digitalized images to be imported into the G.I.S. also as georeferenced images. The locations (or supposed localizations) were modeled as a point vector dataset. The area limits of the mining zones of extraction, reconstructed on the basis of historical documentation, were converted to polygonal vector data. In the same project, the topographic maps used as backdrop images had a scale of 1:10,000. In addition to historical data, a comparison with the geological structure was necessary. Geological data were extracted from the geological dataset of the Umbria Region (https://www.regione.umbria.it/paesaggio-urbanistica/cartografia-geologica (accessed on 12 November 2021)) with a scale of 1:10.0000, integrated with the new Geological Map of Italy, scale 1:50.000, N. 289 "Città di Castello" and N. 299 "Umbertide" (https://www.isprambiente.gov.it/Media/carg/289_CITTA_CASTELLO/Foglio.html (accessed on 12 November 2021); https://www.isprambiente.gov.it/Media/carg/299_UMBERTIDE/Foglio.html (accessed on 12 November 2021)).

The field survey validated the bibliographical information on the geological structure of the study areas. To increase levels of geological information, the point vector dataset edited by the Umbria Region with geognostic surveys was added (https://dati.regione. umbria.it/dataset/banca-dati-indagini-geognostiche-geofisiche (accessed on 14 March 2022)).

The geological data allow us to validate the potential sedimentary units where the lignite deposits could be present.

4. Results

4.1. The Caiperino and Terranera Mines

The Caiperino and Terranera mine area was located along the threshold area of the Aggia and Nestore rivers, both tributaries on the right of the Tiber River (Figure 3). According to the period of exploitation and the administrative boundaries, e.g., for the issuance of claim, we can consider the mining area as two separate mines: the Caiperino mine and the Terranera mine. The Caiperino mine covers the SW part of the cultivation area and is characterized by the presence of a currently hydrographic network that morphologically reshaped the area. The Terranera mine is located northeast of the previous one and has probably been exploited in more recent times.

The beginning of the mining activity dates back to 1866 and lasted for 15 years; the mineralization was identified as a lignite bench that was 1.40 m thick. The area at that time was very isolated, lacking any major road to transport brown coal to Terni (the major steel production city in southern Umbria). In 1901, renewed interest in the mineralized area led to new exploration, and related documents describe the deposit in detail: "the lignite is in layers of blue marl, surmounted by a yellowish limestone sand. The strata have a dip direction equal to 55° N and a dip angle from 15° to 20°". Attached to the report are detailed maps that show the old tunnel entrances and the location of the wells (Figure 4). In 1907, the mining activity was resumed and at least 200 tons of lignite were extracted in the Caiperino area while, in 1916, the area of Terranera produced 500 tons of lignite. Between 1917 and 1918, Austro-Hungarian prisoners present in the Upper Tiber Valley were employed in the mine. After a series of mining activities with a low income, the mine was definitively closed around 1945. There are no traces of the entrances, tracks and tunnels left.

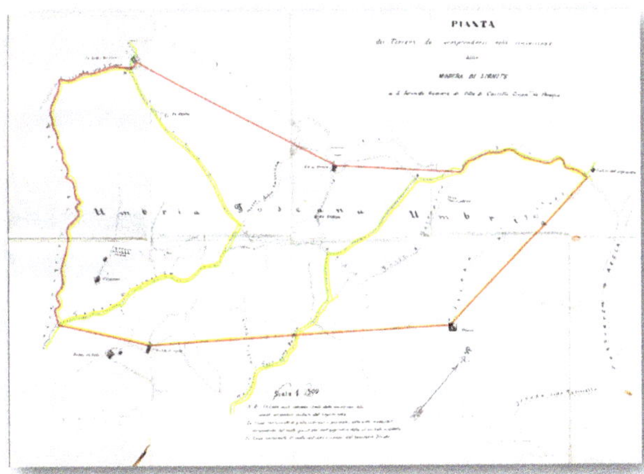

Figure 4. The location map of the different components of the mining area for the Carsuga mine.

Archival research has associated with the Caiperino mine a series of photographs dating back to 1940 that were never catalogued (http://www.archiphoto.it/new/ (accessed on 12 March 2022)). The study was carried out, on a cartographic basis, place names, interviews with local inhabitants and field work to precisely identify the morphological profiles of the places and the locations of the shooting points for each image. The photographic comparison was particularly useful for confirming the location of mine entrances; nevertheless, this process was difficult due to the presence of a dense vegetation cover that has almost completely hidden the hilly and valley profiles (Figure 5). The starting document was a 1:25,000 scale topographic map drawn up between 1925 and 1949 (and updated to 1977) that showed the symbol of a mine (no longer present in the most recent versions of the official topographical cartography).

The georeferencing procedure for the ancient map in G.I.S. returned a good match. The perimeter of the concession area shown on the map overlaps almost perfectly with the reconstructed vector perimeter from archival documents. Analogously, the places indicated on the map coincide with the locations of the current localities. Information on wells and tunnels can therefore be considered as reliable. By using the G.I.S. results, we estimate an areal extension of approximately 62 hectares in the Umbria territory and 42 hectares in Tuscany. These values match the corresponding areas derived from the archival documents.

Figure 5. Comparison between a historical photo (archive G. Tacchini, 1940) and the current landscape in the Caiperino mine. The colored lines highlight the points of reference identified for the comparison between the two images and for defining the placement of the historical photo.

4.2. The Carsuga Mine

The Carsuga mine (also named Pistrino, Citerna or Vingone) started the activity around 1930 and was abandoned in the post-World-War-II period. Although its exploitation was limited in time, the mine left a strong memory in the inhabitants of the immediate surroundings, but, nowadays, both memories and field evidence risk rapidly vanishing. In the current topographic maps, a locality with the toponym 'Mine' is still present. The evidence of a lignite deposit was already present in a map attributed to Leonardo Da Vinci (RL 12676r). The mining activity began in 1934 at a level of 1.20 m thick, classified as mediocre by exploration analysis. For this reason, the extraction was soon abandoned. As for the Caiperino mine, the comparison between the actual landscape and ancient documents (maps and photographs) allowed us to observe the landscape evolution, the activities of the mine and the places where they were carried out. Similarly, the multitemporal and multiscale overlay of maps in G.I.S. evidenced the location of different components of the mining plant (Figure 6).

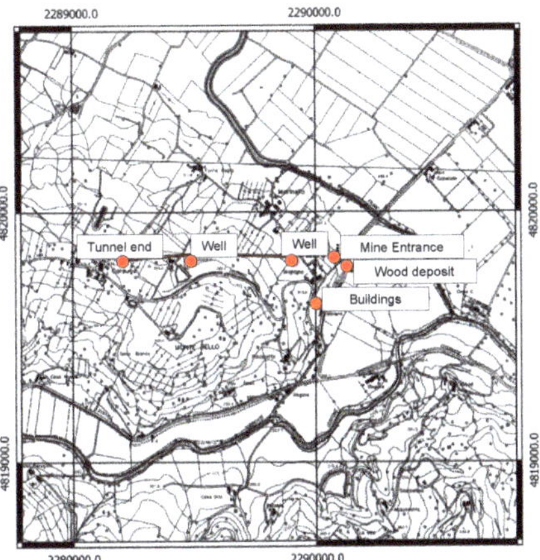

Figure 6. Comparison between a historical photo (archive G. Tacchini, 1940) and the current landscape in the Carsuga mine. The red circle highlights the same reference place identified in both images. In the historical photo, the wood storage area for tunnel covers is present.

5. Discussion

5.1. The Value of Ghost Places in Geoheritage Promotion

Generally, the data used for valuing the geoheritage of a territory are existing landforms, outcrops or samples. Regardless of whether these objects are in situ, such as natural outcrops or features, or ex situ, such as the museum collections, what is deserving of being valued is always a real object and not an abstraction. Moreover, geosites or landscapes with a high geological value that have currently disappeared still retain the memory of an initial spatial planning and the presence of a specific geological arrangement. In other aspects of cultural heritage, different from the geological ones, the so-called 'intangible cultural heritage' is already a well-established value that includes languages, arts, knowledge and social practices and traditional crafts. This kind of cultural heritage is essential for maintaining the cultural diversity and for perpetuating skills and knowledge over generations.

In geoheritage, the memory related to the social aspects of a territory should be considered as being as important as the physical values. If we consider the human presence and activity as a geomorphological modeling agent similar to other forces such as gravity, ice, wind, etc., we should also consider the anthropic landforms related to the geodiversity value. This is particularly true in urban areas where the urban geology and urban geomorphology take into account the anthropic processes and landforms [47,48]. Nevertheless, also in suburban areas, there could be sites, no longer existing, where the geological component was prevalent but coupled with an anthropic contribution. These sites have added geological value to cultural and social values as well, increasing the need to improve and preserve these places.

Mining sites are an excellent example of this heritage that combines abiotic natural values and the action of humans in the modeling of the territory. Abandoned and unrecovered mines are a very fragile heritage. The evolution of and adjustments in the land use are very fast in suburban areas that have already experienced the presence of man. Therefore, it is very likely that, in a few decades, the traces of this geological and social heritage can disappear completely. The conservation and outreach of these 'ghost mines' are not easy to achieve.

5.2. Solution for the Geomining Heritage Promotion

The methodology proposed in this paper is an attempt to store, catalogue and compare historical sources for digitalizing, georeferencing and comparing the present landscape with past scenarios. Even if this technique is common in other disciplines, such as history and archaeology, in geological research, the acquisition of historical data is not so obvious. Moreover, their comparison in a G.I.S. is sometimes underestimated when compared to other quantitative approaches. The promotion of these sites can be developed in several ways. In the Italian territory, the ReMi project (Rete Nazionale dei Parchi e Musei Minerari Italiani - National Network of Italian Mining Parks and Museums) is an official database that is unique for collecting and standardizing information but not very usable for dissemination on the field. A second option could be the realization of geotouristic paths that connect abandoned sites to other points of interest, both cultural and commercial, to increase the attractiveness of the route. However, the disclosure of something that has been completely erased is not immediate. This approach can work in abandoned but still accessible and safe mines but not in areas where the traces of the mines have been completely obliterated.

An efficient alternative may be the use of digital geovisualization technologies with a 3D interactive representation of the different steps for the mining area [49,50]. By using virtual models, the no-longer-existing infrastructures around a mining area can be reproduced. In addition, the virtual models can mimic the original landscape by using digital elevation models derived from ancient maps. These techniques are quite common in urban geomorphology research, particularly for investigating the landscape evolution pre- and post-urbanization. The comparison between the actual elevation model with a DEM derived from altitude values represented in ancient maps, by exploiting the principles of map algebra analysis in G.I.S., resulted in the estimation and location of added or erased volumes of terrain [48]. Finally, if the underground network of tunnels and wells is correctly mapped, the subsurface spatial arrangement can also be modeled and visualized. All these layers can be visualized and navigable on websites using a bird's eye perspective. However, even with this approach, it is not possible to visit the 'lost' mines. If the final aim is the knowledge and use of the territory, efforts should be made to attract people to the real place and interact with them for the most constructive collaboration.

We suggest, as a possible solution, a mobile application for electronic devices [45,51] in order to create an advanced mode of the fruition of naturalistic and artistic sites in rural areas through fusion between itineraries of interest and access to augmented reality content (augmented reality—AR) for the development of immersive and customized touristic experiences. One of the most important goals is integrating society into the processes of scientific research. AR is the optimal solution for visiting ghost mines that no longer exist or are no longer accessible and for displaying a 3D model of structures above and below the topographic surface using virtual tours. Our study has produced the data mining necessary to develop a digital app for the 'ghost lignite mines' of the Upper Tiberin Valley, particularly the historical reconstruction of the two sites of Caiperino and Carsuga mines. The next step will be to select the points of interest (PoIs) along the possible touristic paths. In the linked geodatabase, each PoI is defined by a target image that, in the case of ghost mines, is a detail or a view of the actual landscape. Framing the reference image (image recognition), the digital app will open the multimedia contents that we produced during the field work and data mining as text, drone-fly videos and 3D virtual models that the users can manipulate for an immersive and non-passive experience. With the different app utilities (i.e., 'intelligent paths' based on a deep neural network, links to business and cultural activities in the area), tourists may choose their own visit path and experience the visit at ghost mines to unconventionally deepen their knowledge of the territory.

6. Conclusions

Mining activity has some relevant values: from a scientific point of view, the ore minerals and the geological setting of the area is the first step. Moreover, the geomorphological evolution of the landscape due to the mining activity is another theme, particularly for the anthropic landforms or natural features modified by man. However, knowing mining activity means also discovering the history of men and women who lived at the mine. Thus, the historical, social and cultural values are equally important. Passing on the memory of what has been linked to mining activity means passing on to future generations the complete history of a territory.

The G.I.S. environment allows for the overlaying of different kinds of iconographic documents using a georeferencing procedure, and it is fundamental for limiting the no-longer-existing mining areas. As a result, the memory of the places with human presence is fundamental for the landscape evolution being preserved. Iconographic and textual documents, which would otherwise be lost, were collected in a digital format and in a georeferenced database.

Geoheritage is the joining between geological resources, landscape evolution and human presence; the proposal of a geotouristic path, connecting mines, main towns and villages with cultural value and the naturalistic routes already present in the area appears to be a good solution.

AR technology enhances the visit by correlating different information levels in space (above and under the topographic surface) and in time (the original landscape, the landscape during the mining activity and the actual landscape). The memory of local communities in rural areas is fundamental for avoiding the depopulation and abandonment of extra-urban regions. These areas are not always included in traditional touristic circuits, and cannot benefit from local economic attractors. Their economic recession can be avoided with targeted actions to strengthen geotourism. With the approach presented in this work, 'ghost mines' can be brought back to life as cultural attractions and are embedded in their 'ecosystem' made of both human and geological components.

Author Contributions: Conceptualization, L.M., S.N. and M.P.; methodology, L.M.; formal analysis, M.P.; investigation, M.P.; writing—original draft preparation, L.M.; writing—review and editing, L.M., S.N. and M.P.; funding acquisition, L.M. All authors have read and agreed to the published version of the manuscript.

Funding: This research was funded by he University of Perugia, Department of Physics and Geology, grant number "HUSH (Hiking in Urban Scientific Heritage) UNDERGROUND", RicBAs2018, awarded to Laura Melelli.

Data Availability Statement: Data are available upon request from the authors.

Acknowledgments: Editors and referees are greatly acknowledged.

Conflicts of Interest: The authors declare no conflict of interest.

References

1. Gordon, J.E.; Crofts, R.; Díaz-Martínez, E. Geoheritage conservation and environmental policies: Retrospect and prospect. In *Geoheritage*; Elsevier: Amsterdam, The Netherlands, 2018; pp. 213–235.
2. Gray, M. Geodiversity: The backbone of geoheritage and geoconservation. In *Assessment, Protection, and Management*; Elsevier: Amsterdam, The Netherlands, 2018; pp. 13–25.
3. Reynard, E.; Brilha, J. *Geoheritage: Assessment, Protection, and Management*; Elsevier: Amsterdam, The Netherlands, 2017.
4. Zwoliński, Z.; Najwer, A.; Giardino, M. Methods for assessing geodiversity. In *Geoheritage: Assessment, Protection, and Management*; Elsevier: Amsterdam, The Netherlands, 2018.
5. Brilha, J.; Gray, M.; Pereira, D.I.; Pereira, P. Geodiversity: An integrative review as a contribution to the sustainable management of the whole of nature. *Environ. Sci. Policy* **2018**, *86*, 19–28. [CrossRef]
6. Coratza, P.; Reynard, E.; Zwoliński, Z. Geodiversity and geoheritage: Crossing disciplines and approaches. *Geoheritage* **2018**, *10*, 525–526 [CrossRef]
7. Coratza, P.; Hobléa, F. The specificities of geomorphological heritage. In *Geoheritage: Assessment, Protection, and Management*; Elsevier: Amsterdam, The Netherlands, 2018; pp. 87–106.
8. Melelli, L.; Vergari, F.; Liucci, L.; Del Monte, M. Geomorphodiversity index: Quantifying the diversity of landforms and physical landscape. *Sci. Total. Environ.* **2017**, *584*, 701–714. [CrossRef] [PubMed]
9. Brilha, J. Inventory and quantitative assessment of geosites and geodiversity sites: A review. *Geoheritage* **2016**, *8*, 119–134. [CrossRef]

10. Herrera-Franco, G.; Carrión-Mero, P.; Montalván-Burbano, N.; Caicedo-Potosí, J.; Berrezueta, E. Geoheritage and geosites: A bibliometric analysis and literature review. *Geosciences* **2022**, *12*, 169. [CrossRef]
11. Migoń, P.; Pijet-Migoń, E. Viewpoint geosites—Values, conservation and management issues. *Proc. Geol. Assoc.* **2017**, *128*, 511–522. [CrossRef]
12. Prosser, C.D.; Díaz-Martínez, E.; Larwood, J.G. The conservation of geosites: Principles and practice. In *Geoheritage: Assessment, Protection, and Management*; Elsevier: Amsterdam, The Netherlands, 2018; pp. 193–212.
13. Pelfini, M.; Bollati, I. Landforms and geomorphosites ongoing changes: Concepts and implications for geoheritage promotion. *Quaest. Geogr.* **2014**, *33*, 131–143. [CrossRef]
14. Migoń, P.; Różycka, M. When individual geosites matter less—Challenges to communicate landscape evolution of a complex morphostructure (Orlické–Bystrzyckie Mountains Block, Czechia/Poland, Central Europe). *Geosciences* **2021**, *11*, 100. [CrossRef]
15. Mucivuna, V.C.; Garcia, M.d.G.M.; Reynard, E. Comparing quantitative methods on the evaluation of scientific value in geosites: Analysis from the Itatiaia National Park, Brazil. *Geomorphology* **2022**, *396*, 107988. [CrossRef]
16. Carrión-Mero, P.; Loor-Oporto, O.; Andrade-Ríos, H.; Herrera-Franco, G.; Morante-Carballo, F.; Jaya-Montalvo, M.; Berrezueta, E. Quantitative and Qualitative Assessment of the "El Sexmo" Tourist Gold Mine (Zaruma, Ecuador) as A Geosite and Mining Site. Resources, 9 (3), 28. Chingombe W., 2019. Preliminary geomorphosites assessment along the panorama route of Mpumalanga province, South Africa. *Geoj. Tour. Geosites* **2020**, *27*, 1261–1270.
17. Guijón, R.; Henríquez, F.; Naranjo, J.A. Geological, geographical and legal considerations for the conservation of unique iron oxide and sulphur flows at El Laco and Lastarria volcanic complexes, Central Andes, Northern Chile. *Geoheritage* **2011**, *3*, 299–315. [CrossRef]
18. Ruban, D.A.; Tiess, G.; Sallam, E.S.; Ponedelnik, A.A.; Yashalova, N.N. Combined mineral and geoheritage resources related to kaolin, phosphate, and cement production in Egypt: Conceptualization, assessment, and policy implications. *Sustain. Environ. Res.* **2018**, *28*, 454–461. [CrossRef]
19. Marescotti, P.; Brancucci, G.; Sasso, G.; Solimano, M.; Marin, V.; Muzio, C.; Salmona, P. Geoheritage values and environmental issues of derelict mines: Examples from the sulfide mines of Gromolo and Petronio valleys (Eastern Liguria, Italy). *Minerals* **2018**, *8*, 229. [CrossRef]
20. Preite, M. Il patrimonio minerario e i suoi valori: Perchè e per chi i siti minerari diventano patrimonio? *AIPAI* **2017**, *XI(17)*, 14–21.
21. Prosser, C.D. Communities, quarries and geoheritage—Making the connections. *Geoheritage* **2019**, *11*, 1277–1289. [CrossRef]
22. Mata-Perelló, J.; Carrión, P.; Molina, J.; Villas-Boas, R. Geomining heritage as a tool to promote the social development of rural communities. In *Geoheritage: Assessment, Protection, and Management*; Elsevier: Amsterdam, The Netherlands, 2018; pp. 167–177.
23. Alexandrowicz, Z.; Urban, J.; Miśkiewicz, K. Geological values of selected Polish properties of the UNESCO World Heritage List. *Geoheritage* **2009**, *1*, 43–52. [CrossRef]
24. Carrión, P.; Herrera, G. Proyecto RUMYS: Rutas Minerales y Sostenibilidad. In *Rutas Minerales en el Proyecto RUMYS. Un Factor Integral para el Desarrollo Sostenible de la Sociedad*; Escuela Superior Politecnica del Litoral: Guayaquil, Ecuador, 2009; pp. 7–17.
25. GEMINA. *Ligniti e Torbe dell'Italia Continentale: Indagini Geominerarie Effettuate nel Periodo 1958–1961 dalla Geomineraria Nazionale (GEMINA) di Roma*; Ilte: Torino, Italy, 1962.
26. Argenti, P.; Baldanza A, Chiaverini, R.; Gasperini, A.; Gentili, S. Miniere di lignite in Umbria. In *Quaderni del Laboratorio di Scienze della Terra, Pietre e Terre nel Lavoro Dell'uomo*; Mattioli, B., Ed.; Comune di Spoleto: Spoleto, Italy, 2006; pp. 2–3.
27. Trumbull, J.; Barnes, F.F. *Coal Fields of the United States*; Technical Report; US Geological Survey: Reston, VA, USA, 1960.
28. Bucci, F.; Santangelo, M.; Fongo, L.; Alvioli, M.; Cardinali, M.; Melelli, L.; Marchesini, I. A new digital lithological map of Italy at the 1: 100 000 scale for geomechanical modelling. *Earth Syst. Sci. Data* **2022**, *14*, 4129–4151. [CrossRef]
29. Della Seta, M.; Melelli, L.; Pambianchi, G. Relief, intermontane basins and civilization in the umbria-marche apennines: Origin and life by geological consent. In *Landscapes and Landforms of Italy*; Springer: Cham, Swizerland, 2017; pp. 317–326.
30. Barchi, M.R.; Beltrando, M. The Neogene-Quaternary evolution of the Northern Apennines: Crustal structure, style of deformation and seismicity. *J. Virtual Explor.* **2010**, *36*, 11. [CrossRef]
31. Barchi, M.R.; Ciaccio, M.G. Seismic images of an extensional basin, generated at the hangingwall of a low-angle normal fault: The case of the Sansepolcro basin (Central Italy). *Tectonophysics* **2009**, *479*, 285–293. [CrossRef]
32. Famiani, D.; Brunori, C.A.; Pizzimenti, L.; Cara, F.; Caciagli, M.; Melelli, L.; Mirabella, F.; Barchi, M.R. Geophysical reconstruction of buried geological features and site effects estimation of the Middle Valle Umbra basin (central Italy). *Eng. Geol.* **2020**, *269*, 105543. [CrossRef]
33. Melelli, L.; Pucci, S.; Saccucci, L.; Mirabella, F.; Pazzaglia, F.; Barchi, M. Morphotectonics of the Upper Tiber Valley (Northern Apennines, Italy) through quantitative analysis of drainage and landforms. *Rend. Lincei* **2014**, *25*, 129–138. [CrossRef]
34. Pucci, S.; Mirabella, F.; Pazzaglia, F.; Barchi, M.; Melelli, L.; Tuccimei, P.; Soligo, M.; Saccucci, L. Interaction between regional and local tectonic forcing along a complex Quaternary extensional basin: Upper Tiber Valley, Northern Apennines, Italy. *Quat. Sci. Rev.* **2014**, *102*, 111–132. [CrossRef]
35. Barchi, M.; Brozzetti, F.; Lavecchia, G. Analisi strutturale e geometrica dei bacini della media Valle del Tevere e della Valle Umbra. *Boll. Della Soc. Geol. Ital.* **1991**, *110*, 65–76.
36. Basilici, G. Sedimentary facies in an extensional and deep-lacustrine depositional system: The Pliocene Tiberino Basin, Central Italy. *Sediment. Geol.* **1997**, *109*, 73–94. [CrossRef]

37. Bucci, F.; Mirabella, F.; Santangelo, M.; Cardinali, M.; Guzzetti, F. Photo-geology of the Montefalco Quaternary Basin, Umbria, Central Italy. *J. Maps* **2016**, *12*, 314–322. [CrossRef]

38. Lotti, B. *Descrizione Geologica dell'Umbria*; Istituto Poligrafico e Zecca dello Stato: Rome, Italy, 1926.

39. ISPRA. Città Di Castello, Sheet 289. In *Carta Geologica d'Italia, 1:50,000*; ISPRA: Rome, Italy, 2021.

40. Baldanza, A.; Spirito, A. Dalla classificazione alla musealizzazione: Allestimento di un'esposizione di reperti vegetali fossili presso il "Centro Studi sul Quaternario"(CeSQ) di Sansepolcro. In *Museologia Scientifica Nuova Serie*; Associazione Nazionale Musei Scientifici: Florence, Italy, 2010; Volume 41, pp. 130–136.

41. Camporeale, G. Gli Etruschi: Storia e Civiltà; UTET: Torino, Italy, p. 648. Available online: https://www.utetlibri.it/libri/gli-etruschi-storia-e-civilta/ (accessed on 2 July 2023).

42. Cattuto, C.; Cencetti, C.; Fisauli, M.; Gregori, L. I bacini pleistocenici di Anghiari e Sansepolcro nell'alta valle del Tevere. *Il Quat.* **1995**, *8*, 119–128.

43. Bonali, F.L.; Russo, E.; Vitello, F.; Antoniou, V.; Marchese, F.; Fallati, L.; Bracchi, V.; Corti, N.; Savini, A.; Whitworth, M.; et al. How academics and the public experienced immersive virtual reality for geo-education. *Geosciences* **2021**, *12*, 9. [CrossRef]

44. Janeras, M.; Roca, J.; Gili, J.A.; Pedraza, O.; Magnusson, G.; Núñez-Andrés, M.A.; Franklin, K. Using Mixed Reality for the Visualization and Dissemination of Complex 3D Models in Geosciences—Application to the Montserrat Massif (Spain). *Geosciences* **2022**, *12*, 370. [CrossRef]

45. Melelli, L. "Perugia upside-down": A multimedia exhibition in Umbria (Central Italy) for improving geoheritage and geotourism in urban areas. *Resources* **2019**, *8*, 148. [CrossRef]

46. Luft, J.; Schiewe, J. Automatic content-based georeferencing of historical topographic maps. *Trans. GIS* **2021**, *25*, 2888–2906. [CrossRef]

47. Brandolini, P.; Cappadonia, C.; Luberti, G.M.; Donadio, C.; Stamatopoulos, L.; Di Maggio, C.; Faccini, F.; Stanislao, C.; Vergari, F.; Paliaga, G.; et al. Geomorphology of the Anthropocene in Mediterranean urban areas. *Prog. Phys. Geogr. Earth Environ.* **2020**, *44*, 461–494. [CrossRef]

48. Vergari, F.; Pica, A.; Brandolini, P.; Melelli, L.; Del Monte, M.; et al. Geomorphological classification of the landscape in urban areas. Hints from some study cases in Italy. *Rend. Online Della Soc. Geol. Ital.* **2022**, *57*, 33–39. [CrossRef]

49. Cayla, N.; Martin, S. Digital geovisualisation technologies applied to geoheritage management. In *Geoheritage: Assessment, Protection, and Management*; Elsevier: Amsterdam, The Netherlands, 2018; pp. 289–303.

50. Jiang, W.; Zhang, Y. Application of 3D visualization in landscape design teaching. *Int. J. Emerg. Technol. Learn.* **2019**, *14*, 53. [CrossRef]

51. Melelli, L.; Silvani, F.; Ercoli, M.; Pauselli, C.; Tosi, G.; Radicioni, F. Urban geology for the enhancement of the hypogean geosites: The Perugia underground (Central Italy). *Geoheritage* **2021**, *13*, 18. [CrossRef]

 *geosciences*

Article

Conservation and Education in Spanish Geoparks: Exploratory Analysis of Land Stewardship Experiences and Valuation Proposal through Outdoor Education

Emilio Iranzo-García [1,2,*], Katia Hueso-Kortekaas [3,4] and Ghaleb Fansa-Saleh [2]

[1] Cátedra de Participación Ciudadana y Paisajes Valencianos, Universitat de València, Valencia 46010, Spain
[2] Departament of Geography, Universitat de València, 46010 Valencia, Spain; ghaleb.fansa@uv.es
[3] Department of Mechanical Engineering, ICAI, Comillas Pontifical University, 28015 Madrid, Spain; khueso@comillas.edu
[4] IPAISAL (Institute of Saltscapes and Salt Heritage), 28450 Madrid, Spain
[*] Correspondence: emilio.iranzo-garcia@uv.es

Abstract: Advances in research on environmental problems and public awareness of them have led to renewed concern about the need to establish mechanisms and figures to protect and manage sites so that geoecological processes remain outside the dynamics of anthropic occupation. This research has been approached from an inductive and qualitative perspective based on case studies to examine the articulation of the Spanish geoparks, their dynamics, and the experiences of private valorization in them. Geological heritage is seen as a lever for the promotion of the territory. In all cases, although the geological–geomorphological–paleontological–environmental resources must be significant, this designation aims to enhance the value of all assets, both natural and cultural, conceiving the geopark not as a figure of environmental protection but as a "figure to promote local development". A total of 48 land stewardship initiatives were identified in 11 of the 15 Spanish geoparks. The most significant presence of initiatives was found in the geoparks of Catalonia, followed by the Lanzarote Geopark. No nature-based schools are located within geoparks, except for Wild Me in Central Catalonia. However, the presence of nature-based schools in biosphere reserves (BRs) seems to be more common. Framing alternative proposals, such as nature-based schools in these areas and using land stewardship in their operation, can become an opportunity to protect a region's geological and cultural heritage and improve local communities' quality of life through sustainable and responsible economic and tourism activities. Early-years education in the natural environment facilitates the acquisition of long-term pro-environmental skills, competencies, and behaviors that last into adulthood and act as multipliers for others.

Keywords: Spanish geoparks; land stewardship; outdoor education; forest kindergartens; territorial development

Citation: Iranzo-García, E.; Hueso-Kortekaas, K.; Fansa-Saleh, G. Conservation and Education in Spanish Geoparks: Exploratory Analysis of Land Stewardship Experiences and Valuation Proposal through Outdoor Education. *Geosciences* **2023**, *13*, 276. https://doi.org/10.3390/geosciences13090276

Academic Editors: Hara Drinia, Panagiotis Voudouris, Assimina Antonarakou and Jesus Martinez-Frias

Received: 16 August 2023
Revised: 7 September 2023
Accepted: 10 September 2023
Published: 13 September 2023

1. Introduction

Advances in research on environmental problems and public awareness of them have led to renewed concern about the need to establish mechanisms and figures to protect and manage sites so that geoecological processes remain outside the dynamics of anthropic occupation. Thus, the systematic declaration of protected areas has been consolidated as a strategy of environmental policy [1,2]. However, it should be remembered that nature conservation's origin does not lie in public action. On the contrary, the conservation movement arose thanks to the interest of naturalists. Concerned about biodiversity loss, they began to propose private conservation initiatives by acquiring land and developing the physical protection, management, and regeneration of habitats [3]. But this work required government support and legal regulation, which has materialized at different speeds depending on the state or region, with the establishment of categories of natural protected areas (hereinafter NPAs), such as parks, reserves, marine areas, natural monuments, or protected landscapes, all depending on the assets to be protected and the management objectives to be achieved.

Under the new light of environmental preservation and sustainable development, states, regions, and international organizations have promoted protection strategies through the delimitation of relevant enclaves to ensure the preservation of their geoecological and heritage values. It is understood that they are our natural capital and the basis of environmental sustainability as they are the sources of resources that provide (ecosystem) services that must be understood and unaltered (or resilient) in the most effective way possible. The protection and management of these sites depend on different international, national, or regional legal figures, which are only sometimes adequately coordinated with each other [4]. A new concept that is gaining ground, thanks to UNESCO initiatives, is that of the geopark. It corresponds to an international figure that has been integrated into the legislation of different countries, although its potential has yet to be popularized [5,6].

Since 2015, UNESCO, with the designation of UNESCO Global Geopark, has been recognizing outstanding and unique enclaves from the point of view of their geological structure as habitats of flora and fauna and framework of life of local communities. These sites constitute, together with biosphere reserves and World Heritage Sites, areas with exceptional qualities, whose values must be conserved both to ensure the maintenance of environmental processes and to constitute wealth for society. They bring recognition, contribute to the knowledge of the landscape and its functions, and promote dynamics for the socio-economic development of the sites [6–9]. The network consists of 177 geoparks distributed in 46 countries. The geological richness of Spain, its varied geomorphology, and the landscapes resulting from the anthropic management of these areas explain the recognition of 15 areas as UNESCO Global Geoparks.

The territories delimited as geoparks are not only areas of high geoecological value. They also include intervene (anthropize) landscapes among which, in addition to forest or agricultural resources, are also the communities that manage and exploit them. This is why different management strategies are needed according to the dynamics and character of each geopark, trying to coordinate the policies and legal protection figures linked to these areas. Although geoparks are not a legal figure of protection, the sites or elements within them must necessarily be protected at the local, regional, or state level [10]. In fact, in Spain, in some of the geopark, areas there are other natural protected areas (Figure 1). According to Law 42/2007, of 13 December 2007, on Natural Heritage and Biodiversity, all those natural areas formally designated as per the provisions of international conventions and agreements of which Spain is a party are considered protected areas by international instruments. This regards, in particular, the following: wetlands of international importance of the Ramsar Convention; natural sites on the UNESCO World Heritage List of the Convention Concerning the Protection of the World Cultural and Natural Heritage; protected areas of the Convention for the Protection of the Marine Environment of the Northeast Atlantic (OSPAR); Specially Protected Areas of Mediterranean Importance (SPAMI) of the Convention for the Protection of the Marine Environment and the Coastal Region of the Mediterranean; geoparks, declared by UNESCO; biosphere reserves, also declared by UNESCO; and the Council of Europe's Biogenetic Reserves.

Environmental, outdoor, and nature-based education have a long tradition within NPAs; it is also relevant in the dynamics of geoparks. These educational approaches, which have been gaining prominence in recent decades [11], seek places with a series of environmental values for its development, which makes geoparks an excellent setting for developing educational projects. Their didactic value must be added to the interest of their environmental, cultural, scenic, and recreational values. Although aspects of the relationship between NPAs and ecosystem services, including educational services, that they provide to society have been studied in detail [12,13], there are fewer studies on geoparks [14]. Furthermore, there needs to be more research on the role of citizen initiatives in the safeguarding of these areas. Participatory movements, in defense of specific collective interests, are a fundamental element of modern societies to facilitate the relationship between administrators and the initiatives administered and to promote good practice through effective governance [15]. Protecting and managing these enclaves must inevitably involve the participation of those who own land or live in it. Land stewardship has been shown to be an effective strategy for conserving the natural and cultural values of landscapes. To this end, it uses a series of mechanisms, such as voluntary land transfer agreements; direct acquisition of land ownership by private organizations; technical advice; economic incentives for the implementation of sustainable practices; and environmental education [16].

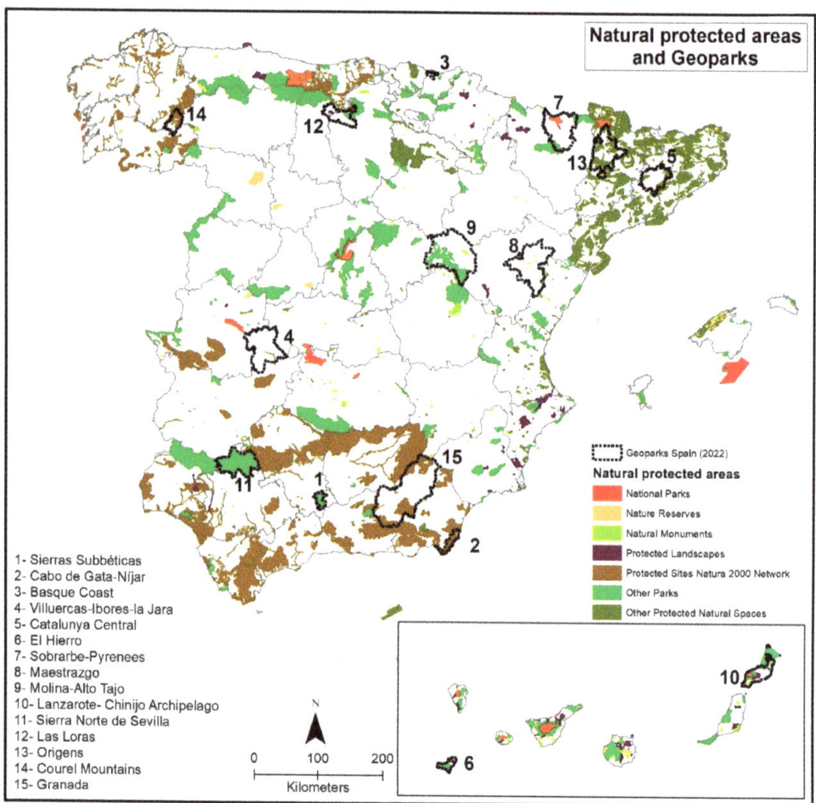

Figure 1. Natural protected areas and geoparks in Spain. Own elaboration.

This paper reflects on the potential of NPAs and geoparks as a framework for educational action, as well as on the role played by civil society in their management through the mechanism of land stewardship. Land stewardship is conceived as a tool for direct intervention, in which groups join forces to maintain specific sites with unique values [17–21]. One of the research questions raised in this paper is whether private conservation–management initiatives (land stewardship) are an attractive option for the management of geoparks. And another is whether, linked or not to land stewardship initiatives, nature-based education projects are identified as an opportunity in the framework of Spanish geoparks. The objectives of this paper are to present the explanatory factors of the articulation of Spanish geoparks and to explore whether there are land stewardship experiences in the field of geoparks in Spain, particularly experiences that are linked to nature education initiatives based on geoecological resources, which value the potential of these areas. As a secondary objective, the paper discusses the opportunities that a nature education project based on land stewardship can generate for the affirmation and projection of a geopark.

Based on case studies, the results are obtained through an exploratory qualitative research design. The most relevant dynamics that help understand the structure and functioning of the Spanish geoparks have been identified. The land stewardship initiatives implemented in their territories are studied and the state of stewardship in these enclaves have been assessed. Nature-based education experiences linked to protected areas are located to point out the possibilities in the interaction between geoparks–land stewardship–nature education.

2. Conceptual Framework

2.1. Spanish Geoparks

Geoparks are geographic areas defined with a three-fold purpose: the conservation of geological, geomorphological, and landscape heritage; scientific study and education; and the sustainable development of their territories. Geoparks seek to conserve geological formations and unique landscapes. But they also promote knowledge and understanding of geology, environmental processes, and human uses. They offer opportunities to learn about a territory's natural and cultural history. In addition, through good sustainable management of their resources, the promotion of educational/cultural activities, product development, and the provision of services can become an engine for local development [5,22,23].

The Spanish geography has four geological domains (Iberian Massif, Alpine Chains, Cenozoic Basins, Volcanic) at the base of a complex and unique geomorphology. Its landscapes are a magnificent territorial resource to promote the reconnection of people with nature, to generate identity and a sense of belonging, and of course, to promote sustainable local development. It is no coincidence that Spain is in second place worldwide, after China, in the ranking of countries with the most geoparks recognized by UNESCO and in first place among European countries [6,24]. The valorization of sites for their geological heritage started in Spain during the first third of the 20th century. It was always linked to the conservation of nature and the scenic values of their landscapes, and not so much for strictly scientific criteria. Only in the last twenty-five years have geosites been identified thanks to participation in international projects [25], and geoconservation proposals have been promoted, mainly thanks to the protection of Law 42/2007 on Natural Heritage and Biodiversity [22].

The origins of the European and global networks for the safeguarding of geological heritage can be found in Spain. The European Geoparks Network (EGN) was created in 2000 by four geoparks from France, Germany, Spain, and Greece, aiming to protect the geological heritage and promote sustainable development in their territories. The EGN Charter remains the basic document that inspires the functioning and development of the European Geoparks Network. In 2001, the network signed an agreement with UNESCO, and in 2004, the Global Geoparks Network (GGN) was founded to implement actions and establish quality standards for the territories involved. The European Geoparks Network was recognized as an official branch of the UNESCO Global Geoparks Network in 2005. In 2014, the GGN became a non-profit organization subject to French law. All in all, it aims to generate and consolidate regional structures, facilitate the exchange of experiences, and develop joint initiatives and projects. The European Geoparks Network is the Regional Network of the GGN and follows its statutes.

Geoparks seek to balance conservation and education with the region's economic development, working closely with local communities to develop sustainable tourism and business activities. In Spain, the 15 UNESCO Geoparks in order of declaration date and province in which they are located are as follows: 1, Sierras Subbéticas (Córdoba), 2006; 2, Cabo de Gata-Níjar (Almería), 2006; 3, Basque Coast (Guipúzcoa), 2011; 4, Villuercas-Ibores-la Jara, (Cáceres), 2011; 5, Catalunya Central (Barcelona), 2012; 6, El Hierro (Tenerife), 2014; 7, Sobrarbe-Pyrenees (Huesca), 2015; 8, Maestrazgo (Teruel), 2015 (although in 2017, it was excluded from the World Network, rejoining in 2020); 9, Molina-Alto Tajo (Guadalajara),

2015; 10—Lanzarote-Chinijo Archipelago (Las Palmas), 2015; 11, Sierra Norte de Sevilla (Seville), 2015; 12, Las Loras (Palencia-Burgos), 2017; 13, Origens (Lérida), 2018; 14, Courel Mountains (Lugo), 2019; and 15, Granada (Granada), 2020.

Geoparks in Spain can be classified according to their geological and landscape characteristics. For example, Cabo de Gata-Níjar, Lanzarote-Chinijo Archipelago, and El Hierro stand out for their coastal character and volcanic geology. The Basque Coast Geopark is also coastal in nature, although geologically, it is characterized by its sedimentary rocks (flysch) and karst processes. Other geoparks in which karst is a protagonist, although located in the interior of the peninsula, are the Sierras Subbéticas, the Loras, or the Maestrazgo with outcrops of Paleozoic rocks, especially rocks belonging to the Mesozoic and Cenozoic eras. The Sobrarbe-Pyrenees Geopark presents an alpine landscape, with glacial and periglacial formations; meanwhile, Origens showcases Pyrenean landscapes of carbonate rocks. And in the Courel Mountains, mountain landscapes with various rock formations stand out: sedimentary rocks, such as slate, quartzite, and sandstone, and igneous rocks, such as granite and rhyolite.

2.2. Private Management Initiatives: Land Stewardship in Spain: Theoretical and Legal Framework

2.2.1. Land Stewardship as a Land Conservation and Private Management Initiative

Consolidating non-regulatory territorial distinctions, legal instruments, and protection figures for enclaves with environmental or heritage values originate in private conservation initiatives [19,26,27]. Although, at present, it is the areas regulated under some legal protection figure that focus the attention of administrations and public opinion, some enclaves are being managed by civil society through complementary conservation mechanisms and strategies, which are successfully contributing to the safeguarding of natural and cultural heritage [28,29]. Public administrations only sometimes have the necessary economic or human resources to address biodiversity and heritage conservation in its entirety [28]. This is when the sense of complementary mechanisms, such as land stewardship, becomes clear.

Land stewardship is defined as the responsible care and management of land to maintain its environmental and cultural values and long-term productivity. Its practice involves the participation of civil society, organized in collectives or entities, in managing land, cultural assets, and landscapes sustainably [19,20], while always considering the environmental, social, and economic impacts of their actions. Land stewardship is a strategic instrument of territorial and environmental management, aimed at conserving biodiversity, enhancing the landscape, and safeguarding natural and cultural heritage [19,20,30]. It is a mechanism through which civil society can become involved in the protection of the environment, collaborating with public action [16] in what has been called shared governance of resources and territory, involving "power" with the administrations [31].

Its practice has a long tradition, especially among Anglo-Saxon countries. It is a strategy that was applied even before some of the state or regional protection instruments and figures appearing in the 20th century. It should be noted that, at the beginning, the promotors of land stewardship did not recognize themselves in their safeguarding work [32]. It was at the end of the 19th century when the first land stewardship entity for safeguarding lands of high ecological interest was founded in the United States. The Trustees of Reservations emerged as the first non-profit organization aimed at the preservation and conservation of enclaves and landscapes in Massachusetts, and soon after, the initiative spread to other countries and continents [32,33]. The first initiatives emerged in Europe in the United Kingdom and the Netherlands. In the United Kingdom, the role played by the National Trust was and is fundamental in stewardship strategies. Currently, public–private collaboration in conservation is very important. In France, the public administration has promoted the implementation of land stewardship strategies. The work of the French Conservatoire du Littoral has been fundamental. In addition, there are also examples of private conservation initiatives, such as the Conservatoires d'Espaces Naturels. Land stewardship experience in Latin American countries is trying to be boosted by the growing number of NGOs involved in private conservation. However, the legal framework needs to be further developed to enable the action of conservation entities [28].

In each location, the terms of agreement have been adjusted to their legal conditions, but the original spirit has been maintained. In Spain, except for isolated experiences, land stewardship took root in the first decade of the 21st century [34]. Land stewardship was introduced in Spain in 1999 [35] and began in Montesquiu, Catalonia [36]. Nature protection and management tasks have fallen mainly on the administrations. Legislation has been passed, and legal and technical instruments have been developed. However, public action presents economic and human limitations to achieve the objectives of resource and habitat conservation. It is for this reason that the participation of citizen groups is necessary in the challenge of safeguarding the environmental and heritage values of the territory [37]. In this context, land stewardship emerges as a valuable and timely tool. Some authors define it as a strategy involving landowners and land users (whether public or private) in conserving habitats and landscapes [38]. Through stewardship, conservation practices are expanded beyond conventional legal agreements.

Land stewardship is implemented through the signing of voluntary agreements or contracts between one or more landowners and one or more stewardship entities. The purpose of this is to carry out conservation projects and actions, to sustainably manage the land and its ecosystem services, and to enhance the value of resources, preferably in places with sensitive habitats, unique landscapes, or heritage sites. Landowners can implement these projects with different levels of involvement: from an indefinite cession of the land to co-participation in the management work with a long-term commitment [39–41]. Some stewardship organizations use revolving funds to acquire land and resell it to other landowners with clauses so that the latter can only use the land in accordance with conservationist principles. In this way, they reinvest, giving continuity to the protection cycle [40].

There is no single method of understanding and implementing land stewardship globally, especially in complex territories [41,42]. Often it is the entities (associations, cooperatives, NGOs, foundations) that directly manage lands ceded by their owners, carrying out safeguarding projects in them that can present different formats according to the entities' objectives. When landowners are directly involved in stewardship initiatives, it is either because they have decided to be part of an agreement, or because they have acquired the property with an agreement already signed. In both cases, the landowners' motivations are similar: environmental sensitivity and a sense of belonging to the land, as well as access to technical support for management and even financial or fiscal incentives obtained in some states or regions by participating in this mechanism [43]. There is research that indicates that satisfaction predominates among landowners who have signed stewardship agreements. However, when they transfer the land to new landowners who must continue the agreement, their satisfaction decreases and may compromise the achievements made [41].

2.2.2. Legal Framework of Land Stewardship in Spain

In Spain there is legislative support for land stewardship mechanisms. Law 42/2007 of 13 December 2007, on Natural Heritage and Biodiversity, calls for the promotion of voluntary agreements between entities concerned with the conservation of environmental, landscape, and cultural values with landowners and owners of cultural assets located on them, within areas declared as specially protected. Thus, this law is committed to incorporating agreements between public and private agents to safeguard natural areas under protection [28]. However, stewardship is not a strategy that is limited exclusively to officially protected sites.

There are numerous projects by private entities aimed at maintaining the ecosystem services of a site, which have been initiated regardless of its level of administrative protection. The spirit of land stewardship is not to replace but to supplement public action and its conservation mechanisms by civil society. The Spanish conservation groups and associations that opted for land stewardship to contribute to safeguarding natural, cultural, and landscape heritage often did not have the technical and economic capacity to achieve the desired objectives. For this reason, groups of entities have been formed to connect, advise, and make them visible to strengthen their collaborative conservation work through more solid structures [44]. In addition, there are forms of cooperation between public administrations and groups through grants and agreements. After the Montesquiu Declaration (2000), a document that formalized the concept and movement of land stewardship in Spain, networks and platforms of stewardship organizations began to be established at regional and supra-regional scales [19]. Examples include the Xarxa de Custòdia del Territori in Catalonia, Avinença in the Valencian Community, the Rede Galega de Custodia do Territorio in Galicia, and the Red Transcantábrica de Custodia del Territorio formed by individuals, companies, organizations, and foundations from Asturias, Cantabria, the Basque Country, and the mountainous north of Castilla y León.

The increase in the number of groups oriented to stewardship and the consolidation of networks of organizations at the regional level led to the creation in 2007 of the Platform for Land Stewardship at the Spanish state level. Its objective as a platform is to support organizations in their projects and to energize the stewardship movement. This platform has been promoted by the Biodiversity Foundation (Ministry for Ecological Transition and Demographic Challenge), founded in 1998 to contribute to the protection of the natural heritage of Spain. In addition, in 2011 the Forum of Networks and Entities of Land Stewardship (FRECT, in its Spanish acronym) was created, a representative entity of the collective of land stewardship entities in Spain whose purpose is to promote stewardship at the institutional, legal, and social level and to ensure its incorporation in land management.

2.3. Nature-Based Education

Everyday use of natural heritage in general, and geological heritage in particular, is educational. This is a sustainable activity because it is based on its intangible value, which does not impact this resource, its educational value, and capacity to raise awareness of its long-term protection. Although there is a long tradition of using natural areas as a setting for environmental education, another educational model stands out as a more recent appearance on a global scale, called nature-based education. It consists of "learning based on regular, direct and permanent contact with the natural environment", which takes advantage of the territory and the resources it offers, always based on respectful relationships between people and the environment [45]. Unlike environmental education applied in informal and non-formal contexts, nature-based education is comparable to formal education, i.e., education that is offered in schools. It occurs in a stable group of children (generally boys and girls in the pre- or primary school stages) who regularly remain in the natural environment, constituting their site of reference and developing their learning daily. The activity, in this educational model, consists fundamentally of (free) play in nature and with the elements offered by nature itself. However, occasionally, there may be few teaching proposals, also with generally natural materials. Given its affinity for formal education, the accompaniment of students, the facilitation of educational processes, the monitoring of children's development and well-being, the recording of evidence of the acquisition of competencies and skills, and tutorial action are essential. All these aspects are very relevant when it comes to understanding the depth and transcendence of the bond children acquire with the natural environment and the consequences that this entails in the short, medium, and long term.

For two reasons, children engaged with regular nature-based education sessions tend to present pro-environmental attitudes in adulthood. On the one hand, because the significant experiences that occur in the natural environment have a tremendous evocative power both in the short and long term, the intense sensations that arise during these experiences generate a strong emotional bond with the place where they have taken place and incite an intimate and genuine desire to protect it. On the other hand, the permanence in nature, i.e., a vast, open, diverse, and changing environment, generally friendly but sometimes hostile, forces the development of autonomy, tolerance, resilience, empathy, and leadership skills, which are more challenging to attain when the educational experience takes place within the confines of a classroom. These skills, in adulthood, can be used to promote activism in favor of the natural environment, as has been repeatedly demonstrated, even in the case of prominent activists who today practice land stewardship [46,47].

There are multiple benefits to staying regularly in nature. In the educational field, it is worth mentioning the acquisition of skills and competencies, such as those mentioned above, as well as others of a social and emotional nature that are so highly valued in today's liquid society, as described by Bauman [48]. Learning in the natural environment occurs on site and is hands-on, with the materials and scenarios in synchrony with the cognitive processes that lead to it. In addition, nature gives countless benefits for the physical and mental health and well-being of children and their educators [13]. Within the natural elements that offer opportunities for learning and well-being, geological heritage constitutes a relevant asset. It is the substrate that gives coherence and identity to the landscape and the living beings that inhabit it, supports the activities that take place on the ground, provides spaces of mystery, refuge, and shelter from inclement weather, and gives rise to countless questions of a philosophical and transcendent nature even among the youngest children ("who put those mountains there?", "how are the stones held together?", pers. obs.). Some geomorphological elements also have incentives for developing fantasy (for their peculiar shapes, textures, and colors), psychomotor skills (to climb them), or aesthetic appreciation. One of the great treasures that children collect in nature and awaken their interest in science is stones and minerals. Stones in general are also excellent companions for games and experimentation, with which they can create characters, build, or even acquire curricular skills in the field of science (arithmetic, geometry, weight, volume) or the arts, either by painting (e.g., storytelling stones) or creating installations with them (mandalas, land art, stacks).

This type of nature-based school has experienced a major global boom in the last decade, with various variations. These include three main models: nature schools, where pupils spend a large part of the day outdoors (nature-based schools sensu stricto); regular outdoor sessions for pupils in mainstream schools; and blended learning systems, where pupils spend part of their time in mainstream schools and part in nature schools. Although it is difficult to provide accurate figures for the number of such schools in Europe, it is estimated that there are about more than 3000 (data from the International Congress of Children's Forest Schools, held in Prague in 2017), with there being about 60 in Spain [49].

Thanks to the regular permanence in the natural environment, nature-based schools maintain a very intense relationship between the educational experience and the space they inhabit. The choice of location is not trivial. Their promotors must strike a balance between offering a natural environment as varied and diverse as possible and one that is located in a practical and accessible place for families who must take their children there daily. On the other hand, they must comply with current regulations on the use and occupation of land (authorized activities in rural areas, requirements for educational settings), safety, compatibility of uses with other activities (hunting, livestock raising, forestry), and education (facilities, curriculum). Given the novelty of the approach, the legislation regulating the creation of schools does not usually facilitate their existence outside urban centers. It requires facilities that will be underutilized (since most of the time, they are outdoors) and obliges teaching in a way that makes adapting to the outdoors difficult. In Spain, most nature-based schools are forced to look for alternative accreditation models to offer contact with nature to children within the framework of quality education [50].

Once the site where the project is to be located has been chosen, it is necessary to agree on the terms of use with the landowners. The most common types of agreements are rental, cession, or simple occupation, in the case of public land. Educational initiatives based on land stewardship agreements are yet to be common, despite the opportunities they can offer, and agreements tend to be short term and without much further commitment to conservation. In the first case, renting generally implies a very high cost for the schools because not only do they need to have the land almost exclusively at their disposal (given the incompatibility with most other uses), but it is also necessary to have a built site that acts as a shelter from inclement weather and as an educational facility as required by the regulations. The legal requirements make it very expensive to rent and maintain the outdoor space, as well as indoor facilities. For this reason, many schools choose to enter into a lease agreement with a public entity (usually a municipality) or a private individual (e.g., former schools, in exchange for minimal maintenance) or landowner (which, for example, they perceive as unproductive). These are informal agreements with some uncertainty in their implementation, conditions, and duration. In the case of the occupation of communal, public, and similar woodlands, the projects are designed without the setup of infrastructures, so the educational project must look for a nearby location to use as a shelter and school.

Land stewardship combines the advantages of some of these options. Establishing a long-term agreement allows for continuity in exchange for a symbolic cost and collaboration in maintaining the ceded or rented property. For the owner, the transferred asset maintains or even increases its value in terms of both the land and the built property it includes. Not many educational projects in nature have availed themselves of this option, although some are already exploring it. This paper presents the case of the Edunat Cooperative in Bunyola, Mallorca (Balearic Islands) as an example of the advantages that a land stewardship case can offer to the agents involved in a nature-based educational project.

3. Materials and Methods

The research was approached from an inductive and qualitative perspective based on case studies to examine the articulation of Spanish geoparks, their dynamics, and the experiences of private valorization in them. It used exploratory analysis to identify conservation, educational, or recreational initiatives not directly linked to public administration and to understand the contextual conditions that are the basis of these experiences.

The identification of experiences in geoparks was carried out by exploiting different sources of information. Firstly, the Land Stewardship Platform of the Biodiversity Foundation (Ministry for Ecological Transition and Demographic Challenge) database was consulted. Secondly, various online thematic directories on nature education were consulted, contrasting the information, when necessary, with the entities identified. Thirdly, interviews were carried out with crucial territorial stakeholders and managers–directors of nature education centers.

Managers and technicians of the geoparks of Villuercas-Ibores-La Jara, Maestrazgo, Granada, Courel Mountains, Origens, Las Loras, and Sobrarbe-Pyrenees were interviewed. Additional interviews were carried out with managers and directors of nature-based schools and members of land stewardship organizations operating within the framework of the geoparks. Sixteen interviews were carried out, ten with geopark managers, four with nature-based school managers, and two with members of land stewardship organizations. The interviews, which lasted approximately 45 min, were recorded and transcribed for processing using CAQDAS (computer-aided qualitative data analysis).

Given the exploratory nature of the research, the interviews that were carried out were of focused or centered types, which allows, with the information obtained, the explanatory factors of the dynamics of the valorization of the resources of the geoparks to be pointed out. They also helped raise new questions and perceptions about the role of these enclaves [51,52].

The information obtained from the interviews were codified by establishing a codification and categorization [53] necessary to analyze the nature of the experiences and the motivations of their promoters. Atlas Ti software (https://atlasti.com/ (accessed on 22 June 2023)) was used for this purpose. To increase the reliability of this exploratory analysis, a triangulation process was finally carried out with experiences identified in well-established natural protected areas, such as national parks or natural parks (regional-scale protection). Part of this analysis was carried out through secondary data (online information from the services of the NPAs, land stewardship entities, and tourist offices). To determine the spatial relationships between geoparks, natural protected areas, biosphere reserves, nature-based schools, and land stewardship projects, spatial overlap analysis was carried out using a geographic information system (GIS). To this end, the necessary layers were obtained or created. Likewise, synthesis cartography was made with the spatial analysis results.

4. Results

4.1. Structural and Alternative Processes for the Dynamization and Territorial Development in Spanish Geoparks: From Conservationism to Participative Management

Analysis of the development dynamics generated in the geoparks was based on the perception of the crucial territorial stakeholders interviewed. One of the ideas expressed is that geoparks arise as initiatives to make geographic areas with unique geological characteristics visible and to promote them socio-economically. Several studies point to the idea that the ultimate purpose of the geopark is the socio-economic revitalization of the reference area [22–24]. This idea was confirmed after the interviews. Based on geological heritage, they are more a figure of territorial socio-economic promotion than of nature conservation. Although each geopark is forged uniquely, the local–rural development groups play a relevant role in its start-up, collaborating with public administrations of scales close to the citizen.

Geological heritage is seen as a lever for the promotion of the territory. In all cases, although the geological–geomorphological–paleontological–environmental resources must be significant, this designation aims to enhance the value of all assets, both natural and cultural, conceiving the geopark not as a figure of environmental protection but as a "figure to promote local development". The conservation of nature is an important issue. However, in this context, a more dynamic and progressive safeguarding approach is chosen to address environmental challenges effectively, with solutions based on the responsible use of territorial resources [23].

Geoparks behave as participatory management entities with an open and integrative structure and a bottom–up approach, in which management does not necessarily depend on a public administration. The role of public administrations in the functioning of the entity varies depending on the geopark, given the differences between them (some geoparks only host 3 municipalities, with a surface area of 578 km^2 and a population of 5406 inhabitants, such as Courel Mountains, while others host 47 municipalities, with a surface area of 4722 km^2 and a population of 100,000 inhabitants, such as Granada). Many of them are taking advantage of pre-existing structures and entities (associations of municipalities, local action groups, local or regional associations of different types). They are committed to joining business associations so that they have the geoparks as a frame of reference. Scientific activity in the geoparks is another of their pillars. Therefore, there is a scientific committee in all of them that proposes research and conservation actions. This is where universities, research centers, scientific societies, and NGOs are involved in the functioning of the geopark. However, conservation is not the primary purpose of geoparks. It is an alternative figure to the NPAs included in European directives and state or regional legislation.

The local territorial stakeholders have observed the figure of the geopark as an alternative to the problems derived from the crisis of the rural environment or the mountain areas. The stakeholders interviewed point out that the geopark has been the most accepted among the possible ways for territorial promotion since it is not based on a restrictive regulation with the activities and uses typical of rural areas. The incorporation of use restrictions to those already existing, and sometimes controversial for the inhabitants of these areas, would imply social disagreement and the proposal's failure. However, the declaration of geoparks, after territorial valorization and acceptance by the population, requires a significant amount of previous, careful, and sensitive work on the part of their promotors. Explaining the opportunities derived from the declaration honestly and convincingly to avoid internal resistance is essential. The geopark cannot be presented as a panacea for the structural problems of a disadvantaged rural area (aging, depopulation, lack of facilities and services). However, it can be presented as a vector for the generation of new businesses, the maintenance of some existing ones, the creation of some quality jobs, etc.

For the management of these geoparks, participatory strategies are proposed for three or four years, which are implemented through annual action plans. The entities that comprise the Geopark Council or Executive Committee seek funding to develop projects and activities using European, regional, or local funds. One of the premises is to take advantage of synergies and only generate a little public expenditure than is strictly necessary. The strategies and lines of action revolve around recurring themes: conservation, research, tourism, environmental education, entrepreneurship, and local empowerment.

Many of the projects proposed revolve around tourism since it is a phenomenon that has a territorial, social, and economic dimension. Tourism becomes an economic engine in the municipalities of the geoparks, generating income for existing businesses (not only in the tourism sector) and encouraging the emergence of new ones. Without entrepreneurs and businesses, generating dynamism and settling inhabitants are difficult. The arrival of visitors who value the resources of the geopark is increasing, which acts as a revulsive, not only economically but also in terms of identity. It generates pride and a sense of belonging that empowers the rural community, which is historically undervalued. Residents have begun to internalize that they live in a territory recognized by UNESCO, which, through these senses of pride and empowerment, serves as an incentive to promote the development of other initiatives. The geopark acts as a catalyst for the territory, a dynamization tool manifested in creating more associations and companies. They have succeeded in initiating structuring processes in communities that were previously less cohesive.

However, the research shows that the involvement of civil society in the dynamics of the geopark is still a challenge. Local communities still need to be better organized and the low population density and especially its aging do not help to involve the residents in the structure and daily management of the area. Some avenues, such as volunteer activities, work camps, and even land stewardship initiatives, are being explored.

In the perception of the geopark managers, analyzed in more detail, the land stewardship mechanism is underutilized, despite its potential. There is agreement that it has a long way to go and needs to be promoted within geoparks for three reasons: 1. The local community is organized around a project, increasing participation and social cohesion. Alternatively, in the worst case, people from outside the territory become involved in the development of initiatives, bringing new human capital to the area; 2. These initiatives are self-managed without needing, a priori, the impulse of the public administration; 3. Abandoned land is recovered for conservation or regeneration, and the landscape and cultural heritage are enhanced.

Land stewardship is presented as an opportunity to support participatory management, considered in the guiding principles of the geoparks. In some cases, such as the Geopark of Granada, although land stewardship is currently barely present, it will be addressed within the framework of the Tourism Sustainability Plan. Funding will be provided for a project whose aims are as follows: to identify plots of land for the implementation of land stewardship initiatives; to establish the most appropriate type of stewardship project in each area; and to identify groups–entities interested in implementing projects to safeguard the environmental and cultural values of the geopark. Other geoparks, such as those in Catalonia, the Canary Islands, or Aragón, already have stewardship experiences. Finally, some geoparks, such as Las Loras or the Basque Coast, do not present stewardship initiatives as such for the moment, but see their development as valuable and are already implementing projects close to land stewardship approaches. For example, in the Loras, there is an agroecological project of the geopark that works with local farmers and ranchers to enhance the value of local organic products. The aim is to develop more sustainable production, distribution, and consumption models while recovering land abandoned by landowners.

However, one of the problems observed is the need for more communication between land stewardship organizations and geopark management bodies. The study has verified how there are stewardship projects within a geopark of which the Geopark management is unaware.

4.2. Identification of Land Stewardship Experiences in Spanish Geoparks and Nature Education: An Opportunity to Be Developed

4.2.1. Inventory of Stewardship Experiences in Spanish Geoparks

The inventory of experiences was carried out by consulting national and regional databases on land stewardship and the information provided by local stakeholders. To increase the reliability of this exploratory analysis, a triangulation process was carried out with experiences identified in consolidated natural protected areas, such as national parks or natural parks (regional-scale protection instruments). Part of this analysis was carried out using secondary data (online information from the services of the NPAs, land stewardship organizations, and tourist offices).

Table 1 shows 48 land stewardship initiatives identified in 11 of the 15 Spanish geoparks (Figure 2). The most significant presence of initiatives is found in the geoparks of Catalonia, which is explained by the following: the tradition of this tool in the region, with a well-established regional stewardship network; the social cohesion existing in Catalonia; and the strong roots of citizens in the territory. The Lanzarote Geopark, with seven initiatives, also stands out, although the conservationist motivation linked to other existing protection figures prevails over the question of identity.

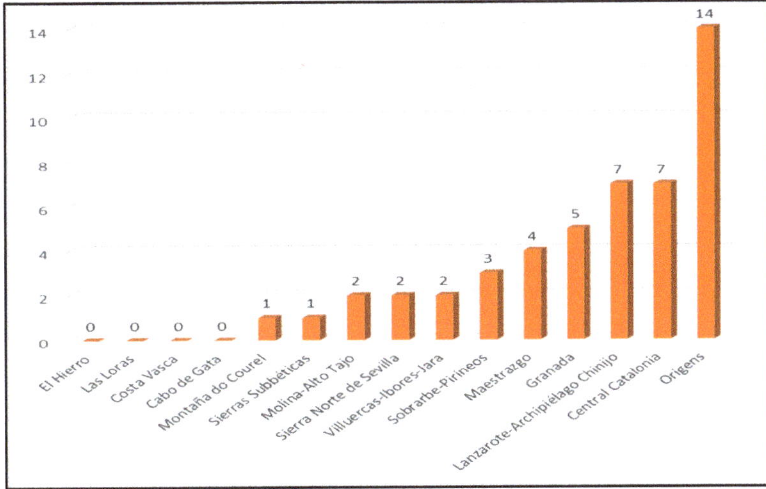

Figure 2. Number of land stewardship agreements per geopark in Spain.

In the rest of the geoparks, the number of initiatives is reduced, and even in Cabo de Gata, Costa Vasca, Las Loras, and El Hierro, no projects have been detected within the framework of the stewardship mechanism.

The 48 land stewardship initiatives depend on 23 stewardship entities, both public (municipalities) and private (associations, foundations, financial institutions, etc.). As shown in Figure 3, private entities, in the form of civil associations, are the ones that have developed or are developing more projects in the Spanish geoparks, followed by foundations and local administrations. This distribution of initiatives according to the type of entity is very similar to that which occurs in Spain as a whole [54].

Table 1. Inventory of stewardship experiences in Spanish geoparks.

Geopark	Name of the Initiative	Municipality	Entity	Goal of the Initiative
Catalunya Central	Coaner	Sant Mateu de Bages	Minyons Escoltes i Guies de Catalunya	Sustainable use of resources
	Camí de la Gavarresa	Avinyó	Associació Hàbitats-Projecte Rius	Habitat restoration and recovery
	Bosc de Cal Cuques	Manresa	Fundació Catalunya la Pedrera	Habitat restoration and recovery
	Custodia fluvial riu Cardener	Manresa	Ayuntamiento de Manresa	Ecological processes
	Aiguamolls de la Bóbila de Santpedor	Santpedor	Obra Social Caixa Catalunya	Habitat conservation
	Bosc de Mas Lluca	Santpedor	Fundació Catalunya la Pedrera	Habitat restoration and recovery
	Món Sant Benet	Sant Fruitòs del Bagés	Fundació Catalunya la Pedrera	Sustainable use of resources
Origens	APC Gavet de la Conca	Gavet de la Conca	Associació Amics Natura 2000 Pirineu	Favouring specific species
	Els Plans d'Aguiró	Aguiró	Associació Marques de Pastor	Conservation of traditional land management
	Tremoluga de Naens—Foradada 1	Naens (Senterada)	Associació Marques de Pastor	Habitat conservation
	Tremoluga de Naens—Foradada 2	Naens (Senterada)	Associació Marques de Pastor	Habitat conservation
	Tremoluga de Naens—Foradada 3	Naens (Senterada)	Associació Marques de Pastor	Habitat conservation
	PAS del Codó	Senterada	Associació Trenca	Favouring specific species
	Les Feixes	La Pobla de Segur	Associació Amics Natura 2000 Pirineu	Ecological processes
	Seixos	Talarn	Ajuntament de Talarn	Ecological processes
	Vedat	Talarn	Ajuntament de Talarn	Ecological processes
	Roqueres 1	Talarn	Ajuntament de Talarn	Ecological processes
	Roqueres 2	Talarn	Ajuntament de Talarn	Ecological processes
	Farratge	Talarn	Ajuntament de Talarn	Ecological processes
	Siall	Isona	Associació Trenca	Ecological processes
	Congost de Mont-rebei	Sant Esteve de la Sarga	Fundació Catalunya la Pedrera	Conservation of traditional land management

Table 1. *Cont.*

Geopark	Name of the Initiative	Municipality	Entity	Goal of the Initiative
Granada	Coto de caza De Castril	Castril	Fundación GYPAETUS	Promotion of the sustainable use of resources
	Viveros Ponce Lajara	Galera	Asociación para la Custodia del Territorio y el Desarrollo Sostenible ACUDE	Conservation of traditional land management
	Los Isidoros	Castillejar	Asociación para la Custodia del Territorio y el Desarrollo Sostenible ACUDE	Conservation of traditional land management
	GRA-1	Gorafe	Fundación de Amigos del Águila Imperial, Lince Ibérico y Espacios Nat de Carácter Privado	Preservation of fauna
	Olivares Vicos +	Guadix	Seo Birdlife	Habitat restoration and recovery
Montaña do Courel	Microrreservas da Serra do Courel	Folgoso do Courel	Asociación Galega de Custodia do Territori	Preservation of flora
Sobrarbe-Pirineos	Estación Biologica Mte. Perdido (EBMP)	Bielsa	Fundación para la Conservación del Quebrantahuesos	Preservation of fauna
	Ayuntamiento de Ainsa-Sobrarbe	Ainsa	Fundación para la Conservación del Quebrantahuesos	Sustainable use of resources
	Refugio Natural de la Peña Montañesa	Pueyo de Araguas	Fundación para la Conservación del Quebrantahuesos	Conservation of traditional land management
Maestrazgo	Resforestación de zonas afectadas por incendio	Ejulve	Asociación para el Desarrollo del Maestrazgo	Habitat restoration and recovery
	Guardianes del territorio	Molinos	Asociación para el Desarrollo del Maestrazgo	Habitat conservation
	Escuela de Actividades de la Naturaleza S. L.	Castellote	Asociación para el Desarrollo del Maestrazgo	Sustainable use of resources
	Mejora de hábitats y recursos turísticos con hides fotográficos.	Montoro de Mezquita	Asociación para el Desarrollo del Maestrazgo	Habitat restoration and recovery
Molina-Alto Tajo	La Huerta Rigüela	Valhermoso	Asociación Nacional MICORRIZA	Others
	La Sarga del Masegar	Pinilla de Molina	Asociación Nacional MICORRIZA	Others
Sierras Subbéticas	Las Quebradillas	Zuheros	Fundación Internacional para la Restauración de Ecosistemas	Conservation of traditional land management

Table 1. *Cont.*

Geopark	Name of the Initiative	Municipality	Entity	Goal of the Initiative
Sierra Norte de Sevilla	SEV-2	Alanís	Fundación de Amigos del Águila Imperial, Lince Ibérico y Espacios Nat de Carácter Privado	Preservation of fauna
	Almadén de la Plata	Almadén de la Plata	SEO/BirdLife, Sociedad Española de Ornitología	Favouring specific species
Villuercas-Ibores-Jara	Conservación integral de valores medioambientales	Navatrasierra	Sociedad Extremeña de Zoología	Preservation of flora
	CC-1	Alia	Fundación de Amigos del Águila Imperial, Lince Ibérico y Espacios Nat de Carácter Privado	Preservation of fauna
Lanzarote-Archipiélago Chinijo	Muladar en PEÑA HUMAR	Teguise	Asociación de defensa medio ambiental vientos del noreste	Favouring specific species
	La Cuestita	Teguise	Asociación Paisajes Atlánticos	Habitat restoration and recovery
	Surte Gorritz	Teguise	Asociación de defensa medio ambiental vientos del noreste	Favouring specific species
	Sembrando para hubaras	Teguise	Asociación de defensa medio ambiental vientos del noreste	Favouring specific species
	El Higueral	Teguise	Asociación Paisajes Atlánticos	Conservation and improvement of landscape
	Vivero Ferrovial	Arrecife	Asociación Ambiental Arrecife Natura	Preservation of flora
	Huerta Vieja	Tinajo	Asociación Cultural y Social Trib-Arte	Conservation of traditional land management

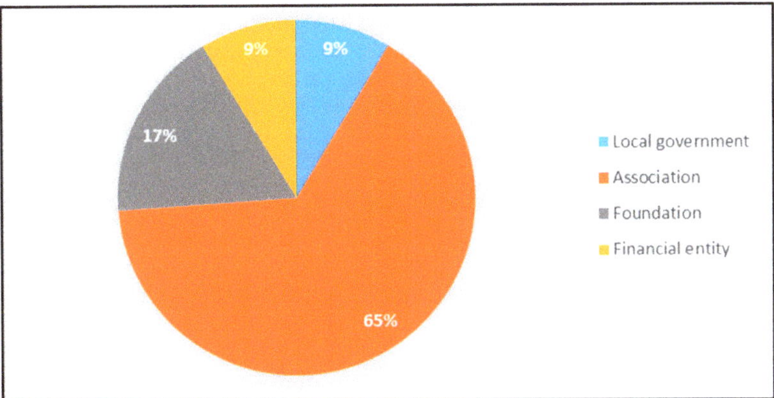

Figure 3. Distribution of agreements in geoparks per type of stewardship entities.

Regarding the purpose of the stewardship initiatives identified in the geoparks (Figure 4), a disparity is observed, although conservationist objectives predominate, such as those related to habitat conservation and restoration, the maintenance of ecological processes, the enhancement of specific species, and wildlife conservation. However, these purposes coexist with others beyond the most classic conservationism. Thus, initiatives linked to conserving traditional uses of the territory and promoting the sustainable use of territorial resources also exist.

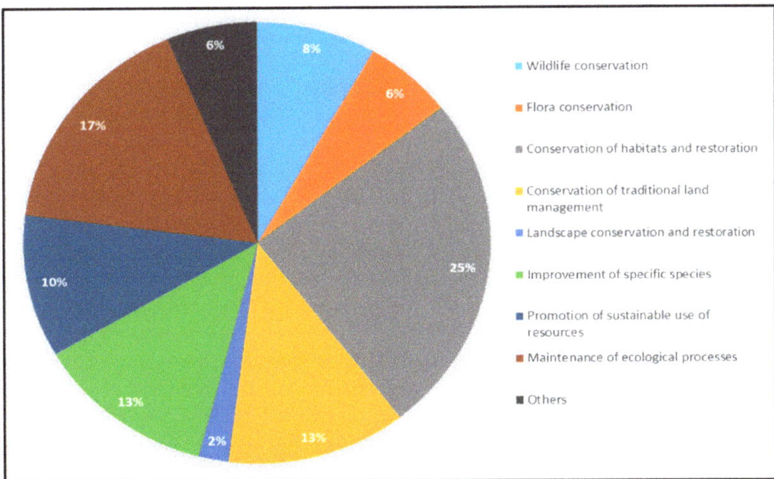

Figure 4. Goal of the stewardship initiatives identified in geoparks.

The land stewardship projects identified in the framework of geoparks do not originate from UNESCO but were initiated by stewardship organizations and landowners (sometimes also municipalities) according to conservationist objectives or to safeguard environmental/cultural values. They are not a direct instrument promoted by the managing entities of the geoparks, but they act as an indirect instrument that contributes to the proper functioning of the geoparks. However, the managers of some geoparks, such as Granada or Villuercas-Ibores-La Jara, are exploring the possibility of directly promoting land stewardship initiatives. Entities that carry out stewardship projects within the geoparks, such as the Foundation for the Conservation of the Bearded Vulture in the Sobrarbe-Pyrenees Geopark or the Spanish Ornithological Society SEO Birdlife in the Granada Geopark, act within them or have joined them as collaborating entities. However, they have yet to sign stewardship agreements directly with the managing bodies of the geoparks. Notwithstanding, the interest of the stewardship initiatives of these entities that have agreements with landowners or municipalities within the geoparks is that they involve the local society. They also mobilize human capital, knowledge, and external economic resources that positively impact geoparks. The associative fabric is improved with collaborations that increase social cohesion. And it contributes to both environmental management and local development.

In the present work, no stewardship projects directly related to safeguarding geological resources have been detected. As can be seen in Figure 4, there are no initiatives related to environmental education or nature education either. Although many of the stewardship proposals indeed incorporate strategies for the dissemination, interpretation of the natural environment, and awareness of the values they protect, these actions are secondary to the primary purpose and are carried out sporadically. Nature-based education, contrary to traditional environmental education, is an activity that requires full-time dedication and would be a protagonist in its execution, something that is not observed here.

4.2.2. Land Stewardship and Nature-Based Education

Figure 5 shows that no nature-based schools are located within geoparks, except for Wild Me in Central Catalonia. However, the presence of nature-based schools in biosphere reserves (BRs) seems to be more common, as is the case of *Amadahi* in the Mariñas Coruñesas e Terras do Mandeo BR, *Nenea* in the Terras do Miño BR, *Grupo de Juego en la Naturaleza Saltamontes* and *Aúlla* in the Cuencas Altas de los ríos Manzanares, Lozoya y Guadarrama BR, or *El Huerto de los Girasoles* and *Laboratorio Lululand* in the Intercontinental Reserve of the Mediterranean, to name a few examples. Interesting is the *Bosque Escola Avelãs* school in Marvão, Portugal, which acts as a transboundary initiative and is located in the international Tagus-Tejo Biosphere Reserve.

While there is some interest in land stewardship initiatives in the geoparks, no nature education proposals exist in them, except in one case. In Castellfollit del Boix, Barcelona, the Wild Me project is located within the Central Catalonia Geopark. During the interview held with the person in charge, it does not seem that the existence of the geopark was a determining factor in the choice of location since they are far from the most characteristic geoecological landmarks of the site and the children move a short distance from their daily meeting point, so they do not get to know them. Instead, in Wild Me, they perceive that the management regulations of natural protected areas restrict public use activities and hinder the implementation of nature-based education initiatives, so it is not usually a criterion for their choice. This perception is shared by other project managers, both those interviewed for this work and those of whose opinion the authors have direct knowledge.

The main difficulty encountered by nature-based education initiatives is that being centers that function as schools, children must attend daily, or at least regularly, throughout the week. This means that the location, while it must be close to or in nature, should be close to where the children live and where their parents or caregivers work. Many protected areas, including geoparks, are found far from towns. However, since nature is the common thread of their pedagogical approach, many are in fact located in or near other natural protected areas.

A case in point is the *Amadahi* school in Oleiros, A Coruña, found within the Costa de Dexo natural monument and in the Mariñas Coruñesas e Terras do Mandeo Biosphere Reserve. Its manager indicates that the school's setting was based on other criteria but that the location in these areas is essential for the quality of the contact with the natural environment. It actively collaborates with the authorities and landowners in the dissemination and good use of its values. *Amadahi* offers training courses to nature education professionals and emphasizes the importance of preserving the natural treasures they enjoy (P. G., pers. comm.). On the other hand, the *Grupo de Juego en la Naturaleza Saltamontes*, in Collado Mediano, Madrid, is located at the gates of the Sierra de Guadarrama National Park within the Cuencas altas de los ríos Manzanares, Lozoya y Guadarrama Biosphere

Reserve. The first NPA type, despite the school being only within its area of socio-economic influence, has been the most attractive for the project, being the second relatively unknown NPA among those interviewed and its educational community. Many families approaching this project are residents who have recently settled in the municipality or even moved there to participate. The main reason for residing in the area is the environmental quality of the surroundings. The presence of an educational project that takes advantage of it is, for them, a very logical step and an added value (EC, pers. comm.).

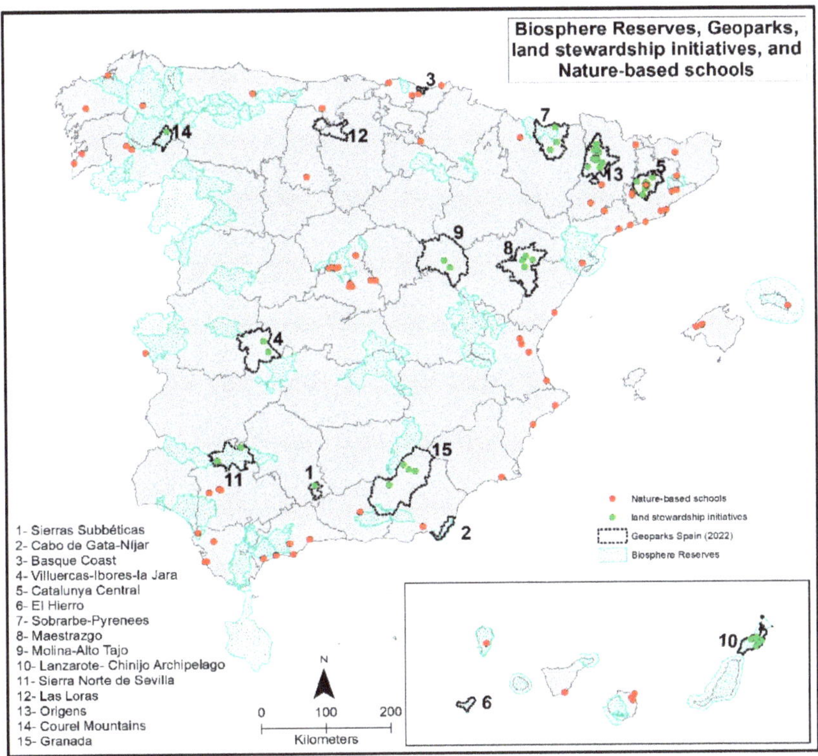

Figure 5. Location of the biosphere reserves, geoparks, land stewardship initiatives, and nature-based schools in Spain.

Nature-based initiatives that wish to obtain formal recognition as schools have a difficult time due to incompatibility between conservation regulations in protected areas, land and urban planning, and education-related regulations. While educational legislation requires a permanent building or construction for the location of classrooms and auxiliary facilities, land-use and urban planning legislation in rural areas and environmental legislation, such as that on the conservation of protected areas, do not allow the installation of buildings for these purposes. In countries such as Germany, the Czech Republic, or Italy, this obstacle is overcome because nature education is recognized as a different (or innovative) educational modality within the system, and what is required, precisely, is that no permanent construction should be used as part of the school facilities. Only mobile structures (such as construction-site-style containers) or ephemeral structures (such as yurts, wagons, or similar) are allowed, as long as they provide the services needed (shelter and storage, mainly) under the health and safety conditions required for any other educational space. This adaptation of the regulations has led to an explosion of nature-based education centers in these countries. Germany has several thousands of them, and the Czech Republic and Italy have over a hundred.

Despite these difficulties, there are two approved nature-based schools in Spain. According to the managers interviewed, this recognition and the acceptance of its implementation often depends on the attitude and flexibility of the official in charge of its processing. This was the case of *Bosquescuela* in Cerceda, Madrid, located in the Cuenca Alta del Manzanares Regional Park, a fact publicly acknowledged by the person in charge of its homologation. In the second case, *Bosqueko Forest School* in Gran Canaria, the school has been approved thanks to strict compliance with current regulations, which has involved oversizing the necessary facilities, given that teaching takes place mainly outdoors. Of the rest of the nature education initiatives in Spain, many have applied for other recognition types (e.g., leisure clubs or similar) that allow them to operate within the current regulations.

4.2.3. The Case of Edunat Cooperative in Bunyola (Mallorca)

Land stewardship is a handy tool for initiatives with a low economic capacity that wish to carry out low-impact projects or positively impact the territory. It is also advantageous for landowners, who obtain maintenance and improvement services for their land in exchange. This is precisely the case of the *Edunat Cooperative* of Bunyola, in Mallorca (Balearic Islands), an entity dedicated to nature-based education on privately owned land, whose owners have signed a stewardship agreement for ten years. The estate has an area of 4.5 hectares and consists of pasture, agricultural land, and Mediterranean forest of holm oaks, carob, and almond trees. There was also some livestock in the past, although the owners were not farmers. It also has a small building, which the cooperative can use. Contrary to what usually happens in the Balearic Islands, the owner family did not wish to sell their property to large investors, generally foreigners, for fear that its rustic use would be distorted towards a more touristic one and that the benefits of this use would not revert to the local community. According to the informant, the property treasures "the memories of the family, the experiences, the laughter of the children [. . .] that continue to resonate there" (G. V., pers. comm.).

For this reason, after long deliberations, the *Edunat Cooperative*'s proposal fell on fertile ground. Their main activity is the educational use of the farm, with children from 2 to 6 years of age visiting the area from Monday to Friday. They also have a playgroup and a bushcraft school for older children, which occur during the afternoons and weekends. In addition, the cooperative manages an organic vegetable garden. The building has been adapted as a shelter in case of bad weather, for the storage of materials and tools, and as an educational and services space (kitchen, toilets). In exchange for using the land and the building, the cooperative maintains the estate, repairs damages, waters the fruit trees, installs nesting boxes and feeders for wild animals, and generally watches over it.

The cooperative's founders have extensive experience with nature-based education, having worked on public land for more than ten years in a similar project in the area. Thanks to this experience, they appreciate the advantages of staying on private land, which allows them to develop their proposal with greater flexibility, adding activities, such as the vegetable garden or the possibility of keeping poultry, planned for the future. It also allows them to collect rainwater, which can be used for irrigation and for recreational and educational purposes. Having control over the space also makes it possible to leave materials, infrastructure, and play elements installed on site without dismantling, moving, or storing them every day. By not coexisting with other users of the space, the possibility of conflict or even accident is minimized in the event of activities incompatible with educational activities (e.g., forestry work, motorized sports, hunting). The area's spaciousness also facilitates the drop-off and pick-up of children, as there is space for parking.

According to the informant, the experience and seriousness of the proposal have been considered in the custody agreement, which includes the guarantee of compliance with regulations, especially concerning insurance and employee contracts. On the other hand, unambiguous agreements have been established for conflict prevention, communication, and transparency between the parties. They are very grateful to the owners for their open and interested attitude and the possibility of developing their activity in exchange for maintenance tasks. The economic conditions are, therefore, favorable, but they require a large investment in time and effort. The cooperative must not only develop its activity daily but must also meet the commitments made with the property. It is an investment in time and effort that must be considered before signing such a custodial agreement.

5. Discussion and Conclusions

According to Rosskamp and Valdati (2023), geoparks are a tool for local development, supported by concepts such as geoconservation, geotourism, and geoeducation [55]. Beyond differentiating geographic areas with significant geological heritage, they play an essential role in promoting sustainable tourism, the conservation of natural and cultural heritage, and the promotion of economic and social development of local communities [56,57]. After studying how the managers of eight of the fifteen Spanish geoparks perceive the internal dynamics, we observed that, in the face of different casuistry related to their territorial features, different surface area, population, and number of settlements, they share approaches, organizational structures, and even problems. Geoparks act as a distinguishing mark; taking advantage of geological and geomorphological characteristics, this instrument makes disadvantaged rural territories visible. Their declaration encourages the stimulation of local economic activity, the promotion of sustainable tourism, the improvement of essential services and connectivity, the rooting of the population, and the revitalization of community life [57,58].

Assuming geoparks are considered not as a figure of protection of geological heritage, but as an instrument of territorial promotion, implementing the land stewardship mechanism is seen as a stimulating option to help achieve both conservation and development objectives. Land stewardship promotes sustainable land use by involving civil society in land-use management and decision making [16]. Stewardship projects promote the conservation of environmental resources and the practice of activities that respect the health of ecosystems, seeking collaboration and implementing the most appropriate actions for each context [16,19,35].

So far in Spain, land stewardship projects to protect the geological heritage of geoparks, promoted by civil society, still need to be explored. For example, in four of the fifteen geoparks, no experiences have been identified; in two, only one has been detected; and in three, only a couple have been identified. However, as seen in the results, there are some experiences, and it is necessary to recognize the work that specific conservation organizations and groups are implementing through the mechanism of land stewardship to complement the conservation of these enclaves' geological heritage, biodiversity, and culture. However, these are projects that are born outside the geoparks' management. In other words, they are private stewardship entities that work within the scope of the geoparks even before their existence. The initiatives analyzed in this paper show that there are alternatives to managing nature conservation. These are more flexible, participatory, and cooperative formulas that incorporate strategies for socio-economic development and identity strengthening for these territories.

But on the other hand, land stewardship presents challenges and problems that hinder its implementation. There are still land managers and landowners who are unfamiliar with this mechanism, which is a drawback for its use. In addition, the legal and administrative framework still needs to be improved, which can generate bureaucratic obstacles and a lack of confidence regarding the implications related to ownership. Another area for improvement is linked to project financing and human capital since stewardship is promoted by private entities (generally non-profit associations based on volunteerism) that do not have high economic resources to sustain the proposals over time [42,59].

Land stewardship is a tool that is not equally rooted throughout the Spanish territory. In some regions, this mechanism has been more widely used than in others. This is the case of Catalonia, the region with the highest number of land stewardship entities in the state (63) and the highest number of agreements (710), according to the Report of the 6th Inventory of Land Stewardship Initiatives in Spain [54]. This is also reflected within the geoparks. Origens and Central Catalonia are the areas where more stewardship projects have been identified. The reason for this is to be found in the region's strong territorial identity and the solid associative network present. Associations in Catalonia are a vital characteristic of the region's civil society. It promotes cultural and territorial identity, the defense of rights, citizen participation, and collaboration to develop projects and activities for the benefit of the community. In addition, Catalonia has consolidated a network of organizations and institutions that participate in the stewardship process, in which an organization, such as the Xarxa per a la Conservació de la Natura (XCN), contributes to promoting the mechanism, weaving alliances between actors and providing technical assistance and support to associations [42].

Geoparks are spaces of interest for learning about environmental processes and transmitting territorial values. They have the necessary ingredients to propose new educational methods based on direct contact with nature. Environmental education activities are carried out in most of the Spanish geoparks. There are different educational materials and projects of different formats through which schoolchildren approach these enclaves' geological and landscape values and socio-economic reality. These experiences are closely linked to what is known as environmental education. However, only one experience has been detected within the geoparks of what is called an outdoor or nature-based school. It has been shown that these schools have an indirect relationship with the presence of natural protected areas, but this is limited to those that protect their biotic values. Geoparks, or geological values in general, are not usually considered when choosing a site. However, these elements are very important in shaping the landscapes and, therefore, the identity of the residents. Geoparks can be excellent scenarios for education and connection with the natural environment through tools such as land stewardship. Given the flexibility in their management, installing equipment for educational use, as required by sectoral regulations for this type of initiative, should be fine. Perhaps the most significant weakness lies in the sensitivity of the entities managing these educational projects, whose training and priority is linked to education (they are mainly people with degrees in teaching, early childhood education, social work, and psychology) and not so much to geology, biology, or rural development. They may likely lack the necessary information about the geoecological values and the opportunities for public use provided by the geoparks, and it may be necessary to disseminate this information to this group.

Another area for improvement of the nature-based schools is their managerial structure. They are small projects, usually managed by associations or cooperatives, with little financial power and a high staff turnover. They need help to finance and manage the rental or purchase of land to carry out their activities. For this reason, due to its flexibility and adaptability to different social agent profiles, scenarios, and situations, land stewardship is a valuable tool that can be applied to nature-based education initiatives in geoparks.

In summary, encouraging the articulation of civil society through structures that use land stewardship mechanisms can contribute to consolidating the triple purpose of any geopark: nature conservation, education, and sustainable development. Framing alternative proposals, such as nature-based schools in these areas and using land stewardship in their operation, can become an opportunity to protect a region's geological and cultural heritage and improve local communities' quality of life through sustainable and responsible economic and tourism activities. Early-years education in the natural environment facilitates the acquisition of long-term pro-environmental skills, competencies, and behaviors that last into adulthood and act as multipliers for others. This, in turn, would reinforce the objectives of the conservation and dissemination of the values of the geoparks from one generation to the next.

Author Contributions: Conceptualization, E.I.-G. and K.H.-K.; methodology, E.I.-G., K.H.-K. and G.F.-S.; validation, E.I.-G., K.H.-K. and G.F.-S.; formal analysis, E.I.-G., K.H.-K. and G.F.-S.; investigation, E.I.-G., K.H.-K. and G.F.-S.; writing—original draft preparation, E.I.-G. and K.H.-K.; writing—review and editing, E.I.-G., K.H.-K. and G.F.-S.; visualization, E.I.-G., K.H.-K. and G.F.-S. All authors have read and agreed to the published version of the manuscript.

Funding: This research received no external funding.

Data Availability Statement: The data is contained within the article.

Conflicts of Interest: The authors declare no conflict of interest.

References

1. Adams, V.M.; Mills, M.; Weeks, R.; Segan, D.B.; Pressey, R.L.; Gurney, G.G.; Groves, C.; Davis, F.W.; Álvarez-Romero, J.G. Implementation strategies for systematic conservation planning. *Ambio* **2019**, *48*, 139–152. [CrossRef] [PubMed]
2. Tiebel, M.; Mölder, A.; Plieninger, T. Small-scale private forest owners and the European Natura 2000 conservation network: Perceived ecosystem services, management practices, and nature conservation attitudes. *Eur. J. For. Res.* **2021**, *140*, 1515–1531. [CrossRef]
3. Sayer, J.A. Science and international nature conservation. *CIFOR Occas. Pap.* **1995**, *4*, 14.
4. Florido, G.; Lozano, P.J. Las figuras de protección de los espacios naturales en las comunidades autónomas españolas: Una puesta al día. *Boletín Asoc. Geógrafos Españoles* **2005**, *40*, 57–81.
5. Nikolova, V.; Sinnyovsky, D. Geoparks in the legal framework of the EU countries. *Tour. Manag. Perspect.* **2019**, *29*, 141–147. [CrossRef]
6. Pérez-Calderón, E.; Prieto-Ballester, J.M.; Miguel-Barrado, V. Perceived Rural Development in UNESCO Global Geoparks in Spain. *Land* **2022**, *11*, 1086. [CrossRef]
7. Farsani, N.T.; Coelho, C.; Costa, C. Geotourism and Geoparks as Gateways to Socio-cultural Sustainability in Qeshm Rural Areas, Iran. *Asia Pac. J. Tour. Res.* **2012**, *17*, 30–48. [CrossRef]
8. De Castro, E.; Loureiro, F.; Patrocínio, F.; Gomes, H.; Castel-Branco, J.; Cezar, L.; Fernandes, M.; Azevedo, P. The Estrela UNESCO Global Geopark Territorial Development Strategy: A Holistic Vision for the Twenty-First Century. In *Economics and Management of Geotourism Tourism, Hospitality & Event Management*, 1st ed.; Braga, V., Duarte, A., Marques, C.S., Eds.; Springer: Cham, Germany, 2022; pp. 19–46.
9. UNESCO. UNESCO Global Geoparks (UGGp). Available online: https://en.unesco.org/global-geoparks (accessed on 22 June 2023).
10. Pásková, M.; Zelenka, J. Sustainability management of Unesco Global Geoparks. *Sustain. Geosci. Geotourism* **2018**, *2*, 44–64. [CrossRef]
11. Peris Reig, L. *Outdoor Education. Una forma de Aprendizaje Significativo*, 1st ed.; Punto Rojo Libros: Sevilla, Spain, 2017; p. 138.
12. Troitiño Vinuesa, M.A. Espacios Naturales Protegidos y Desarrollo Rural: Una Relación Territorial Conflictiva. *Boletín Asoc. Geógrafos Españoles* **1995**, *20*, 23–37.
13. Hueso-Kortekaas, K. *Educar en la Naturaleza*, 1st ed.; Plataforma Editorial: Barcelona, Spain, 2021; p. 256.
14. Hueso-Kortekaas, K.; Iranzo-García, E. Salinas and "Saltscape" as a Geological Heritage with a Strong Potential for Tourism and Geoeducation. *Geosciences* **2022**, *12*, 141. [CrossRef]
15. Sending, O.J.; Neumann, I.B. Governance to governmentality: Analyzing NGOs, states, and power. *Int. Stud. Q.* **2006**, *50*, 651–672. [CrossRef]
16. Brown, J.; Mitchell, B. The stewardship approach and its relevance for protected landscapes. *Georg. Wright Forum* **2000**, *17*, 70–79.
17. Mitsos, M.; Ringgold, P.C. Testing Stewardship Concepts on Federal Land. *J. Sustain. For.* **2001**, *13*, 305–320. [CrossRef]
18. Mumaw, L. Transforming urban gardeners into land stewards. *J. Environ. Psychol.* **2017**, *52*, 92–103. [CrossRef]
19. Basora Roca, X.; Sabaté i Rotés, X. *Custodia del Territorio en la Práctica. Manual de Introducción a una Nueva estrategia Participativa de Conservación de la Naturaleza y el Paisaje*, 1st ed.; Xarxa de Custodia del territorio, Fundació Territori i Paisatge—Obra Social Caixa Catalunya: Barcelona, Spain, 2006; p. 80.
20. Pietx, J. Custòdia del territori a Catalunya. *Medi Ambient Tecnol. Cult.* **2008**, *42*, 34–43.
21. Iranzo-García, E.; Puig Lanas, M.L. La custodia del territorio: Un instrumento para la puesta en valor colaborativa del paisaje: El caso de Castielfabib. In *Objectius de Desenvolupament Sostenible en el Territori Valencià: Trobada 2020*; Universitat de València: Valencia, Spain, 2020; pp. 151–163.
22. Carcavilla, L.; Durán, J.J.; García-Cortés, A.; López-Martínez, J. Geological heritage and geoconservation in Spain: Past, Present, and Future. *Geoheritage* **2009**, *1*, 75–91. [CrossRef]
23. Brilha, J. Geoheritage and geoparks. In *Geoheritage: Assessment, Protection and Management*; Reynard, E., Brilha, J., Eds.; Elsevier: Amsterdam, The Netherlands, 2018; pp. 323–335.
24. Hilario, A.; Carcavilla, L. Twenty Years of Spanish Geoparks: Analysis and Future Prospects. *Geoheritage* **2020**, *12*, 87. [CrossRef]
25. García-Cortés, A. Spanish Geological Frameworks. In *An Approach to Spanish Geological Heritage of Internacional Relevance*; Instituto Geológico y Minero de España: Madrid, Spain, 2009; p. 215.

26. Álvarez, S.; Hernández, S. La custodia del territorio como instrumento complementario para la protección de espacios naturales. *Rev. Catalana Dret Ambient.* **2011**, *2*, 1–22. [CrossRef]

27. Jackson, S.F.; Gaston, K.J. Incorporating private lands in conservation planning: Protected areas in Britain. *Ecol. Appl.* **2008**, *18*, 1050–1060. [CrossRef]

28. Barreira, A.; Rodríguez-Guerra, M.; Puig, I.; Brufao, P. *Estudio Jurídico Sobre la Custodia del Territorio*; Plataforma de Custodia del Territorio de la Fundación Biodiversidad: Madrid, Spain, 2010; p. 277.

29. Capdepón, M.; Durá, C.J. Introducción al concepto de la conservación privada: «nuevas» herramientas para la conservación de la biodiversidad. *Ciudad Territorio. Estud. Territ.* **2019**, *51*, 27–42.

30. Brown, J.; Mitchell, B. Extending the reach of national parks and protected areas: Local stewardship initiatives. In *National Parks and Protected Areas: Keystones to Conservation and Sustainable Development*, 1st ed.; Nelson, J.G., Serafin, R., Eds.; Springer: Berlin/Heidelberg, Germany, 1997; Volume 40, pp. 103–116.

31. Borrini, G.; Farvar, M.T.; Kothari, A.; Pimbert, M.; Renard, Y. *Partager le Pouvoir: Cogestion des Ressources Naturelles et Gouvernance Partagée de par le Monde*; International Union for Nature Conservation (IUCN): Gland, Switzerland, 2009; p. 548.

32. Fagan, S.M. An Analysis of the Evolution of Theory and Management in the Trustees of Reservations. Master's Thesis, University of Pennsylvania, Philadelphia, PA, USA, 2008.

33. Brewer, R. *Conservancy: The Land Trust Movement in America*; Dartmouth College Press: Lebanon, PA, USA, 2003.

34. Durá Alemañ, C.J. La Custodia del Territorio y sus Nuevas Técnicas Para la Conservación del Patrimonio Natural, el Paisaje y la Biodiversidad: Un Invento Norteamericano y su Expansión al Resto del Mundo. Ph.D. Thesis, University of Alcalá de Henares, Alcalá de Henares, Spain, September 2013.

35. Pietx, J. Custodia del territorio, una nueva estrategia de conservación. *Quercus* **2000**, *169*, 20–23.

36. Declaración de Montesquiu. 2000. Available online: https://www.custodia-territorio.es/content/declaraci%C3%B3n-de-montesquiu-2000 (accessed on 8 March 2023).

37. Ruiz, A.; Navarro, A.; Sánchez, A. *Libro Blanco Construyamos el Futuro de la Custodia del Territorio*; Foro de Redes y Entidades de Custodia del Territorio: Madrid, Spain, 2018.

38. Racinska, I.; Barratt, L.; Marouli, C. LIFE and Land Stewardship. Current Status, Challenges and Opportunities. Report to the European Commission. 2015. Available online: https://www.landconservationnetwork.org/node/9535 (accessed on 17 January 2023).

39. Kamal, S.; Grodzińska-Jurczak, M.; Brown, G. Conservation on private land: A review of global strategies with a proposed classification system. *J. Environ. Plan. Manag.* **2015**, *58*, 576–597. [CrossRef]

40. Hardy, M.J.; Bekessy, S.A.; Fitzsimons, J.A.; Mata, L.; Cook, C.; Nankivell, A.; Smillie, K.; Gordon, A. Protecting nature on private land using revolving funds: Assessing property suitability. *Biol. Conserv.* **2018**, *220*, 84–93. [CrossRef]

41. Groce, J.E.; Cook, C.N. Maintaining landholder satisfaction and management of private protected areas established under conservation agreements. *J. Environ. Manag.* **2022**, *305*, 114355. [CrossRef] [PubMed]

42. Quer, B.; Asensio, N.; Codina, J.; Chalard, L.; Pietx, J.; Rodríguez, P. *Study of the Development and Implementation of Land Stewardship in the Different Participation Regions*; LANDLIFE: Girona, Spain, 2012; p. 125.

43. Horton, K.; Knight, H.; Galvin, K.A.; Goldstein, J.H.; Herrington, J. An evaluation of landowners' conservation easements on their livelihoods and well-being. *Biol. Conserv.* **2017**, *209*, 62–67. [CrossRef]

44. Ruiz, A.; Arias, F.; Navarro, A. *El Tercer Sector Ambiental: Un Enfoque de las Entidades Ambientales no Lucrativas*; Colección Cuaderno de Campo; Asociación de Fundaciones para la Conservación de la Naturaleza: Madrid, Spain, 2016; Volume 1, p. 20.

45. Asociación Nacional EDNA. Guía de escuelas en la naturaleza. In *Información Práctica Sobre la Vida y Organización de Experiencias en la Naturaleza en España*, 1st ed.; La Traviesa Ediciones: Málaga, Spain, 2020; p. 112.

46. Chawla, L. Childhood experiences associated with care for the natural world: A theoretical framework for empirical results. *Child. Youth Environ.* **2007**, *17*, 144–170. [CrossRef]

47. Matsuba, M.K.; Pratt, M.W. The Making of an Environmental Activist: A Developmental Psychological Perspective. *New Dir. Child Adolesc. Dev.* **2013**, *142*, 59–74. [CrossRef]

48. Bauman, Z. Educational challenges of the liquid-modern era. *Diogenes* **2003**, *50*, 15–26. [CrossRef]

49. In Natura. Available online: https://escuelainnatura.com/directorio-escuelas-en-la-naturaleza-espana/ (accessed on 5 February 2023).

50. Hueso-Kortekaas, K. A critical view on the strengths and challenges of outdoor preschools and nature-based education. In *Regenerative Desing in the Latin-Mediterranean Context. New Visions, Emerging Practices and Perspectives*; Cobreros, C., Giorgi, E., Cattaneo, T., Eds.; Springer Nature: Cham: Switzerland, 2023; *in press*.

51. González Río, M.J. *Metodología de la Investigación Social*; Aguaclara: Alicante, Spain, 1997; p. 300.

52. Saldana, J. *Thinking Qualitatively: Methods of Mind*, 1st ed.; SAGE Publications, Inc.: Thousand Oaks, CA, USA, 2014; p. 240.

53. Miles, M.B.; Huberman, A.M.; Saldana, J. *Methods of Predicting Qualitative Data Analysis: A Methods Sourcebook*, 4th ed.; SAGE Publications, Inc.: Thousand Oaks, CA, USA, 2014; p. 408.

54. Prada, O. *Informe del 6° Inventario de Iniciativas de Custodia del Territorio en España*; Fundación Biodiversidad, Ministerio para la Transición Ecológica y el Reto Demográfico: Madrid, Spain, 2019; p. 223.

55. Eder, E.W.; Patzak, M. Geoparks-geological attractions: A tool for public education, recreation and sustainable economic development. *Episodes* **2004**, *3*, 162–164. [CrossRef] [PubMed]

56. McKeever, P.; Zouros, N.; Patzak, M. The UNESCO global network of national geoparks. In *Geotourism: The Tourism of Geology and Landscape*; Newsome, D., Dowling, R., Eds.; Goodfellow Publishers: Oxford, UK, 2010; pp. 221–230.
57. Rosskamp, D.; Valdati, J. Geoparks and Sustainable Development: Systematic Review. *Geoheritage* **2023**, *15*, 6. [CrossRef]
58. Zouros, N. The European Geoparks Network. *Episodes* **2004**, *27*, 165–171. [CrossRef]
59. Sabaté, X.; Basora, X.; O'Neill, C.; Mitchell, B. *Caring Together for Nature. Manual on Land Stewardship as a Tool to Promote Social Involvement with the Natural Environment in Europe*, 1st ed.; xct in partnership with Eurosite, CEN L-R, Legambiente and Prysma: Barcelona, Spain, 2013.

MDPI
St. Alban-Anlage 66
4052 Basel
Switzerland
www.mdpi.com

Geosciences Editorial Office
E-mail: geosciences@mdpi.com
www.mdpi.com/journal/geosciences